THE END OF THE SOUL

THE END OF THE SOUL

Scientific Modernity, Atheism, and Anthropology in France

Jennifer Michael Hecht

Columbia University Press
New York

Columbia University Press
Publishers Since 1893
NewYork   Chichester,West Sussex

Library of Congress Cataloging-in-Publication Data
Hecht, Jennifer Michael, 1965–
    The end of the soul : scientific modernity, atheism, and anthropology in France /
Jennifer Michael Hecht.
       p. cm.
    Includes bibliographical references and index.
    ISBN 0–231–12846–0 (cloth : alk. paper)
       1. Atheism—France—History. 2. Religion and science—France—History. 3.
Anthropology—Religious aspects—History. 4. Anthropology—France—History. 5.
France—Religion.   I. Title.

BL2765.F8 H43 2003
211'.8'094409034—dc21   2002035093

Columbia University Press books are
printed on permanent and durable acid-free paper.
Printed in the United States of America
c 10 9 8 7 6 5 4 3 2 1

To my sister and brother

Contents

∞

ILLUSTRATIONS

∞

Illustrations are set following p. 210.

Acknowledgments

∞

Many people helped this book come into being. I would first like to thank Robert Owen Paxton, who guided my study of French history when I was a doctoral student at Columbia University, and Nancy Leys Stepan, who introduced me to the history of science there, especially the history of scientific claims about human difference. While I was engaged in the early research for this project in France, Maurice Agulhon at the Collège de France discussed the work with me and brought to my attention a study of French ethnology recently published by Nélia Dias. Nélia and I met in New York a few years later, and since then I have been delighted to share with her our odd familiarity with the same otherwise half-forgotten band. Historian and race theorist Pierre-André Taguieff discussed Lapouge with me for several hours on a rainy Paris afternoon and later helped me gain access to the archives in Montpellier. Claude Blanckaert shared his expertise of the Société d'anthropologie de Paris and its archives. A small, international crew of scholars had variously decided to take notes in the Musée de l'homme every day and go to Professor Blanckaert's seminar once a week at the Muséum d'histoire naturelle, after which he would join us at a café to drink espresso and wine and argue about Broca. I learned a lot about the mood of nineteenth-century European natural science in those sessions. I would also like to thank the librarians at the Paris Musée de l'homme for their gracious and informed assistance. Claude Blanckaert invited me to take part in the CNRS conference "La race: Idées dans les sciences et dans l'histoire" in 1993, where my study of Lapouge benefited from participants' observations.

Alexander Alland, Atina Grossmann, and Vera Zolberg read parts of the manuscript at various stages and offered helpful comments and much appreciated encouragement. I must also thank the anonymous readers of *French Historical Studies*, the *Journal of the History of Ideas*, *Isis: Journal of the History of Science Society*, and the *Journal of the History of the Behavioral Sciences* for their often exceedingly helpful criticism and questions. Similarly heartfelt thanks go to the anonymous reviewers who offered comments on the present work as it made its way to press. After the book was finished, two scholarly events in which I took part offered new things to think about: At the 2001 History of Science Society conference, Ian Dowbiggin, Mike Hawkins, Richard Weikart, and I delivered papers for a session entitled "The Life Sciences and the Crisis of Ethics," chaired by Edward Larson. It was good to compare notes from France, England, Germany and the United States. Evelynn Hammonds's "Race and Science" workshop at MIT (February 2002) brought together twenty scholars working on that subject. The lively weekend of talk influenced some of my final thoughts here and redoubled my enthusiasm to share this curious chapter in the history of anthropology.

Thanks to Mary Keller, as always, for all sorts of advice, insight, inspiration, and encouragement. Thanks to Steven Hull for going to the Library of Congress with me so many times while I researched this book, and for many long conversations. Thanks, too, to the Ginocchio family for their camaraderie and hospitality during my several stays in Paris. The support of my parents, Carolyn and Gene, has been inestimable.

I thank my husband, John Chaneski, for his considerable help in preparing the manuscript for print, his ebullient companionship in all things, and his surprising height. Finally, over the years my siblings, Amy Allison and Jamey, read the manuscript in its entirety and in parts and engaged me in my extended reveries on French anthropology, autopsies, and free thought. They're also funny. This book is dedicated to them both. Finally, I would like to sample notions from the acknowledgments pages of the first books of my two advisers at Columbia. One noted the role of the Meade's pear tree that she could see from her window as she wrote; I am happily reminded to give witness to the stained glass of the Eglise St. Médard that faced my Paris apartment and to the low red brick and high metal spires of New York City. The other's disclaimer seemed so right to me that I am compelled to borrow it: "These friends and helpers are in no way responsible, of course, for whatever errors, perversities of judgment, or downright idiosyncrasies may appear in this book. Those belong altogether to the author."

The End of the Soul

The End of the Soul

This book is about atheism and its relationship to science, especially the science of people—of race, gender, class, and nation—at the end of the nineteenth century and the beginning of the twentieth. I started researching this topic about ten years ago because I read of the existence of a Society of Mutual Autopsy, and I wanted to know more. The French anthropologists that created it dominated the Society of Anthropology of Paris in the last decades of the century and championed an outspoken, overt mixture of science and anticlerical politics. The history of the Society of Anthropology of Paris in this period had been explored in several works, most notably, in French, in several articles by Claude Blanckaert and a book by Nélia Dias and, in English, in an article by Michael Hammond and the dissertations of Joy Harvey and Elizabeth Williams.[1] These answered a lot of questions, and the present work is much indebted to them. But these studies were all primarily concerned with tracing the interaction between political ideology and the development of particular lines of scientific theory; most often, this had to do with more or less subtle assumptions of human hierarchy. I wanted to know more about some of these anthropologists' outspoken defense of an entirely unsubtle, politicized science and their zealous campaign against belief in God. I soon found that a distinct group had first come together as freethinkers—atheists—in the period of the conservative Second Empire and had then entered into anthropology as an intact group, with the explicit intention of using the young science against religion, God, and, specifically, the Catholic Church.[2]

By the late nineteenth century, French culture was dominated by the notion that tradition, church, monarchy, and dogma were naturally and inextricably united in a struggle against change, freedom, democracy, and science. The confrontation between science and religion was a constant theme. But before the publication of Darwin's *Origin of Species* in 1859, even most left-wing intellectuals felt that some deistic notion was necessary to account for human existence. Evolutionary theory did not cause anticlericalism or atheism, but it was a great and encouraging windfall for those who were already in opposition to the church. Some forms of anticlericalism reach back before the Revolution, but under Napoleon III's Second Empire, political repression and the privileges of the Catholic Church wildly exacerbated the hostility between Catholicism and republicanism. As Theodore Zeldin has put it, "Clericalism and anticlericalism became probably the most fundamental cause of division among Frenchmen."[3] Indeed, Zeldin credits the anticlerical passion of this period with the creation of a two-party political system in France: without much agreement on anything else, everyone in France was either for or against the clergy (p. 1027). In this great division, there were several sites of the anticlerical avant-garde—or those who saw themselves as such—and this book is the history of one of them.

By the time the Third Republic was instituted in 1871, a community of left-wing atheists were using anthropology to argue against religion and, more surprisingly, using the rituals of this new science to cope with the distress and alienation occasioned by the loss of God and of church community. These freethinking anthropologists of Paris were as intent on their freethinking mission as on their anthropology and saw themselves as central figures in a great project of transforming France into a scientistic, antireligious—indeed, atheist—country. They were jubilant in this dechristianizing project, but they were also somewhat agonized over the end of the soul and its consequences for humanity. In interesting ways, they managed this agony through the invention of various anthropological ideas and practices.

Throughout its investigations, this book concentrates on relationships and behaviors as much as it does on the ideas of science. Through behaviors as well as ideas, the freethinking anthropologists created a purposeful cult, whose central gesture was a somewhat fantastic translation project by which the entire context of public and private discourse was to be changed from basically religious to basically scientific. Even in the world of the new, secularizing Third Republic, the freethinking anthropologists were extremists—by most contemporary estimations, they took antitheism and antiphilosophical metaphysics a bit too far—but now they found a niche. For pragmatic secularists who saw the Catholic Church and political conservatism as their real enemies, they were useful allies, and the government and the general public supported them by funding and flocking to a variety of their anthropological activities—books, schools, journals, conferences, and tours. For the much smaller group of French men and women who truly agreed with the freethinking anthropologists on questions of God and naturalism, they offered a more dramatic service. They came to provide a kind of replacement cult, complete with death rites in the form of an autopsy society and a variety of other services that paralleled Catholic ritual. Thus the community had common rites of a most serious nature, as well as an eschatological vision in which the triumph of science over faith coincided with a worldly utopia of equality, democracy, and self-fulfillment.

Because the freethinking anthropologists saw themselves as egalitarians before they were anthropologists, it is not surprising that they argued that anthropology was inherently emancipatory and egalitarian in its conclusions. Nevertheless, their philosophical materialism and brash hostility to all metaphysics caused them to flatten the human experience into that which could be weighed and measured. Numbers rarely meant much in the anthropological theories they generated, but the freethinking anthropologists carried out a tremendous amount of measuring anyway and proselytized the truth and antimythic purity of facts expressed in weights and numbers. This is particularly important because some of the central techniques by which governments have come to understand their populations through biological measurements and statistics were created by students of the freethinking anthropologists. When these former students went on to become influential scientists and politicians, they were generally far less focused on evangelizing atheism than their anthropologist teachers had been, but they remained animated by the freethinkers' sense of materialism, naturalism, and measuring. Borrowing ideas, language, techniques, and behavior from the freethinking anthropologists, they turned these toward measuring bodies for criminal identification, or trying to control the national birthrate, or theorizing mass exterminations in order to "correct" the population. In these and other endeavors—literary as well as political—some key French men and women passionately connected their work to the freethinking anthropologists and their particular version of materialism. Through discussions of physical anthropology, these students of humankind manipulated, "proved," and publicized a host of deeply private and broadly public concerns—and generated some very troubling doctrines. In late-nineteenth-century France, these anthropologists were by no means the only people replacing interest in the soul with interest in the body, but their particular variation on this theme had some significant consequences.

The first part of this book is about a self-identified group and its lay following. The book then follows that story from the extraordinarily zealous, leftwing freethinking anthropologists and their conclusions about humanity to a second generation of body measurers, some of whom, such as the Bertillon brothers, successfully brought these numerical techniques to the modern state as systems for gaining usable, if problematic, information about the populace. By contrast, some of these students of the Paris anthropologists, most notably the scientific racist Vacher de Lapouge, brought biological reductionism to a level that could not be supported by the republican regime. Finally, I will follow the early-twentieth-century dismissal of the more racist and sexist of these doctrines from within anthropology, as well as from philosophy and from Émile Durkheim's sociology. In the context I examine, leftist, secularist concern with the body (instead of the soul or the spirit) helped the modern state come to *see* its population and try to ameliorate its troubles, but it also generated an attitude toward humanity that was not compatible with leftist ideology. This forced an explosive confrontation over the issues of atheism, religion, morality, racism, sexism, and equality, and neither the political left nor the political right has ever been the same since. The left had seen itself as the keeper of science, but when it became clear that the peculiar authority of science could be used to create dogma and false hierarchies as well as to dismantle them, some preferred equality to science. While science remained in the arsenal of the politically progressive, its numbers and laws were suddenly and vividly understood as a potential enemy of equality and as a possible support for any given social hierarchy. The final part of the book offers a revision of the common narrative of the history of racism, which held that scientific racism went relatively unchallenged until the horrors of Nazism made clear its dangers.

This book, then, is the story of a leftist, atheist movement and the fascinating experience of being an atheist in France when it was both the absolute cutting edge and a wild bit of the fringe. Following key students of this group, the book examines how the atheist anthropologist's turn from the soul to the body helped to generate several theories of biological determinism. Finally, it demonstrates how some of the more moderate of the irreligious were able to catch the error, dramatically reject those theories, and revise the left's antimetaphyscial, scientific ideals in order to defend its moral vision.

The atheism of the anthropologists had a variety of interesting consequences and influences that will show up throughout this book. Though the freethinking anthropologists had hundreds of devoted followers who identified themselves as atheists, nationally this was still a small group, and though they had the attention and even the ardent sympathy of many important figures in the French public world, they were rarely matched in their antitheist zeal by even their own body-measuring students. A host of better-known figures encountered the freethinking anthropologists, however, and let it be known that the dogmatic, passionate materialism of these anthropologists was crucial to their own artistic, political, intellectual, and even personal lives. Thus, as fascinating as they are on their own, much of the importance of the freethinking anthropologists lies in this network of associations. Much of the significance of the present study is that it examines the figures in this network—Emile Zola, Jules Ferry, Arthur Conan Doyle, Maria Montessori, Margaret Sanger, Paul Valéry, Paul Verlaine, Hamlin (Hannibal) Garland, and Bram Stoker are among the better known—in a new light, that is, in a context made visible by the freethinking anthropologists' attack on theism and their attempt to replace it with anthropology.

The ardor of these anthropologists and their followers is perhaps best characterized by the Society of Mutual Autopsy—a club in which one waited for one's friends and fellow members to die and then dissected them—unless they got to you first. When the great republican statesman Léon Gambetta died in 1882, only his heart was laid to rest in the Panthéon, surrounded by the more complete tombs of other national heroes. Gambetta had believed in science with such conviction that he had willed his brain to its most outrageously dedicated disciples, the freethinking anthropologists, who had promised that brain autopsies would yield scientific advancement and, through it, social progress. This event also characterizes the anthropologists' profound failure: next to nothing was learned from Gambetta's autopsied cerebrum. Instead, the society's great success was its ability to lend a sense of meaning and purpose to a death otherwise experienced as meaningless. As this book will show, anthropology served not only to provide hypotheses for questions of morality and mortality but also to alleviate the fears surrounding these questions and to provide a community of hope and enthusiasm for those who had explicitly rejected the spiritual.

∞

CHAPTER ONE

The Society of Mutual Autopsy
and the Liturgy of Death

On October 4, 1889, the Prefecture of Police of Sables-d'Olonne authorized the exhumation of the remains of Eugène Victor Véron so that they might be shipped to Paris for examination and preservation.[1] There was no suspicion of foul play, and this was by no means a fresh corpse: Véron had died on May 23.[2] Though Véron's death certificate called him a journalist, it was his anthropological associations that led to his rather odd posthumous adventure in late 1889: years earlier, on October 19, 1876, in Paris, Véron and eighteen other men had pledged to dissect one another's brains.

This pledge was the birth of the Société d'autopsie mutuelle—the Society of Mutual Autopsy. The society acquired over a hundred additional members in its first few years, including many notable political figures of the left and far left. From its heyday in the last two decades of the century until just before World War II, the society carried out many encephalic autopsies, the results of which were periodically published in scientific journals. This published material alerted historians to the formation of this unusual group, and works by Michael Hammond, Elizabeth Williams, and Joy Harvey all comment briefly on the society's existence.[3] An essay by Nélia Dias is the only analysis to extend beyond a few lines, and though it relies on published sources, it offers an insightful sketch of the society's publicized anthropological, political, and freethinking concerns.[4] What the society's archives reveal, from their dusty box in the basement of the Paris Musée de l'homme (down a very dark spiral stairwell—one brings a flashlight) is a more tender and fascinating business.[5] While founders and members all described their endeavor as profoundly secular, the society's autopsies and ancillary rituals were modeled on religious be-

6

haviors. Indeed, the founders created a confessional, liturgical memorial system, and, in surprisingly self-conscious ways, members embraced this devotional system as a replacement for a spurned Catholicism. The anthropologists who created the Society of Mutual Autopsy were self-proclaimed atheists—freethinkers—who explicitly hoped that science could replace religion. In founding the society, the freethinking anthropologists were constructing an arena for atheist proclamations and celebrations, creating active, science-oriented rituals for a community that was otherwise united only by a rejection of metaphysics and a refusal to take part in the ceremonies of faith.

DEATH IN A SECULAR WORLD

It is a commonplace that for atheists the significance of life is greatly increased by the disappearance of an afterlife: the absence of an eternal life allows mortal life to bloom in importance. Since the Enlightenment, scientific progress has been imagined as a replacement for religious eschatology, with worldly utopia replacing heavenly bliss. The understanding of human existence maintained its narrative format but was given a new ending, and this time the whole thing took place on earth, among the living. In a very similar fashion, the various assumptions of inevitable progress inherent in the modern theories of history (specifically those of Voltaire, Condorcet, Hegel, Marx, and Comte) can be understood as reconfigurations of Christian eschatology.[6] The Christian model is especially notable in Marxian ideology, because unlike more Fabian versions of gradual progress, the revolutionary event provides a parallel to Judgment Day. Atheist historical narratives give meaning to individual lives by making them part of a progressive march toward earthly paradise. In this schema, mortal life is not a mere test to get through on one's way to paradise, because there is no paradise unless human lives are spent creating it. Yet if life is more meaningful, the end of a life certainly loses meaning in the new configuration.

For those who wanted God and all brands of metaphysics to be declared dead, death suddenly came alive as the most significant human problem.[7] Even the most utopian notions of human history cannot fully replace the promise of a spiritual eternity in which all the faithful, irrespective of life span, take part in the eventual glory. Faithful Marxists or other utopians get to build a future paradise, but if they die before it is realized, they never have the chance to participate in its marvels. Worse yet, in anthropological terms, the true mode of progress was evolutionary, and paradise was at least partially conceived of as eugenic in its origins and its results. But evolution is an exceedingly gradual process that, in Darwinian terms, cannot be greatly altered by individuals. If a person is worthy, he or she can assist the progress of humanity by marrying well and reproducing prodigiously; having done so, he or she is not of much more use to the project. If a person is not particularly gifted or has an overriding heritable problem, then his or her contribution to evolution is to die childless and get out of the way. Since most random mutations are either inconsequential or negative (to any given schema, whether survival ability or some human standard of improvement), many individuals would find themselves in this compromised position. Even were evolution considered to be necessarily progressive, one would be hard-pressed to imagine a more wasteful system of improvement: thousands upon thousands of creatures are generated, and most of them are useless because they carry no valuable, heritable mutation.

That is why this was such a difficult doctrine to uphold. Whether a secularized, scientific paradise or the messianic expectations of modern historical narrative, utopian progress ends for you when you die. Even if you manage to add some genetic or social benefit to the human project, death still ends your part in it. The central project of the Society of Mutual Autopsy was to connect the individual's death to progress and thereby to eternity, aggressively confronting the tragic nature of progress thus conceived. For people who rejected religion so strenuously that they saw burial as an abhorrent, cultish ritual but could not bear utterly disappearing, the society provided great comfort. There were never very many of them—a few hundred—but the Paris anthropologists eased the fears of solitary atheists in the provinces and gave succor to those French men and women who, in severing ties with the church, had lost their only confessor and their only friend in death.

While this chapter's central concern is to elucidate this materialist reinvention of Christian last rites and liturgy, it also describes some of the more contentious intellectual and political issues of the early Third Republic. The history of the Society of Mutual Autopsy highlights the problems of early republican secularism as they were negotiated between citizens and scientists, men and women, government and academic institutions, and, last, healthy people, safe in their convictions, and those same people, later, on the brink of the abyss.

FORMATION OF THE SOCIETY

The foundation of the Society of Mutual Autopsy was suggested by the medical doctor Auguste Coudereau on October 19, 1876, at a meeting of the illustrious Société d'anthropologie de Paris, an institution widely acknowledged as the international center of anthropological studies. The pioneering anthropologist Paul Broca had created this society and added to it a laboratory, a school, a museum, and a library. It was within this institution that the Society of Mutual Autopsy was conceived. Along with Coudereau, the founding members included Louis Asseline, Yves Guyot, Louis-Adolphe Bertillon, Abel Hovelacque, Gabriel de Mortillet, Henri Thulié, Charles Letourneau, and Eugène Véron. The latter six were all professors at the Ecole d'anthropologie. Mortillet was also current president of the Société d'anthropologie, and several of the others would or had served as such. All served, by turns, as the chief officers of that body. Asseline was a man of letters, Thulié was a medical doctor, and Guyot was an important economist, while Bertillon, Hovelacque, Mortillet, Letourneau, and Véron were accomplished, prolific anthropologists with significant reputations: Bertillon was a famous and formative investigator of demography. Hovelacque published and professed linguistics. Mortillet founded the first archaeological journal and created a nomenclature for that science that is still in use today. Letourneau wrote copiously on "anthropological sociology," through which he hoped to describe humanity in its essence by comparing attitudes and behaviors across cultures and across time. Véron specialized in the anthropological study of art and aesthetics. This prestigious company was marked not only by the anthropological accomplishments of these men but also by their politics: they were deeply anticlerical members of the political left wing. They advocated feminism and socialism and frequently invoked the notion that science would help deliver society from priests and dogma, from the inferior status of women, and from general inequality. Within days of Coudereau's proposition, the statutes and membership of the new society were published in the *Revue scientifique*. Soon after, they appeared in the medical journal *Tribune médicale* and in the politically republican journals *Les droits de l'homme* and *Le bien public*.[8] Attention from the press would continue throughout the project, becoming especially heavy when the society got hold of a particularly famous brain, such as that of Gambetta.

In public and private letters to potential members, the society leaders adopted a brash proselytizing style, asserting that "without question," autopsies on brains were the soundest way of increasing knowledge about the functioning of the mind and the physical location of particular abilities and characteristics.[9] The project was not often described as a revival of phrenology, but most people involved in it did not deny the connection—with good reason. Phrenology had been invented around 1800, by Franz Joseph Gall, an Austrian neuroanatomist who wanted to remove the metaphysical from psychology and ground it on a more material basis. In medical school Gall had somehow concluded that the smarter students generally had prominent, bulging eyes, and on the basis of this odd observation, he tried to figure out what else might be discernible from features of a person's head. With his disciple, Johann Spurzheim, Gall decided that feeling the bumps on a skull could give information about some thirty-seven human attributes. As the two popularized the science on either side of the Atlantic, phrenology came to be associated with left-wing reform: phrenologists were for temperance, against corsets, and at least mildly feminist. Many people also associated practitioners with irreligion; indeed, because phrenology sited the mind's functions in a material location, the Scottish philosopher Sir William Hamilton dramatically claimed, "phrenology is implicit atheism."[10]

Gall's science had been resoundingly rejected by midcentury, but the Paris anthropologists believed the ambitions of phrenology to have been replaced by a still more empirical interpretation of mind-brain relationships founded on Paul Broca's work on aphasia. Broca was a doctor and the founder of French anthropology. Through clinical study and postmortem examination, he had famously established that the "third left frontal circumvolution" was the area of the brain that controlled speech; a lesion there produces effects on speech that are still called Broca's Aphasia today. Mathias Duval, professor at the Paris Faculty of Medicine, argued that the principles of the Society of Mutual Autopsy were "perfectly in accord . . . with the order of study that, since Broca, we have come to represent under the title of 'aphasia.' "[11]

Autopsies were routine in France by the late nineteenth century (in contrast to the contemporaneous situation in England) but were not concerned with relating brain morphology to human characteristics.[12] In its publications, the society explained that it was possible to perform research-oriented autopsies only on the "poor and unattached" elements of society—those that end up nameless and without resources, dying alone at the charity hospitals. The freethinking anthropologists believed this was a double tragedy: first, because only members of "the disinherited section of the population" were being autopsied and studied as examples of humanity; and, second, because the personalities of these specimens were unknown, making it impossible to find connections between mind and brain morphology.[13] The solution was as simple as it was radical: the nineteen men donated themselves to one another and set out to recruit future corpses into the fold.

At first glance, Coudereau's idea was dangerously contrary to his political beliefs, flattering society's elite in suggesting that they were more worthy of dissection than were the unclaimed bodies at the charity hospital. Indeed, there is a tendency to equate theories of biological determinism with social conservatism, and in general this has been a historically accurate association. The fact that these radical social progressivists were so interested in the body is partially explained by the French biological theory that the function makes the organ, as well as the related Lamarckian concept of the inheritance of acquired characteristics. Both of these suggest that the life society forces you to lead might have a tremendous impact on your own physical morphology and on that of your children. Coudereau explained that the "disinherited part of the population" was only less interesting because the "defects of our social organization had not given them the means to develop the cerebral aptitudes that they possess in 'germ.'"[14] He believed that it would be easier to find direct relations between brain areas and specific human abilities in the "cultivated class . . . well-known people valued as scholars, writers, industrialists, and politicians, etc." (2).[15]

This was the primary explicit goal of the society, but there was another, often articulated goal: the identification of hereditary diseases in the interest of protecting future generations. It was never directly explained how the postmortem identification of such illnesses (and this assumes they were not fully apparent during the subject's life) was to "safeguard against their development" in future generations (2). While a eugenical project was implied, it was not stated. The assertion was merely that doctors ought to be informed about the diseases identified in an older, deceased generation of a patient's family.[16] Coudereau insisted, in inflammatory tones, that a family had an "incontestable right" to the autopsies of their departed members. He argued against the "numerous prejudices, which ha[d] their source in unthought-out sentimentality" and had created the general opposition to autopsies.[17] Clearly, the whole enterprise was designed to be outrageous. The members were attempting to attract attention to their endeavor, and they were as eager to offend as they were to convert. But this was not mere provocation: they were deciding the fate of their own bodies. There is no place like the deathbed for a scorned religion to be refound; even the contemplation of the event is likely to give the fair-weather atheist pause. That the anthropologists were willing to go so far exemplifies the depth of their freethinking convictions, and most civilians who joined the society referred to themselves as avid freethinkers as well. New members were each required to draw up a will leaving their brains to the society—they generally offered their bodies as well—and agreed to pay annual dues to cover the costs of running the society and performing their eventual postmortems. These dues also served to keep track of distant members and occasioned a yearly reconfirmation of their commitment. Members were also required to write a short essay detailing their physical health throughout their lives, as well as their intellect, character, sensations, and abilities. Such was the project. The first member to die was Louis Asseline, in 1878. Broca performed the autopsy with help from Drs. Coudereau and Thulié.[18] The civil burial was something of an event, at which the entire Paris "scientific intelligentsia" was present and André Lefèvre delivered an ardent speech.[19] Broca's own brain came into the society's hands two years later. The autopsies were all done in the Laboratory of Anthropology, which, for decades after Broca passed, was nominally under the control of the famous but generally absent doctor Jean-Baptiste Vincent Laborde and was actually run by the extraordinary young scientist Léonce Manouvrier.

Coudereau was dissected in 1882.[20] Gambetta's brain was autopsied that same year.[21] Most of the society's founders and many involved members were eventually dissected, the results of which sometimes—as in the case of Gambetta—drew a great deal of popular interest. It is crucial to keep in mind that the anthropologists and some of the lay members were friends and colleagues. They worked, socialized, and even vacationed together, along with their families. Some of them *were* family: the demographer Louis-Adolphe Bertillon was followed into the society by his son Jacques, also a demographer and one of the most powerful forces behind the pronatalist depopulation scare that enveloped France from 1870 through most of the twentieth century. Many extrafamilial relationships will become clear later in the book; here, it is enough to note that these people were friends, and when they died they cut open each other's heads and investigated the brains inside. This is uncommon behavior in modern Europeans and suggests that the anthropologists were seeking to maintain the nervous instability of their existential position, regularly stoking their own crisis. There could be no more direct way to contemplate the weird connection between the material self, on the one hand, and life, consciousness, feeling, and thought, on the other.

THE REPUBLICAN PUBLIC DONATES ITS BRAINS

The public was made aware of the society through a great number of articles that appeared over the years in the scientific and republican, nonscientific press, many of which were published soon after the journals' editors became members of the society (though this was never mentioned in the articles).[22] Many members referred to these articles in their letters of application; for instance, Aline Ducros, a Parisian woman, joined the society after reading an article in *L'homme libre* in 1877.[23] Another major source of publicity was the Paris World's Fair of 1889, where the society held a detailed exhibit showing plaster casts of brains as well as charts, graphs, and attestations of the founders' political positions and cultural contributions. Bursts of popularity were also brought on by the membership of such important political figures as Stéphen Pichon, one of

Clemenceau's close friends and the editor of the political daily *La révolution française*. Pichon later served lengthy terms as deputy, senator, and minister of foreign affairs.[24] Another important member was General Léon Faidherbe, who began his outspoken republican career under the Empire and remained both colorful and politically committed throughout his life. The general was a charismatic figure, and a number of popular books were written about him from 1871 to 1932.[25] When he joined the society, Faidherbe explained that his own corpse would be worthy of study because "I will furnish, when the time comes, the most beautiful case of ataxia that one could ever hope to see."[26] For over a year after Faidherbe's death, new adherents indicated that they had heard about the group via reports on the late republican general.[27]

Most journals, such as *L'echo de Paris*, reported Faidherbe's membership in the Society of Mutual Autopsy with relative equanimity. Some, however, like *Le temps*, were rather critical, and *Le siècle* stated that the information offered by the society was "singularly vague" and that the conclusions it proffered were "singularly arbitrary."[28] Politics guided the various reactions to this scientific society, and strongly republican journals were generally very positive; but positive or negative, the press brought attention to the Society of Mutual Autopsy. Furthermore, the freethinking anthropologists published extensively, creating their own public image. They collaborated on several journals and largely controlled the *Bulletin de la Société d'anthropologie*. They also published a plethora of books with a variety of publishers. In many of these, mention was made of the society. For example, in Eugène Véron's study of aesthetics, a lengthy footnote was devoted to the society's work, promising that "such an institution cannot fail to furnish very useful data."[29]

Once interested, the potential members wrote to the society and received a template will and testament for their application. This stated that "the undersigned" desired to be of use, after death, to "the scientific idea" that he or she had upheld during life and therefore would donate his or her cadaver to the society. Later, an optional passage for freethinkers was added and published in several journal articles. It ran as follows: "The goal that I pursued during my life, and that I desire to contribute to after my death, is above all else scientific. All religion is, in its essence, extrascientific and hostile to the development of science. I therefore demand, as a logical consequence of my convictions, that the burial of the parts of my body that the laboratory does not keep for its studies will be done without any religious ritual and that the ceremony be purely civil."[30] Most new members, like most of the group's leaders, availed themselves of the model, spicing it liberally with their own opinions and experiences. Many also included a short section on why they had turned to the Society of Mutual Autopsy and why they had turned away from the Catholic Church. These documents lend tremendous insight into the distress of atheist French men and women—and there were more than a few women involved in this project—as they contemplated the meaning of death.

The testament of André Lefèvre is an excellent example. Lefèvre was a professor at the Ecole d'anthropologie and a founding member of the society. The essential passage of his will reads: "Freethinker, faithful to scientific materialism and to the radical Republic, I intend to die without the interference of any priest or any church. I leave to the Ecole d'anthropologie de Paris my head—face, skull, brain, and more, if it is useful. . . . The rest of me should be incinerated."[31] Many members of the society made a sharp distinction between the value they placed on parts of their body that could be of use to science and those that could not be of use. The virulence of their disdain for the latter, especially in comparison to their high esteem for the former, is only comprehensible when taken within the context of the issues involved. In leaving their bodies to the society, members were rejecting the power of religion to invest meaning in their death. At the same time, they were attempting to make an analogous investment of their own. Members negotiated this distinction by being harshly derisive of the nonuseful parts of their bodies (which might be buried or burned even if other parts were preserved), while requesting that the "scientifically useful" parts of their bodies not only be examined and discussed but also preserved and publicly displayed.

Claudius Chaptal, one of the first laymen to join the society, initially wrote to the group in April 1878 to say that he was "convinced that many of the singular events in the life of a man permanently mark the cerebral organ" and that he would like his brain to be examined in light of this notion.[32] Along with his testament, Chaptal included a vita detailing his education and work as a mathematician and physicist. He also sent a list of publications so that the society would have an idea of his aptitudes and worth.[33] In the testament itself, Chaptal instructed the society to use what parts of him they wanted and to send the rest to a medical school.[34] It was rather common for professors who joined the society to express the desire that some part of their corpse be given, as a pedagogical device, to the school at which they had taught. Chaptal thus well represented his peers when he hoped that his skeleton might hang at the Lycée de Nîmes, where he had been a student and later served as a professor. He also mused that the lycée might make use of his heart, liver, and intestines for anatomy demonstrations.

Claudius Chaptal was also representative of many adherents in his denigration of his own corpse outside its scientific usefulness, ardently repudiating the traditional Catholic notion upholding the sanctity of the body after death. From religious training and tracts, as well as from contemporary public debates over civic funeral laws, nineteenth-century French men and women were quite familiar with the Catholic doctrine of material continuity.[35] This doctrine held that the resurrected self would be a translated version of actual bodily remains. Destruction of one's corpse was not simply sinful; it was self-annihilating. Given this, there is profound commitment in Chaptal's statement that his unwanted body parts should not even be buried, because he "attach[ed] no importance whatever to such rotting garbage."[36] Georges Laguerre's request was along the same lines. Laguerre joined the society in 1883, perhaps largely motivated by his far-left politics; several years later he was elected to the Chamber of Deputies and soon after became one of the eleven members of the general staff of General Boulanger's "national party" (all but three of whom had come from the radical or socialist extreme left). Laguerre specifically requested that the scientifically interesting parts of his body be placed on public display at the Musée d'anthropologie and consigned the rest of his body to any convenient, casual disposal.[37] A more virulently anticlerical expression of this is to be found in the testament of Eugène Véron. Anthropologist, journalist, and founding member of the society, Véron asked that there be no ceremony after his death and that his remains not be buried. Wrote Véron, "I attach no type of importance at all to that assemblage of decomposing matter which has lost the ability to feel and to think and of which the elements now do nothing but increasingly disassociate from each other." Véron provided for the possibility that he would be buried despite his request by appending the instructions that any such burial should be extremely simple. "I do not want," he explained, "after my death, to contribute, even a little, to the accumulation of the wealth of the clergy, against which I have combated all my life and that never ceases to do to France and to the Republic all the evil in its power."[38]

Paul Robin also expressed anticlericalism through derision of his future corpse, writing that if for any reason his dissection was impossible, he wanted "to be put into a hole, naked or in a cloth or a basket; 'to be buried like a dog' following the charming expression of the priests."[39] Robin elsewhere wrote an impassioned letter to the society, asserting that people have no control of their own bodies during life, citing "military service, industrial service and marriage" as his examples, and arguing that French citizens had no control over their own bodies after death, either, citing "funeral rites, still under control of the Catholic clergy, even in the City of Light."[40] Robin was a freethinker and an anarchist who would become quite well known for founding the Ligue de la régénération humaine, a group dedicated to the instruction of birth-control practices.[41] The euphemism of the day was "neo-Malthusianism."[42] It should not be surprising that pronatalist Jacques Bertillon and neo-Malthusian Paul

Robin joined the same society: they were both concerned with bodies and with translating the pastoral duties of the church into concerns of science and the state.[43] Like any other new member of the Society of Mutual Autopsy who had a favorite cause, Robin used his last will and testament to promulgate his beliefs:

> As for those people who would come by affection or by routine to take part in the spectacle of a burial of the contemporary fashion—obstructive for masses of passersby, terrifying for the simpleminded, and grotesque for thinkers—with its waste of flowers and crowns, I ask you to please usefully consecrate the time and money that you would have wasted. Spend that time, instead, on the propaganda and on the practical undertaking of the humanitarian ideas and works that are dear to me: good birth, which is to say not produced by chance but obtained by scientific selection, by liberated mothers, reasoned and voluntary; good integral education (I have created a specimen and many times explained the principles); and good social organization (easy to create and to maintain by and for people who are well born and well educated). My idea on this last point can be summed up in these words: society without money or masters.

While most testament writers did not express such extensive political platforms, most did mention their love of science and the Republic and their hostility toward superstition and religion. For some members, participation in the society was their only opportunity to proclaim unbelief and to express elaborate convictions in place of lost Christian catechisms. Yet to fathom these new convictions required ritual and practice, just as the old had. While Catholics and cosmopolitan scientists could proclaim, practice, and act on their beliefs, the solitary atheist of a family or rural community was stuck in a position of silence and inaction. The Society of Mutual Autopsy helped to define atheism by the things that one did, rather than the things that one refused to do. It was the group's extremism—from their provocative name, to their general public demeanor—that attracted its members, and in reading their testaments one senses that these men and women each harbored a ferocious desire to demonstrate their convictions actively.

Many people felt unworthy of scientific interest and sought to justify themselves as valid specimens. The young Paul Robin was relatively sure of himself, writing that "since my mental development is a bit removed from average banality, the study of my brain might be of interest to anthropologists."[44] But his need to make this justification was shared by many others who were less confident. Léonce Harmignies, a twenty-six-year-old Parisian, wrote plaintively,

"I hope, sir, that this unknown who you so courteously welcome will return one day this favor through works dignified of the scientific idea which you inspire."[45] Barbe Nikitine, a writer for the journal *La justice*, voiced a common concern when he wrote that he had hesitated in joining the society only because he "surmised from the names of its founders and its first members that only remarkable brains of an extremely well established value were deserving of study and investigation." What had emboldened him to send in his testament and dues in 1883 was the greatly increased membership of the society, along with the decision that there would be interest in the "cerebral organization of all men and women who, born and raised in the milieu of our old society, break away from it so much as to enter onto the path of an intellectual and social revolution."[46] The autopsy testament was thus a site for proclaiming one's republicanism, and at the same time one's republicanism became the justification for the autopsy. Many new members mirrored Nikitine in presenting themselves as worthy of dissection in their "quality as a humble champion of the grand cause of human emancipation" (2).

The one exception to this was Georges Vacher de Lapouge, the famous antirepublican anthropologist and the founder of scientific racism in France. Lapouge spent the majority of his life conceiving and proselytizing a version of biological determinism that divided the human race into two basic racial groups based on head shapes. The dolichocephalic, or long-headed, race was Aryan and superior, the brachycephalic, or round-headed, race was hardworking and good but inferior. Modernity had stirred up the proper social roles, and now the brachies (as he called them) had too much power and were ruining everything. Worse, Lapouge thought Jews were a venal version of the higher race who might at any time dupe the brachies and, disastrously, take the helm of civilization. He even predicted that the twentieth century would see vast exterminations conducted in the name of racial dominance. But he, too, donated his brain to the deeply republican, egalitarian Society of Mutual Autopsy. In April 1897 he contracted typhoid fever. Quite sure that he was near death, Lapouge wrote to the professors at the Ecole d'anthropologie in Paris to offer them his brain for dissection after his death.[47] This is not surprising if one considers that Lapouge based his whole racist ideology on skull measurements. He believed profoundly in the direct relationship between brain morphology and human characteristics, he was deeply interested in his own mind, and there was no other autopsy group to which he could turn. These ideological enemies were so dedicated to their separate agendas (the school desiring a brain to dissect, Lapouge desiring his brain to be thus honored) that they rallied to work together. Only Lapouge's physician fought against the arrangement, taking umbrage at the implication that he could not save his patient.[48]

The society asked for frequent reports on the patient's progress and were eventually informed, to their reserved pleasure, that Lapouge would indeed pull through.

Many people were simply excited to have an arena in which to share their beliefs. A brief selection from a rather long poem illustrates the point well. Victor Chevalier included this piece of verse along with his testament in 1889. He wrote it just after visiting the Society of Mutual Autopsy's anthropological exhibit at the Paris World's Fair of 1889. The following translation renders into English the general (rather lamentable) cadence of the original, though the extraordinary enthusiasm is difficult to recapture.

UNDER THE DOME OF THE FOYER
OF INSTRUCTION—SITE OF THE AUTOPSY EXHIBIT

prehistoric man emerged from natural selection
humanity was excusable, it looked for its path
it remained partly animal, there was no trick for its election
man kept, alas, his cruel instincts intact

but Anthropology is instructing us, man is marching toward progress
its slow work that is marked off by the centuries
the future race will march toward justice and reason
under the guidance of science our perfect goal we will address
where beings are equal in the universal formation

. . . . .

my goal is liberty, equality, fraternity
in the eternal, just, and reasoned love of nature, the only divinity.[49]

Calling nature "the only divinity" was strong stuff, but it was understood that the Society of Mutual Autopsy stood not for deism or agnosticism, but for atheism. A newspaper article on the society, unsigned but written by one of the founders, claimed that much had been learned from the brains of the illustrious donors. "We would have really liked to have the cranium of Victor Hugo," he added. "The society did all that it could to get it. But Victor Hugo was a deist!"[50] There had been no previous mention in the article that one had to be an atheist to be interested; it was assumed.

For most members, the Society of Mutual Autopsy was their only link to a specific doctrine of utopian progress. The society gave people a chance to connect their death, and not only their life, with eternity. The anthropologist

Charles Letourneau's obituary included a quotation from his own work which exemplifies this notion. It reads:

> This perspective of unlimited progress is the modern faith: and to our advantage this new belief replaces the mirage of a lost paradise; it sustains and consoles us in our public and private trials. Encouraged by it, we regard ourselves as laborers in an always unfinished work, but a work to which all men great and small, obscure and celebrated, can and may lend their hands. As cruel as may be the miseries, injustices, and calamities of the present, we may regard them as but mere accidents in the long voyage of humanity toward a better life and accept them with patience, all the while seeking remedies.[51]

The stretch of time into the future is what seems most impressive here—the "perspective of unlimited progress," the "long voyage of humanity toward a better life"—but it is wonderfully bittersweet: humanity needs to be sustained and consoled, we suffer public and private trials, our labor toward a better world is "always unfinished," the present is overrun with cruel miseries, injustices, and calamities, and we must accept them with patience while seeking remedies. It is this bittersweet, brave resignation that distinguishes these free-thinkers from the stereotype of the self-satisfied, almost patronizingly calm scientist who dismisses religion and never gives it another thought. The Society of Mutual Autopsy was an arena for aggressively confronting formerly religions questions and for further translating these questions into the secular world—with something of their mood intact.

ANTHROPOLOGIST AS CONFESSOR

A testament-donation to the Society of Mutual Autopsy was, however, more than a platform for self-flattery, political propaganda, the venting of anger against Catholicism, or a pledge of faith in science. As a considerable number of letters and wills attest, the relationship between lay adherents of the society and its anthropologist leaders reproduced a priest-parishioner relationship in some unexpected ways. The scientist was already conceptually connected to the priest because, in many senses, the one had taken over the authority of the other on matters of truth, human origins, the origins and age of the universe, the meanings of illness and health, and so on. But, except in the case of physicians, this conceptual link did not often produce a transference of responsibilities. For many members of the society, however, the notion that this group would be handling their bodies after death was profoundly meaningful. The anthropologists of the society had convinced their adherents that an autopsy would be truly revealing, and this belief led members to write to the anthropologists with great candor. Also, the project demanded valid personal information, and if one did not have a particularly accomplished brain to offer, one could at least excel in intimacy, telling all in order to merit inclusion. They wrote, it would seem, both because they wanted to confess and because they expected that their suffering would somehow be marked on their brains.

Johann Joyeux, who joined the Society of Mutual Autopsy after visiting its display at the 1889 Exposition Universelle, wrote repeatedly to the society's leaders in the years that followed. His testament made it clear that he had repudiated all aspects of Catholicism. He asked that any part of his body not useful to science should be thrown into a communal grave and hoped that his skeleton might hang in the Collège Rollin.[52] In 1892 he wrote to the society lamenting his beloved wife's alcoholism, saying, "I have shown you my completely naked soul." His long missive gave wrenching details of his wife's transformation from "the sweetest and best of women" to "a violently maddened fury" and from the "most honest person" into "a miserable wretch who doesn't know the difference between mine and yours." Joyeux expressed all of this scientifically, writing that his wife was the "daughter of an alcoholic father who died ruined, almost crazy with chagrin. My poor wife had in her the fatal germ of an ignoble and terrible passion."[53] After Joyeux's death, Manouvrier decided that his skeleton was too decrepit to hang at the Collège Rollin and kept it, instead, in the Laboratory of Anthropology.[54]

Such sorrowful stories as that of Joyeux, explained to the anthropologists with confessional ardor, were not rare. In another example, on April 6, 1881, Paul Monnot, a Parisian civil engineer, wrote a brief note to the Society of Mutual Autopsy saying, "I am going to die tonight by asphyxiating myself in the little closet of the antechamber of my apartment."[55] Along with this note, Monnot sent several confidential pages to the anthropologists, one of which told the long and detailed story of a wife who no longer loved him (so much so that when she lost a suit of separation she "was still obstinate and went to live in a convent") and a son whose military career had taken him far away. In the same envelope, Monnot sent a résumé of the story of his life, the range of his aptitudes, and the history of his health—including everything from scarlet fever at ten years of age, to typhoid fever at eighteen, and ending with death, "by his own will," at forty-six. Monnot was successful in his suicide attempt. The society obtained his brain for study, but not before sending a telegram to Monnot's son informing him that his father proscribed any religious ceremony. The telegram also threatened the young man, explaining that according to Monnot's testament, if this aspect of his will were contested, half the estate would be given to the city of Paris.[56]

Although the confessional relationship was generally enacted between the society's leadership and a deeply troubled or dying society member, there were instances when it occurred between the anthropologists and a member of the deceased's family. This was the case with Jeanne Véron, the wife of Eugène Victor Véron. Véron had vehemently requested that his corpse be autopsied, that his remains not be buried, and that there be no ceremony after his death. The best-laid plans, however, are difficult to carry out after one's own demise. Véron died, far from Paris, in May 1889. Because his death had left his wife and daughter "absolutely without resources," the widow Véron did not have the funds required to convey his whole body to Paris.[57] Deeply regretful that she would not be able to carry out the entirety of her departed husband's wishes, the widow Véron arranged, after "many formalities and numerous difficulties," to have a brief autopsy performed locally and for the doctor who performed this service to remove her late husband's brain, put it in alcohol, and send it to the anthropologists in Paris.[58] Months later, she obtained official authorization for the exhumation of her husband's skeleton, which she also shipped to Paris.[59] She then waited for someone, the local doctor or the Paris anthropologists, to tell her something about her husband. But no one had very much to say. The doctor who had performed the cursory autopsy, Dr. Gaudin, offered no information on his own. When the widow Véron requested an autopsy report, he wrote that he would only discuss the results of the autopsy ("description of the tumor, among other things") with another doctor, and he requested that he be put in touch with one of the doctors of the Paris autopsy society.[60] At this point, Véron began writing what was to become a long series of plaintive letters to the "venerable anthropologists," letters that were never satisfactorily answered because, of course, the anthropologists had promised much more than they could deliver. The widow Véron's confidence and interest, however, did not wane significantly. In 1891 she herself joined the Société d'anthropologie de Paris, and though she did not live in Paris, she kept up a lively correspondence with that society's leadership.[61] Véron was one of the first women in the Société d'anthropologie de Paris. The Society of Mutual Autopsy, however, included female members from its beginnings.

Many articles on the society commented on the group's female members. As reported by the *Morning News*, the Society of Mutual Autopsy included "ten or twelve women, a princess, and the wife of a senator who himself is a member."[62] Beyond the generally feminist attitude of the freethinking anthropologists, one significant reason for the presence of women in the society was the participation of the well-known female anthropologist Clémence Royer.[63]

Royer had gained a wide reputation for the very atheistic, Lamarckian, socio-biological essay with which she prefaced her translation of Darwin's *On the Origin of Species*. She had written a great deal on the subject since. She participated actively in the meetings of the Société d'anthropologie de Paris, represented it at international scientific conferences, and published frequently in its journal. In 1880 she and the prominent philosopher Charles Renouvier were named the French representatives to the Fédération Internationale de la Libre Pensée founded in Brussels (Charles Bradlaugh and Herbert Spencer were the representatives from England).[64] Her work was path-breaking and inspirational, as is clear from the widow Véron's letter of condolence at Royer's death:

> I did not have the honor to know Mme. Clémence Royer personally. But one does not need to be in direct contact with someone to appreciate their moral value and their high intellectual faculties. But beyond that, as a woman, I believe that Mme. Clémence Royer has rendered an invaluable service to our sex. She demonstrated with a remarkable talent, conscience, and high-mindedness that, contrary to common prejudice, woman can—by a rational culture and through the influence of a clear-minded milieu, along with native predisposition—usefully approach the collection of questions that had been envisioned, up to then, as the exclusive domain of men.[65]

Given the ardor of this sentiment, it would seem likely that Royer's membership in the Société of Mutual Autopsy diminished the social barriers that otherwise discouraged women from joining such enterprises.

Some women were courted to join the society. Jacques Bertillon wrote to the society in 1878 stating that the princess de la Tour d'Auvergne had expressed the opinion that the autopsy of all French men and women ought to be enforced by law. What her reasons for this were Bertillon did not say, but he did counsel the society against stressing materialism, as the princess "is not devoted to it, nor to monarchism, nor to republicanism."[66] The princess, who was reported to be fatally ill, joined soon after. Such courtships were rare, however. Most women were surprised and delighted to be admitted.

It certainly cannot be said that women joined the society along with their husbands, as a kind of default way to spend eternity together—on a shelf if not in heaven—for married memberships were extremely rare. Often, however, the women involved were the widowed wives of men who had been involved in anthropology. Their husbands had not become members—perhaps they were not interested in doing so, or perhaps they had died before the society was founded. In either case, it seems likely that the husbands' interest in anthropology had brought the appropriate journals into the wives' homes, and, widowed, the wives continued to read these journals. In any case, women seem to have joined the society "in the interests of science" and with the same basic notions as the male members. Some men followed their materialist wives to the civil grave. Ten years before the Society of Mutual Autopsy was founded, Zoé Bertillon, wife of Louis-Adolphe, mother of Jacques, caught a fever she could not shake and died at thirty-four. She had been "a great reader of sociology" and was "actively concerned with a group of young Parisian women who interested themselves in education, for the sake of both knowledge and citizenship. They were regarded as progressive and even dangerously so."[67] When she died, her husband followed her instructions and saw to it that she had a civil burial. These were prominent citizens, and when the famously proactive (though moderate among his peers), bishop of Orléans, Dupanloup, heard of the burial arrangements, he "wrote an Episcopal charge concerning the scandal and social dangers of burial without the benefit of the clergy" and sent it to Zoé's bereaved husband. Louis-Adolphe responded as follows: "I do not wish to give offense, much less to scandalize anyone. But my wife was a liberal and had ceased to believe in the Catholic faith. The scandal would have been if I had allowed her to be buried with religious ceremonies she did not believe in" (42). Zoé Bertillon's granddaughter, Jacques's daughter, Suzanne, would later write that Zoé came to her marriage young, having received a solid religious instruction. As a result, she was a practicing Catholic, but her family was progressive and republican and after she was married, Zoé "was quickly won over by the liberal and philosophical theories of her husband and henceforth stopped practicing her religion." Indeed, "for the period, she professed very advanced ideas."[68] In another discussion, Suzanne Bertillon wrote of Zoé Bertillon's feminism and noted that it was "judged audacious and provocative" but that "Louis-Adolphe shared all the opinions of his wife and often supported them with the weight of his erudition and his statistics" (33). Her parents and siblings supported her being given a civil ceremony; the radical opinions between Zoé and Louis-Adolphe surely passed in both directions (36).

The widows who joined the autopsy society seem to have had an interest in anthropology while their husbands were still alive, but most of them joined no societies until they were widowed. This does not necessarily mean that their husbands had impeded their prior membership, though that is certainly one explanation. More likely, they saw no need to join on their own until their husbands' deaths broke their contact with a society's proceedings and publications. That was certainly the case for Jeanne Véron, who did not join the Société d'anthropologie until two years after her husband's death. The widow

Véron finally joined the Society of Mutual Autopsy in 1906, when a distressed budget forced her to quit the rather expensive Société d'anthropologie. Her biographical essay and her testament both displayed extraordinary commitment to the cause.[69] Véron explained that although her mother had received a decent education, her father was barely literate and did not believe education was necessary for girls. Still, at what she described as a "very young age," she demonstrated "a taste for the natural sciences," and in adolescence she occupied herself by writing impassioned speeches "in favor of the victims of monarchical oppression." After her marriage, at twenty-two, she learned some Latin and some Greek from her husband and later wrote articles and book reviews for the *Musée universel* and other journals. In the early eighties she wrote a large number of zoological books for young people as well as some general children's fiction, all of which the anthropologists put out with one of their publishing ventures: the Librairie d'éducation laïque (Library for laic education).[70] "Social, political, and religious questions," she explained, interested her as passionately as "the discoveries of chemists, biologists, ethnographers, and anthropologists." Like several other members, she detailed her abilities with some embarrassment, stating that she had been complimented on her writing many times but that she mentioned this "only as information for your researches . . . and because you will not read these notes until I have ceased to live."[71] Instead of copying the model testament that the society provided, Véron wrote her own:

> Desiring to contribute to the diffusion of scientific ideas which I have professed for the last thirty-five years, I leave my body, and especially my brain, to the Société d'anthropologie de Paris. I insist, thus, that my autopsy will immediately follow my death, if I die either in Paris or in the Department of the Seine, and that the cost of this will be paid by my inheritors.
>
> If my death takes place in another area, then I charge my daughter-in-law, Eugènie Véron, to accomplish my will by arranging for a doctor (preferably the one that assisted me during my final illness) to extract my brain, place it in alcohol, and send it to the rue de l'Ecole de médecine—the location of the society. Later on, an exhumation can be effectuated in order to remove my skeleton.
>
> This posthumous gift is made so that one or several members of the society can make a comparative examination of my brain and my bones, in order to make a progressive and exact (if possible) determination of the location of the mental faculties and their relationship to the anatomical constitution of the human being.[72]

For the widow Jeanne Véron, as for so many others, the Society of Mutual Autopsy performed a wide range of functions. It offered a political platform, an opportunity to write her own eulogy, a forum for the demonstration of her faith in science, a means for her to preserve herself, physically, for posterity, and a chance to contribute to the utopian scientific project. Perhaps most important, through the society Jeanne Véron found a functional replacement for the confession, viaticum, and extreme unction of Catholic last rites. Echoing the detail of liturgical speech, she was able to list what would happen at her death with precision and solemnity.

THE AUTOPSY REPORTS

It seems that autopsies were performed on the cadavers of all members in good standing. For the most part, however, the anthropologists only wrote up and published reports of the autopsies that had been effectuated on one of their own. This was another of the crucial lines between the priestly anthropologist and the lay members of the group. Lay members had access to the anthropologists as experts, leaders, and as somewhat anonymous confessors. Here, the anthropologists served and did not receive. Their role was sacerdotal, regulating the details, issuing text, creating doctrine and liturgy, and, of course, preserving relics. Their rewards were sacred if not beatific: when their autopsies and eulogies were published, they each entered into the canonical text of their cult.

The highly symbolic nature of the autopsy and attendant published report becomes clear in the progression of the reports from the bold interpretive pronouncements of the early years to a later style of noninvestigative description. Consider Louis Asseline's autopsy. In the original report, published in the *Bulletin de la Société d'anthropologie de Paris* in 1878 and written by Henri Thulié, there was a great deal of biographical detail stressing Asseline's leftist politics.[73] The report stated that for a man of his age Asseline possessed a particularly heavy brain. Thulié explained, however, that "it [was] not a delicate brain; the circumvolutions are thick, almost fat." For Thulié, this was "a remarkable thing, because what most characterizes the intelligence of Asseline was an exquisite delicateness, to the point of subtlety" (164). The remark demonstrates that some of the society's members had an almost unbelievably simplistic understanding of the kind of connections to be found between mentality and brain morphology, but it also demonstrates a commitment to derive some inductive conclusions from these procedures. Several years later, in 1883, the doctor

Mathias Duval opened another discussion of Asseline's brain after having presented a paper to the Société d'anthropologie describing his further studies.[74] Duval was a member of the society, taught at several Parisian schools, and two years later would be appointed to the extremely prestigious chair of histology at the Paris Faculty of Medicine. The 1883 paper did not insist on a direct linkage between morphological features of the brain and character features of Asseline's mind. Duval did, however, mention that Asseline's brain had certain simian features. This drew an attack from one of the members of the Société d'anthropologie, a M. Foley, who protested that, after hearing Duval's essay, "there is no need for any other argument against such a society and its immorality" (273). Foley went on to say that he could not condone a society that made "these kinds of discoveries." The anthropologist and medical doctor Eugène Dally defended Duval's analogy, saying that further recognition of the kinship of animals and humans might stop the exceedingly prevalent cruelty to animals. Another interesting reply came from Paul Topinard, an anthropologist who considered himself Broca's primary disciple and who had expected to take over the helm of the Société d'anthropologie when Broca died. Topinard was a generalist; his best-known work was simply called *Anthropology*, and though he is better known in America today than the freethinking anthropologists, that is largely because he lived in New York for periods of his life, both as a child and later. He thus had social and professional ties here and a facility with English uncommon among his French peers. In France in the 1880s he was a bit out of favor—partially because he disapproved of the increasingly political climate of the society. "We seek the truth and nothing else," he insisted. "It does not much matter to us if we grow more like animals, or less like them. The only way to become less like animals, however, is to see clearly and not to nourish illusions and preconceived ideas. The Société d'anthropologie sides with no sect, neither in one sense nor in another" (274). In any case, the discussion did not much pertain to the society's stated intentions. Later still, in 1889, Asseline's brain was featured at the society's exhibit at the Paris World's Fair.[75] The literature at the fair made absolutely no claims regarding the relationship between Asseline's personality and the morphology of his brain. Indeed, it conceded that very few such discoveries had been made at all (104).

Somewhat stymied by the task they had set for themselves, the anthropologists soon began merely to describe the brains. Those writing the autopsy report on Coudereau's brain (the brain that had devised the entire scheme) barely mentioned his character.[76] Certainly, much scientific work is but description and data. In this case, however, the descriptive texts were functioning as fully realized monuments, a final contribution to the journal. The endeavor had become more ritual than investigation; more a sectarian memorial service than a science. This pattern was only broken in 1886, when the society was given the brain of the great French statesman Léon Gambetta. In this case, a much wider public was interested in the society's findings, and so some findings had to be offered. Also, Gambetta's brain posed a troublesome conundrum that set the anthropologists to rethinking one of the most fundamental tenets of their discipline.

We do not know exactly how Gambetta came to donate his brain to the Society of Mutual Autopsy, but it is not too odd that he did. Gambetta was essentially the head of political anticlericalism in France in these years. His famous proclamation, "Clericalism, that's the enemy!" of 1877 represented the culmination of decades of republican frustration, and it launched decades of active struggle against the church.[77] Gambetta's speeches hailed France as the country that "inaugurated free thought in the world," and he cursed clericalism as "the enemy of all independence, all light, and all stability," because it was the enemy of all the healthiest and most beneficent aspects of modern society (180, 177). He spoke in political terms—of progress, social justice, and effective education—but he also spoke of truth. Since the rise of Napoleon, announced Gambetta, there had been a struggle between two groups: "those who pretend to know everything through revelation, in an immutable manner, and those who march, thinking and progressing, to the suggestions of science, which every day accomplishes progress and which pushes back the boundaries of human knowledge" (179). It was, in his words, "a civil war." Did he see the new, secular world in terms of anthropology? In a private letter, Gambetta once wrote, "Could it be for peoples, as for animal races, the struggle for existence and authority periodically brings the disappearance of the weakest, most ignorant, most heedless, by the aggression of the most strong, most learned, most wise? Could it be that politics is only a branch of human physiology? Perhaps."[78] He must have strongly indicated to someone, at some point, that when he came to die he wanted to continue to serve the anticlerical campaign.

The anthropologists extracted Gambetta's brain at Ville-d'Avray, where he had died, and carried it by train to Paris, holding it aloft "carefully suspended from the hand," and then took it by car from the Saint-Lazare railroad station to the Laboratory of Anthropology.[79] Their first report on the autopsied cerebellum was considerably longer than any previous report on a brain, encompassing some twenty-three pages and covering every aspect of the organ— except its weight.[80] This lacuna was remarked upon during the discussion following the presentation, but the report's author, Dr. Duval, refused to comment, dramatically withholding the information for his special report on the subject: "The Weight of Gambetta's Brain."[81] It was here that Duval let it be known that the great man's brain weighed in at only 1,160 grams—light enough to convince the scientists who had weighed it, Jean-Baptiste Vincent Laborde and Paul Bert, that an error had been made.

It should here be mentioned that Laborde and Bert were extremely important, well-known figures of the Third Republic: Laborde headed up the department of physiology at the Faculty of Medicine and was editor in chief of *La tribune médicale*; Bert was professor of physiology at the Sorbonne, edited the science section of the left-wing newspaper *La république française*, and served jointly as minister of public instruction and minister of religion in Gambetta's government. In his political capacity, Bert made rabble-rousing speeches against the church and in particular against Catholic education in France, which, he said, "understood nothing of natural laws" and saw the world as dominated by the "caprice of supernatural powers," while "laic education encourages a man to work and gives him confidence in his own abilities, in personal progress, social progress, [and] humanitarian progress."[82] Religion, he asserted, was a mess of "grotesque superstitions" that denied that people had an inner ability to make moral decisions, saw marriage as an inferior state, and had no place in the country of Molière, Rabelais, and Voltaire. Both Laborde and Bert were members of the Société d'anthropologie. Though neither had officially joined the autopsy society, they clearly wanted to be part of the project and were willing to lend their names to it. Especially in Bert's case, because he worked so closely with Gambetta, it is notable that they were willing to hold, weigh, feel, poke, and dissect the brain of their recently deceased friend and colleague.

According to Duval, when Laborde and Bert found such a low weight for Gambetta's brain they assumed they had erred. A second try gave the even more lamentable result of 1,150 grams. Duval went on to offer a plethora of reasons as to why this exemplary man's brain could have been so light. Most of them stressed that, had the brain been handled in some different way, it would have retained more of its water. There followed a number of complicated calculations, at the end of which Gambetta's brain weight was altered to a more respectable heft. In the end, Duval listed a number of other great men with light brains and threw the subject open for discussion.

Léonce Manouvrier, who elsewhere waged an energetic and often successful campaign against biological determinism, here cautioned that brain weight did determine some aspects of intelligence but not all of them.[83] Manouvrier suggested that the talents with which Gambetta had been most gifted were perhaps unrelated to brain weight.[84] This implied, he realized, that Gambetta had some deficiencies, but, "after all, one can be a lawyer, orator, governmental minister, patriot, skillful man, etc., without being perfection itself" (409). Nevertheless, most newspaper reports maintained that Gambetta's brain was considerably heavier than indicated by the first published numbers.[85] The confusion did not stop certain anthropologists from writing a great deal about the subject. For example, in 1898 Dr. Laborde published a full-length book on Gambetta's brain, dramatically claiming that Gambetta's donation had rendered a tremendous service to science. Laborde wrote that Gambetta's brain gave scientists the final proof they needed about the relationship between the brain organ and function, for Gambetta, "one of the greatest orators, if not the very greatest," also had a "third left frontal circumvolution" (Broca's speech area) that was "the most developed and most complete that had ever been witnessed."[86] Laborde may have exaggerated a bit to make this more dramatic, but he did not claim to have found any new relationships between brain morphology and function.

By the time they were done with Laborde's own brain in 1903, most of the anthropologists no longer believed there existed any relationship between brain weight and intelligence, a shift that seems to have been caused in great part by what they referred to as the "embarrassing case" of Gambetta's light brain, coupled now with the "embarrassment" of Laborde's case—for his brain was light, too.[87] Here, again, they fell back on the third left frontal circumvolution, stating that, "Dr. Laborde possessed a faculty of elocution that, without reaching the eloquence of Gambetta, was nonetheless one of the dominant characteristics of his personality. . . . This very interesting result gives proof of the invaluable service the autopsy society can offer to Science" (423).

In truth, nothing substantive had been added to Broca's original findings, but that did not seem to bother anyone very much. The fact that they would joyfully confirm the significance of Broca's aphasia and the "third left frontal circumvolution" for thirty years would be staggering, in fact, if their project had been limited to discovering links between the mind and the brain. Instead, it would seem that the "invaluable service" they had provided was in the realm of memorial, revelation, judgment, and immortality. At the funeral of Louis-Adolphe Bertillon, whose brain had been removed for study and whose whole skeleton would soon hang in the society museum, Charles Letourneau gave an oration on behalf of the Society of Mutual Autopsy.[88] Beginning his speech with a discussion of the society, Letourneau said that sometimes, in order to serve progress, "one must go against our customs and, to a certain degree, our laws" (188). "For Bertillon," continued Letourneau, "to brave those who are prejudiced for the sake of a superior interest was a commonplace, and to devote himself to science and to social progress was a necessity." After a lengthy discussion of Bertillon's work, Letourneau returned to the subject of the society, explaining that Bertillon had been ill for some time, had known very well that he was dying, and had never wavered in his resolve:

One night . . . in one of those moments of physical anguish wherein most men almost cease to think, Bertillon found the strength to renew, by letter, his adhesion to the Society of Mutual Autopsy. He did this in several touching lines that began as follows: "To be useful has always seemed to me to be the most beautiful goal of life," and finished with these words, "It is one o'clock in the morning, and I believe I am on the point of death." . . . Along with his friends, who I have the honor to represent, Bertillon did not entertain the least illusion on the subject of the hereafter. For him, as for them, the only possible survival was that which resulted from acts and from works. Like us, he knew that the only means of not dying entirely was to disperse to the four winds all that one could of the fire of one's heart and the light of one's mind.

(191–192)

The speech was published in the same journal that would later carry Bertillon's autopsy report, and its importance lies in the strikingly self-conscious way in which it refers to the search for a "means of not dying entirely." Equally significant, however, is the liturgical structure of Bertillon's final attestation. Letourneau praised Bertillon for his steadfast commitment to materialism even in the face of physical anguish. This was not just any physical anguish—it was the kind in which most people "almost cease to think," the kind of pain that makes rational creatures abandon logic and fall into a sweet, emotional piety. Why would Bertillon gather his waning strength in order to renew his membership—and why would Letourneau make so much of it? The dying man had already made out his will, kept up his dues, and furthermore (in contrast to many members who knew their relatives would fight a sacrilegious will) Bertillon's own son was also a member of the society. The act can only be understood as a purely symbolic gesture of commitment, based squarely on the Christian deathbed scene and the public description of a "good death" that followed such a scene. The final pronouncement of faith was the central feature of the Christian death ritual, serving as a crucial sign of peace and certainty for the dying and for those about to mourn. Bertillon and Letourneau surely witnessed these rites many times throughout their lives, and, despite nominal iconoclasm, their solace depended on the tradition of ritual. It should also be noted that for a member whose family was religious, failure to issue a final materialist attestation might mean a religious burial. As discussed above, many members placed threatening clauses in their wills in order to ensure that their wishes would be respected after death. The deathbed scene carried with it a liturgical context, re-created because it offered comfort but also because of the solemnity and inviolable covenant implicit in liturgical speech.[89]

There was also a concern that one might betray one's own convictions at the fearsome moment of death. Tales of a good materialist death helped to make the appropriate behavior into a second nature—just as religious authors recommended the frequent reading of illustrative descriptions of a good Catholic death. Indeed, members of the Society of Mutual Autopsy present a fascinating inversion of the classic distress. Traditionally, it is a weak believer who agonizes over sins committed and desperately hopes that a life of faithlessness will be redeemed in a pious deathbed scene. Life's final moments were crucial. According to Catholic doctrine, sinners and saints could both change their status on the basis of a good or a bad death. John McManners's study of death in Enlightenment France highlights the importance accorded to one's frame of mind at the moment of death—and the intense anxiety that this produced among the faithful.[90] However, in the late nineteenth century, many atheists were haunted by the notion that they might convert on their deathbeds, thus invalidating a lifetime of materialist conviction. As Thomas Kselman has pointed out, priests in this period were feared and reviled for the pressure they brought to bear on the unrepentant dying—and for the taunting public spectacle they made of any well-known atheist's deathbed conversion.[91] One thus finds men and women desperately hoping that they would have the presence of mind to die a bad death (in the eyes of the church). This late-nineteenth-century anxiety was based on how one would be remembered rather than on fear of damnation, but it was no less intense for that. Whether faithful to Catholicism or materialism, people hoped for sobriety at the capricious hour of death.

Letourneau's speech also gives insight into the cathectic purpose of these reports and eulogies. Letourneau expansively honored the whole group, including himself, by showcasing Bertillon's purity of commitment: he "did not entertain the least illusion." This was an inspirational tale for the reader of the journal who was not likely to find such praise elsewhere, but, especially for the anthropologists, there was also a cautionary subtext here. Continued dedication to the group's doctrine was supported by communal pressure, which each member propagated and received. This pressure was surely private in many of its guises, but for the anthropologists it was exacerbated by the promise (and fear) of publication and the sense of permanence and remembrance that publication represented in an otherwise ephemeral world.

SCIENCE, CHURCH, AND STATE

The Society of Mutual Autopsy's struggle to obtain state authorization nicely demonstrates the way in which the government mediated, rather haphazardly, between religious and antireligious forces in France. The application for official authorization first went to the prefect of police of Paris in 1880. Feeling that the subject matter in question lay "outside the limits of [his] competence," the prefect wrote to Jules Ferry, then the minister of public instruction, asking him to make the decision.[92] For his part, the prefect of police added, he thought it would be all right to authorize this odd association. The group was, the prefect commented, "occupied at this moment with the study of Broca's brain." He further added that, "though [the society] is composed in large part of men dedicated to politics, it seems to me that such questions are not raised in the meetings of the society, of which the object is purely scientific." Despite his republican, anticlerical politics (he would soon radically secularize the French educational system), Ferry was understandably a bit hesitant about granting a group of politicians the authorization to dissect one another. He decided to get some insider opinions. In September 1880, Ferry wrote to Jules Gavarret, inspector general of medical instruction, asking his opinion.

Jules Gavarret had become well known after his youthful rise to the post of professor of medical physics at the Paris Faculty of Medicine in 1843. His work in both medicine and physics had earned him membership in the Academy of Medicine in 1858, and he would become its president in 1882.[93] He was a member of the Société d'anthropologie as well, had served a short term as its president, and for a while directed the *Revue d'anthropologie*. It was in light of his high standing in the medical and scientific community and in particular in his capacity as inspector general of medical instruction that Gavarret was asked to judge the Society of Mutual Autopsy.[94] Gavarret's reply was a harsh critique of the society and a recommendation that it not be awarded any kind of official recognition.[95] He reported that authorized representatives of the biological sciences had refused to join the society, and he provided a breakdown by profession of the original membership: only eleven were medical doctors, five were men of letters, there were two municipal councillors, one civil engineer, one employee of public assistance, and one archaeologist. Whether these men considered themselves anthropologists above all else, Gavarret did not mention, but it is hard to fault him for recognizing the political tendencies of the group. He railed: "Among these founding members there exists, in reality, nothing other than a *community of very advanced political opinions*" (emphasis in original). He also reported that the general membership (which he estimated as numbering over a hundred), had been "recruited from among artists, men of the world, and even women" (2). Gavarret warned that the society was bound to upset people, stating that the "privileges" conferred on its members in the statutes were of a nature that would "hurt feelings and trouble the peace of families." He also claimed that the nine members of the Society of Mutual Autopsy who were also members of the Société d'anthropologie had "often created serious difficulties in this association of men dedicated to the cult of science." Furthermore, he declared, their goals were purely political and "not at all to work for the advancement of science" (3). In fact, he concluded his evaluation with the assertion that the Society of Mutual Autopsy should not be given authorization because to do so would somehow jeopardize the venerable Société d'anthropologie.

It should be noted that Gavarret was by no means an arch conservative. As we learn from Joy Harvey's intellectual biography of Clemence Royer, in the late 1860s Gavarret invited Royer to give a talk on evolution to his class at the Paris School of Medicine. Madeleine Brès, the first French woman to be admitted to the school and to become a doctor, was in the audience and remembered Royer's impact vividly—both for the ideas of Darwinism she presented and for the fact that her classmates applauded her lecture, showing that they were able to "render homage to all that is good, beautiful, and true, even when it concerns a woman."[96] Gavarret had also been one of the four men who sponsored Royer's membership into the Société d'anthropologie de Paris. So Gavarret's concern seems to have been localized to this group of freethinkers rather than prompted by its general ideals. He must have changed his mind, because, in the years that followed, Gavarret came to work quite closely with the freethinking anthropologists on a number of projects, including the Society of Mutual Autopsy, though there is no testament in his name.

Ferry then wrote to Armand de Quatrefages, professor of anthropology at the Muséum d'histoire naturelle de Paris, asking about possible friction between the new Society of Mutual Autopsy and the long-standing Société d'anthropologie, and also asking if there was reason to doubt "the scientific authority and perhaps the morality" of the group involved (2). Quatrefages was arguably the most established and conservative anthropologist in France, holding what was the oldest and most esteemed position in anthropology, namely, the chair at the museum. He published extensively on race, evolution, and anthropology in general, his best-known works being *L'unité de l'espèce humaine* and *Rapport sur le progrès de l'anthropologie*.[97] He had played conservative foil to Broca's avant-garde notions of evolution and polygenism and was quite well known internationally. Explaining that he had heard a good deal about the society and felt qualified to give an opinion on it, he also made it quite clear that the Society of Mutual Autopsy "could be understood from many different points of view" and that he would only concern himself with "scientific considerations." As far as science went, Quatrefages thought there would be "incontestable" scientific merit in the dissection of the brains of "intelligent and cultivated" men, "gifted with diverse aptitudes" that had been for years appreciated by those who would do the dissecting. Quatrefages assured the minister that such work "could lead to positive results of a very great scientific importance." That is, he entirely agreed with the scientific premise of the society that knowledge about a personality could help one to locate that personality's characteristics in the physical structure of the brain and that those characteristics would be more recognizable or interesting if the subject had been educated, talented, or accomplished. Quatrefages assured the minister that such work "could lead to positive results of a very great scientific importance," warning only that the new society should not be annexed to the Société d'anthropologie in any way, lest the "great and legitimate authority" of the "older sister" association be jeopardized by the new group, which had not yet proven itself. Indeed, he rejected any solidarity between the two groups "with all his strength" and insisted that they be "absolutely distinct."[98] Quatrefages understood that some tension was building in the Société d'anthropologie as a result of the freethinking anthropologists' increasingly outspoken politics. It is important that despite his awareness of the situation, he still gave his support to the freethinking anthropologists' autopsy project. Ferry took Quatrefages's caution against linking the two societies to heart. But he also noted Quatrefages's appreciation of the scientific merit of the adventure along with Gavarret's complete dismissal of it.

Confused, and perhaps overburdened (while remaining minister of public instruction, he had ascended to the position of prime minister of France only four days after writing to Gavarret), Ferry sent the whole file to the minister of the interior. Repeating, in his cover letter, Gavarret's description of the society as a *"community of very advanced politics,"* Ferry seemed most concerned with protecting the older Société d'anthropologie, which he described as a glorious institution, "an entirely French creation" and "unique in the world."[99] Ferry had good reasons to believe this. A few months earlier, the Ministry of Foreign Affairs had written to Ferry recounting for him the events at a recent anthropological conference in Moscow. The letter included several published reports on the conference, one of which, appearing in the *Télégraphe*, reads as follows:

> While most States lack even one chair of Anthropology, Paris possesses a veritable Faculty of Anthropology, composed not of *one* chair, but of a great number of diverse courses that are all about anthropology. This is the result of the work of the creator of the Ecole d'anthropologie, the illustrious Doctor Broca and his colleagues Bertillon, de Mortillet, Topinard, Hovelacque, Dally, Bordier, etc. One would have to have seen the esteem in which the Russian scholars hold French Anthropological instruction in order to really realize the value of what we have in Paris.[100]

So it is not surprising to find Ferry both interested and concerned as he described the situation to his colleague. "We must be proud of this entirely French creation," he told his minister of the interior, "and protect it against all attacks"[101] Interestingly, this did not convince the minister of the interior. Ferry had become the prime minister of the Republic on September 23, 1880, because the previous prime minister, Freycinet, had resigned along with two of his ministers. The actions of his minister of the interior, Jean Constans, had prompted the resignation. It had been Freycinet's policy to negotiate amicably with the clergy and to enforce anticlerical laws with some moderation. Constans, however, was an intransigent anticleric, who would later be responsible for making seminarists serve a year in the military, just like other students and anyone who wanted to teach. Here, as Gambetta had encouraged him to do, he had issued a decree stating that all unauthorized religious orders now had to ask for state authorization and further requiring the dissolution and expulsion of the Society of Jesus.[102] As a result, the government changed hands—Freycinet resigned and Ferry took over—but Constans remained minister of the interior. So it was to Constans that Ferry had sent the file, an extreme anticleric who did not look unkindly on the Society of Mutual Autopsy. Thus, despite the threats posed by the new society—as a rival to the Société d'anthropologie, which was lauded as a source of national pride, as very likely to disrupt family peace, as possibly being incompetent in practice, and, most of all, as being an extremist political group—it was authorized on December 29, 1880.[103]

The difficult time Ferry had deciding on the character of this group could only be redoubled by considering their own description of the project:

> They concede to us that physics and chemistry render perfect explanations of the functions of all the organs. Meanwhile, they say, it is not the same for the brain, which the soul commands. Religion and philosophy prohibit us from studying it as the location of thought. Ah well! Unbelievable! The majority obeys this prohibition. It took the audacity of the anthropologists to attempt several timid experiments on the physiology of the brain hemispheres. How much more audacious were the sensualist and materialist philosophers of the eighteenth century! If they had had our knowledge about the nervous system, the question of the soul would have been resolved a long time ago.[104]

At the beginning of this statement, the anthropologists seem devoted to science, while priests and philosophers seem overly defensive. By the end, it is clear that the anthropologists have an impassioned agenda of their own: to prove the nonexistence of the soul.

The French government again confronted the society when civil burial laws came under discussion in the Chamber of Deputies.[105] In March 1886, Yves Guyot and Gabriel de Mortillet spoke before the chamber, of which they were both members, proposing an amendment to the law on the liberty of funerals.[106]Their amendment was a mandate for the free disposition of the body. The Catholic Church noted the threat and issued a counterattack. In 1886 cremation was specifically and completely prohibited by the church. The free-thinking anthropologists republished portions of this prohibition in their most political journal, *L'homme*, concentrating attention on the prohibition against membership in any "society of which the goal is to spread the practice of the incineration of cadavers."[107]The war was afoot. Guyot and Mortillet were not successful in adjoining their amendment, but the law itself was passed in 1887, despite its being characterized by Archbishop Freppel, a deputy of the chamber, as representing "a materialist paganism that no longer recognizes in the human body the abode of an immortal soul."[108] In 1889 another law was passed, extending secular burial rights so that they allowed for cremation. Guyot soon changed the instructions in his testament from a request to be buried in a common grave to a request for cremation.[109] One of the unique insights betrayed by the Society of Mutual Autopsy is that the desire for a civil burial was not merely a last swipe at the Catholic clergy. Rather, French men and women had internalized the church to such a degree that they could not merely ignore its injunctions. The Catholic Church claimed that bodies were sacred and so, in a furious rejection that nonetheless took quite seriously the equation between church and body, these ex-Christians begged for the right to have their own bodies mutilated, discarded, and burned. Infantilized by religion and enfranchised by science, they hacked at the church within.

A FINAL GASP

For a brief period, the autopsy society was directed by an anthropologist from outside the immediate ranks of the freethinkers, namely, Jean-Baptiste Vincent Laborde.[110] Laborde was director of the Laboratory of Anthropology and may well have permitted the performance of autopsies at the lab contingent on his being named president of the society. As Nélia Dias pointed out in her essay on the society, this marked an important shift.[111] Laborde served as director from 1892 to 1903, and during this time the degree of political fervor in the stated goals of the society dropped appreciably, though the membership continued to represent the more extreme anticlerical republicans. On the occasion of his ascendancy to the presidency, Laborde called for a revision of the society's statutes. Some of the more inflammatory passages were excised, such as the identification of various religious ideas as "numerous prejudices that have their source in an unconsidered sentimentality."[112] It was decided that the rather provocative word *mutuelle* would be dropped from the society's name, giving it a significantly more scientific air.[113] In another change from the original statutes, French society was nowhere accused of "disinheriting its poor" or of allowing them to be so profoundly underdeveloped that they were useless as scientific specimens. The justification for the Society of Autopsy was now simply that, without it, almost all brain autopsies were performed on people about whom the scientists knew nothing, which made it impossible to find correlations between brain morphology and the deceased's former aptitudes.

When Laborde died in 1903, the society returned to the hands of its freethinking founders. Thulié was elected president. In an attempt to revitalize the society, three thousand circulars were sent out to the members of the Société d'anthropologie, encouraging them to join the Society of Autopsy. The circulars demonstrate that the general concerns of the autopsy society had overtly returned to those of the early years: radical, atheist, and distinctly political. Designed specifically to attract anthropologists into the fold, the circulars lament that the society had not yet managed to recruit substantially from its colleagues at the Société d'anthropologie. Essentially, the members of the Société d'anthropologie were taken to task for their "absurd prejudices, which inhibit them from leaving their own brains to the very laboratory where they have examined the brains of others. Their acts are in contradiction to the science that they cultivate and give moral support to religious societies that jeopardize science by propagating the idea that autopsies stamp the dead with a degrading mark." In a further reflection of the return to overt antireligious proclamations, the 1903 circular defined the intention of the Society of Mutual Autopsy as being "to contribute to social education . . . by destroying the prejudices and the superstitions that are still very prevalent."[114] The officers of the society discussed the idea of rewriting the statutes again or of just returning to the original wording as conceived in 1876. "The present printed statutes," read one meeting's minutes (there are very few extant), "are not in accord with those that had motivated the authorization of the society."[115] But nothing came of this.

Instead, in 1904 the society printed up personal history questionnaires for its members, in order to standardize their information. In this way, the society's leaders intended to create a collection of "psycho-physiological documents" that would be "unique in the world."[116] The questionnaire included such queries as: "Do you have any special aptitudes (music, drawing, poetry, eloquence, etc.)?; Are you happy or sad, calm or violent?; At what age did you learn to walk?; Can you voluntarily move your scalp, your ears, your scrotum?; Do you have a good memory of smells and tastes?; What is your religion?; Carefully indicate all the maladies that you have had, both during and after childhood: rickets, infantile convulsions, serious fevers, short- or long-term paralysis, nervousness, etc."[117] A copy was published in the *Revue scientifique* in 1905.[118] Completed questionnaires were returned to the society with some regularity until 1914, and a few continued to trickle in until 1930.

The answers to questions on religion generally included a rejection of Catholicism. Felix Blachette wrote, "Catholic by my parents, baptized at seven years old, and a nonbeliever since the age of ten because the chaplain at school divulged my confession."[119] The simple strangeness of losing God because of the minor betrayal of one priest, made stranger by the young age reported, suggests a mythic and oft-repeated story; the intellectual, social, and political issues subsumed in the experience of hurt feelings and having another option. That was in 1905. By 1913 some, like Clara Chérin, simply noted that they were Catholic. Others made it clear that they wanted an autopsy to be sure they would not be buried alive—a strangely common fear of the period. As always, some members were brief with their information, others were exhaustive and confessional. Dr. Regnauth Perrier, for example, aptly titled his personal health history, "a complete museum of pathologies," while Dr. Alfred Guède wrote only that having lived by the "republican formula" he had maintained an imperturbable health.[120] By this late period in the life of the society, however, the fact that so many detailed questions were being posed was much more intriguing than the answers. Despite the total lack of meaningful discoveries over thirty years of operation, the society was still publicly championing the ambitious project that it had set for itself in 1876.

STRANGE ENDINGS

Things rarely turn out as planned. Georges Papillault had proclaimed, "My corpse can be used as is desired—conserve, dissect, hang up my skeleton, place it in a public area, museum, etc. But I refuse any religious ceremony before, during, or after these manipulations."[121] Georges Montandon, who did the autopsy, however, decided that the skeleton of Papillault was "of limited anatomical interest."[122] Clémence Royer was buried intact.[123] Her will explicitly forbade the removal of any part of her body, including head or skull, but it also included an express refusal of a deathbed conversion. The nuns who were looking after her in her final illness had been counseled that Royer wanted to die with her freethinking beliefs undisturbed, and friends stood by to protect and console her. Among her last words, Royer murmured, "No conversion, not Catholic."[124] As requested, she was buried in a wooden box, lined with sand. Vacher de Lapouge survived his battle with typhoid fever in 1897 and managed to outlive the entire society. He died in 1936, pleased with the "selectionist" revolution taking place in Nazi Germany. Oddly enough, he was given a full Catholic burial. The brain of Laborde, who had often impinged on the time and workspace of his nominal underling, the brilliant Léonce Manouvrier, was tastefully insulted by the latter in the scientific press.[125] Whatever may have happened to Jeanne Véron, her bold testament and proud yet modest biographical notes (including her instructions for the disposition of her body) remained sealed in the society's archives until they were opened in my research for this book.

Paul Robin, the freethinking neo-Malthusian who had used his testament as a political platform, revoked his will in 1904. In doing so, he attested that he had joined the society "in the interests of science" and in the "battle against religion." When he learned that freethinking anthropologist Dr. Papillault had been initiated into the Freemasons (June 1903) and had "married the daughter of his colleague, Dr. Hervé, vice-president of the Ligue des droits de l'homme . . . at the Church of the Trinity!"[126] it drove Robin from the ranks. Jacques Bertillon took exception to the description of his father's brain and personal characteristics that was published in the Bulletin de la Société d'anthropologie de Paris. Letourneau had suggested that the late Bertillon's speech difficulties had verged on aphasia. He later retracted the statement, as "requested by the family," but the damage was done. Jacques left the Society of Mutual Autopsy soon after.[127] Perhaps the spectacle of his father's skeleton hanging in the society's museum was a disturbing and determining factor. As for Gambetta, the autopsy on his brain significantly added to the corporeal disfigurement of his death. There was not enough left of him to be buried properly in the Pantheon—the French burial shrine for national heroes. That is why, among the full tombs of illustrious compatriots, a simple monument there reads, "Here lies Gambetta's heart."

Toward the end of its active existence, the Society of Mutual Autopsy set up what the members referred to as a museum but was, in fact, no more than several glass display cabinets within the Musée d'anthropologie maintained by the Société d'anthropologie.[128] In this shrine was a bust of Gillet-Vital and a cast of a death mask of Louis-Adolphe Bertillon. There were brains in jars, including those of Coudereau, Asseline, and Bertillon, whose jars were labeled "intellectual"—these were the saints of the scientific cult. Bertillon's skeleton was there, too, along with a large collection of brain casts and skulls, including those of Gambetta, Bertillon, Coudereau, and Eugène Véron. At this writing, there is a collection of skulls on display in the Muséum d'histoire naturelle de Paris whose only identification is a label that reads "intellectuals." I suspect it is them.

OUR UNDERSTANDING OF ANTICLERICALISM may suffer from our own complacent reaction to atheism. The intention of this study of the Society of Mutual Autopsy has been to demonstrate the emotional distress experienced by late-nineteenth-century men and women who had made the decision to live without belief in God. Surrounded as they were by believers, these early atheists looked on processions, prayers, priests, and rituals and felt bereft. When the Society of Mutual Autopsy proposed an alternate, secular version of these comforts, they embraced it with zeal. The raving virulence of the anticlerics becomes comprehensible when antireligion is understood as an alternative ritualized belief system—especially one powerful enough to steward lives into death. In the beginning, the founders believed that "the intellectual future of humanity depends entirely on our possession of more or less precise information on the cerebral functions and on the localization of the diverse faculties."[129] This conviction may not have been shared completely by all those who joined in the endeavor, but clearly all wanted to believe in something—as opposed to spending their lives ardently not-believing. The something in which they came to believe became, over time, increasingly untenable. The attendant rituals, however, remained useful, and so, for several decades, the society endured.

These anthropologists attacked the very notion of religion and attempted to take over one of its most crucial roles in human life: the rites of death. Anthropology served as a devotional system entrusted to sort out and preserve the "useful" parts of the body and to dispose properly of the rest, the "rotting garbage" that came to represent the Catholic Church. It created textual and material monuments, kept relics, and listened to the confessions of the afflicted. It forced moments of existential contemplation by creating a ritual in which people actually poked around in the brains of men and women who had only lately been their friends, colleagues, and correspondents. In these endeavors and in the society's general evangelical manner, nominally antireligious behavior was based on religious models. Though political figures of the far left were disproportionately represented in the society, membership was not a status symbol: members did not tend to brag of their involvement. Instead, it was a very personal association built on a complex reaction to modernity: fear of meaninglessness and faith in the human ability to create meaning; a passionate belief that rationality, not passion, is the means to all good ends. The group was not concerned with decadence or degeneration but rather with progress. It spoke of neither power nor despair but rather of "being of use," and of "living on." In its vision for the individual members, it was both mundane—offering eternity in the guise of a brief report and a collection of specimens—and wildly exotic, allowing the individual to climb up onto the altar of science, and suggesting that this act might change the world.

∞

CHAPTER TWO

Evangelical Atheism and
the Rise of French Anthropology

The Society of Mutual Autopsy was not the only quasi-religious project enacted by the freethinking anthropologists of Paris. There were many, and each
can be characterized as a translation of traditionally holy objects, events, ideas,
and gestures into a scientific, materialist frame of meaning. The freethinking
anthropologists translated not only funerals from the sacred to the profane but
also human sexuality and reproduction, city buildings, street names, plots of
land, government personnel, ritual feasts, holidays, animal and human remains, and every conceptual aspect of human culture, from aesthetics to marriage laws, from economics to a philosophy of mind. First, a brief survey of the
religious and political world of France at the end of the nineteenth century
will help situate their mission. Then a very quick tour of the history of anthropology in France will culminate with the strange and surprising manner
in which the freethinkers came to anthropology and redesigned it to fit their
needs.

ANTICLERICALISM AND ATHEISM

By the last three decades of the nineteenth century, it seemed to many that
church, dogma, authoritarianism, and social hierarchy were locked in a permanent battle against science, equality, republicanism, and progress. People
felt that the battle was at least as old as the Revolution, but it was more complicated than that. In the revolutions of 1789 and 1848 the Catholic Church
was seen as allied to the monarchy, but some clerics had also joined the rebels

in both these uprisings and, in memorable and dramatic ways, had helped to dismantle ancient privilege. Catholicism has within it the conservative energies of the established church but also the emancipatory and egalitarian energies of the righteous preacher or prophet, and both these Christian responses to authority were in evidence in the great republican revolutions. Still, it is safe to say that the French Revolution began the separation of church and state: in 1792 the Legislative Assembly pointedly set itself up as a "civil" body and that same year authorized civil marriages, well in advance of most European countries.[1] Even if the church had fully allied itself with republicanism early on and stuck with it, a process of state laicization would have been on the agenda, in defense of the principle of religious freedom for other faiths of the republic. The issue, however, might have stopped there.[2] In any case, the church did not consistently ally itself to the republic. In the Restoration, it was offered a position of prestige and control and took it, and again under the Second Empire the church emerged as a very privileged and socially conservative body, disdainful of the pleasures and pursuits of the common people and the bourgeois alike and cursing the revolutions. Bishops and priests railed against modernity, change, progress, and science, preached of hell and damnation, and, perhaps to satisfy their core audience of arch believers, they often refused communion to "former sinners."[3] When anticlerics listed their complaints against the church, it was common enough to find a refusal of this sort in the litany. Maurice Agulhon's *The Republic in the Village* tells of the detailed civil ceremony that was arranged by the town council for anyone of the village to whom the church denied a religious burial: there were floral crowns and busts of Marianne, and one sung "The Marseillaise," genuflecting at the last two lines.[4]

Napoleon III gave the church leave to expand its educational system enormously, and in this and other arenas the clergy used its power to silence and persecute its enemies. Though its place was secure, the church condemned modern society and held up the world of the Middle Ages as an ideal. The clergy responded to ideals of pluralism and secularism by growing defensive and panicky, trying to repress change rather than find a place within the changing world; in these years, clerical power seemed smirking and cruel to secular republicans, but there was also fear in it. At the same time, the republicans' other grievances were many: everything from having been humiliated or beaten by Jesuit teachers, to being taught myth instead of science because one was female, to watching the clergy grow fat in pomp and circumstance while one's family scrimped and saved, to being censored in public speech and in the press, to cursing users of contraception at a moment when the modern trend toward smaller families had clearly begun. These resentments mingled with a political objection to the large and privileged role the church had in government and education, such that many who were not really irreligious became fiercely anticlerical. The church took up the battle and for the most part, rather than seeking to address the religious needs of the nation, came to behave as a scolding political force, casting scorn on its enemies, cultivating ignorance of modern concerns, and demanding a great deal from the remaining faithful.

By the end of Napoleon III's tenure, the sense that democracy and religion were deeply and inherently opposed to one another was well established and violently emotional. Now and again, under the Empire, there were individual attempts to form a leftist movement based on the revolutionary spirit of faith, and sometimes these drew a burst of support, but they never lasted long; there was just too much anger between church and republicans. Then, in the early Third Republic, the Catholic bishops and priests openly allied themselves with the legitimist party, who hoped to overthrow the republic and reinstate the monarchy. Meanwhile these clergymen actively and effectively participated in the parliamentary elections. Republicans were shocked to see the church willing to use all the democratic tools available in the new government—for the purpose of dismantling it. And once again, it seemed that though they had a good deal of control, the clergy's strength did not calm them. Proudhon complained of the "tyranny of the priests" whose "avowed plan is to kill science, to snuff out all liberty and all enlightenment. Their anger increases in proportion to their power."[5] The clergy also tried to get the young republic to risk its life defending the pope against the forces of unification in Italy. This seemed wrong to republicans for a host of reasons, but perhaps the most important consequence was that it made the church seem hawkish in French government, and the association stuck.

The intellectual side of the struggle over religion and irreligion in French society may be the best known. Throughout the nineteenth century, philosophers and literary authors wrestled with the question, and many came down against religion in favor of science and rationalism; Auguste Comte, Hippolyte Taine, and Ernest Renan were among the most significant. Because science was so successful in so many ways in the nineteenth century, it can appear that its rise was indeed a large impetus toward the fervid rejection of religion. Yet serious scholars of anticlericalism have stressed that the attitude was fundamentally political: its proponents wanted clergy out of the government and out of the schools because the church had repeatedly proven that it was against the republic and its ideals. As Owen Chadwick wrote in his classic *The Secularization of the European Mind in the Nineteenth Century*, "This onslaught upon Christianity owed its force . . . not at all to the science of the nineteenth century. Its basis was ethical; its instrument the ethical criticisms of the eighteenth century. It attacked Christian Churches not in the name of knowledge but in the name of

justice and freedom."[6] The present study is one of the histories of this drama—
though it is precisely interested in the relationship between "the onslaught
upon Christianity" and "the science of the nineteenth century." It agrees that
the latter did not cause the former but asks what their relationship was and
argues that in some ways it was science that owed its energies to anticlericalism and atheism. Science was desperately embraced by people who had already
lost their faith in religion. For many French men and women, political and
moral grievances served to vilify the Catholic—and eventually the theistic—
cosmology long before they had access to a persuasive alternative conceptual
universe.

Again and again, throughout the period, there are conversion stories
wherein atheist, republican men and women "find" science and then vehemently adopt its explanations. Theodore Zeldin, too, has written that the gist
of the matter was political, not scientific. "There was . . . certainly more
positive disbelief of religion in France than in either the USA or Britain. Why
should the French, who saw less of the wonders of science than the Americans
or the British, have been more convinced that science had disproved religions?
Perhaps for that very reason, but perhaps also they were not convinced by science as might be supposed."[7] Some were shaken in the specifics of their faith
because of scientific discoveries, but mostly science was the adopted doctrine
of people who were already at odds with Christianity. Zeldin continues, "What
was perhaps peculiar to France was that there was a whole combination of
grievances against the Church, and therefore much more radical argument.
The cause of science was eagerly seized on, as a result, by the enemies of
Christianity." Though stressing political, economic, and social reasons for irreligion, Zeldin also attests to the fact that many of the French men and
women who left the Church in the nineteenth-century were "profoundly interested by religion, tormented by uncertainty, and sometimes heretics more
than unbelievers" (1029). How that interest was manifested, and how that torment was handled, is a more problematic question. The constant and vociferous denial of religion is itself a profound interest in religion, and torment may
not be due to uncertainty but to the consequences of an atheism that is itself
quite certain. Though such extremists often denied interest in religion, they
could not take their minds off the matter and fiercely wrestled with just those
questions that had always fascinated the religious.

The political origin of anticlerical wrath is true for the freethinking anthropologists, too, and for their peers, but it is not the whole truth. Many
people felt that in their youth the church had misled them about the real nature of the world and the real basis of morality, and now they knew better and
were angry. Again, historians of anticlericalism have worked to describe the

political, social, and economic sources of anticlericalism in order to correct the wrong impression one is prone to get at first glance: that powerful intellectual writers and scientists secularized the population and made it militant. But, in asides, these historians have also noted that within this broader world of anticlericalism, some people were very much animated by ideas and by real philosophical problems about existence. To continue with Zeldin's comment that some early freethinkers were profoundly interested in religion: "The attitude of such men differed from that of the simply cynical, or of the wits who poured scorn on religion in what they believed to be the tradition of Voltaire, or of M. Homais, the village pharmacist, who had no doubts at all and derived considerable satisfaction from his position as the local philosopher. The party of Voltaire, when it comes to be analyzed, will be seen to have infinite variations in it" (1029).

M. Homais, the priest-baiting pharmacist of Flaubert's *Madame Bovary*, is famous for having thought he had all the answers, and we can learn something from attending to what he actually said. Homais's most direct attestation of his anticlericalism was this: "I do have a religion, my religion, and I have rather more than that lot with their jiggery-pokery. I'm the one who worships God! I believe in the Supreme Being, in a Creator, whoever he may be, I care not who has put us on this earth to do our duty as citizens and fathers; but I don't need to go into a church and kiss a lot of silver plate, paying out for a bunch of clowns who eat better than we do!" There were many who were both anticlerical and irreligious yet not really atheist; the "party of Voltaire" was witty or cynical about church and dogma but believed in a God of some sort and did not believe that jettisoning the church would require a profound overhaul in how human beings saw themselves or the universe. This book studies the experience of those who felt that a profound overhaul in how we saw ourselves was necessary. A bit more from Flaubert's M. Homais shows that even he only barely fits his own category of dispassionate cynic:

"You honor him just as well in the woods, in a field, or even contemplating the ethereal vault, like the ancients. My God is the God of Socrates, of Franklin, Voltaire and Béranger! I'm one for the creed of the Savoyard Curaet and the immortal principles of '89! I cannot, therefore, abide an old fogey of a God who walks round his garden with a stick in his hand, lodges his friends in the bellies of whales, dies with a loud cry and comes back to life three days later: things absurd in themselves and completely opposed, what is more, to every law of physics; it all shows, incidentally, that the priests have always wallowed in squalid ignorance, doing their utmost to engulf the population along with them."

He paused, looking around him for an audience, for, in his effervescence, the pharmacist had briefly fancied himself at a full meeting of the municipal council. But the landlady had stopped listening to him; her ear was straining after a distant rumbling.[8]

There were others—the freethinking anthropologists, for instance—who cared very much about the question of "who has put us on the earth" and what it was all for and who were interested in replacing religion in a much more vigorous fashion. Yet even M. Homais was emotional about the intellectual problems of religions and truth. It is also of some significance, as I shall explore below, that the village woman to whom he was speaking so casually allowed her attention to wander back to the physical world.

The scientism of the important figures of the highest rungs of the intellectual world flourished a few decades in advance of popular agreement with that viewpoint. Auguste Comte had systematized a philosophical position called "positivism" that celebrated dogmatic faith in science, inevitable progress, and the rejection of all unempirical knowledge. Long before there was a stable republic, Comtian positivism was widely touted as the worldview of republicanism. But eventually Comte tried to fashion a religion of his own, believing that human beings have spiritual needs that must be met; he thought Catholicism did a good job but was annoyed at the mythic aspect. There is also the case of Ernest Renan, who wrote The Future of Science in 1848, claiming that science must be the source of all truth, but did not publish it until 1890, by which point he was not so sure. His readers were ready for the midcentury message; they missed the caution in his late-century introduction to the book. But, with clearly dampened expectations, Renan's 1890 introduction still tells us something about the intellectual facet of irreligion: "Science saves us from errors more than it gives us the truth, but it is already something not to be a dupe."[9] Politics and social progress were both important, but so were science and ideas. Part of irreligion in France was about not being a dupe. Renan also wrote the profoundly influential Life of Jesus, a secular history of the origins of Christianity and the first of its kind. It was written in an accessible, novelistic prose and was widely read and debated. To get a hint of the secular tone here: Renan's translator tells us in a preface to the book, "A young French lady put down the Vie de Jesus with the remark: 'What a pity it does not end with a marriage!'"[10] The book was answered by Dupanloup, bishop of Orleans—the same who scolded Louis-Adolphe Bertillon for the secular nature of Zoé Bertillon's funeral—but Dupanloup's Life of Jesus was a rehash of traditional quotations exhorting Christians to patience and resignation.[11] It had little impact. Still, even here we are not far from the political: Bishop Dupanloup was a senator under Napoleon III and a deputy in the early Third Republic.

Amid these intellectual revsions, there was a drift away from the church by ordinary people, the result of many factors of modernity, not least the rise of capitalism, the new variety in entertainment and leisure, and the increasing mobility of the population. Even before the French Revolution there had been a falling off of church attendance. By the second half of the nineteenth century it was still true that the vast majority of French citizens were baptized, and most seem to have believed in some kind of God and some kind of afterlife, but that was where agreement stopped.[12] In the 1860s through the 1890s there was a general dechristianization; baptism fell off a bit, and civil marriages and burials became more common. There were regions of France where religious practice was piously enacted, but for most of the country the mood was often indifference, sometimes hostility. The Paris Basin was particularly lax—urban workers, in general, tended to be hostile to the church—and there is much evidence that churches in rural France were increasingly ill attended in these years. John McManners has explained that official statistics of the 1870s record 35,000,000 people as Catholics and only 600,000 Protestants, 80,000 freethinkers, and 50,000 Jews; but, "convenient as these figure were for apologists, they had little significance. Many of the 35,000,000 accepted no obligation beyond making their Easter communion, many merely attended mass occasionally, or came to church to be married or were brought there to be buried. . . . The church's influence was exercised over only a minority of citizens in some areas with the good will of the rest, in others (as, for example the Aube, the Corrèze, and part of the Creuse) there was a preponderance of anticlericals and unbelievers."[13] Much of France was never more than nominally converted to Catholicism, subscribing instead to the local flowering of superstition, astrology, and folktales and merely mouthing the words commanded by the priests.[14] Jacques Gadille has shown that some French anticlericalism was generated by Christian believers who did not agree with the church's interpretation of the religion.[15]

At the beginning of the new century several groups had considerable success championing the idea of Christianity as the spirit of equality, rebellious social justice, and generosity—Marc Sangnier's Sillon movement stands out— but these experiments in Christian democracy did not hold up long against the continued sense on the left that religion was anathema to justice, and on the right that religion ought to support established authority. The bourgeoisie were inclined to be indifferent or hostile toward the church in favor of Enlightenment suspicion and the rationalism of modernity. Many were also disappointed with Pope Pius IX's "Syllabus of Errors," which specifically condemned "progress, liberalism, and modern civilization." In fact, active anticlericalism, as opposed to mere lack of interest in the church, was so firmly associated with the middle class that some socialists argued that anticlericalism was being offered as a new opiate for the people; the real enemies were not those in robes and collars but those in furs and jewels.[16] More often, like M. Homais's landlady, workers simply did not see their interests represented in this conversation, and they did not attend to it.

Overall, religiosity was becoming increasingly feminine, because girls were generally given a much more religious and less scientific education than were their brothers. There is ample evidence that women did resent the church's positions on such issues as birth control and abortion, and republican women often expressed disdain for the pretense of priestly authority, but the generalization still holds. Indeed, the common understanding of why, in this cradle of democracy, women did not obtain the vote until the end of World War II, is that republican men and women feared that the great bulk of Frenchwomen were Catholic monarchists and would vote the republic out of existence if given the chance. The relationship can be overstated, for churchgoing declined across the genders. Still, the effect of convent schools for girls and lay schools for boys was that denunciations of French priests almost always mentioned their destruction of family peace; among other things, priests counseled wives to insist that sex be a matter confined to efforts of procreation. In 1845 Jules Michelet wrote of confessors "seducing" wives, "flagellating" them with "spiritual rods," and insisting that they confess details of their private experiences that they would never think of telling their husbands.[17] It is understandable that Michelet might be jealous: the popularity the church had with women in this period has been understood, at least in part, as an effort by women to liberate themselves from the tyranny of their husbands.[18]

It must also be added that throughout the second half of the century, in a kind of answer to the thinning congregations, a significant miracle cult grew up around the Virgin Mary and the saints. This miracle cult was even more pronouncedly a women's movement. The central gesture of the cult was making pilgrimage to miracle-producing shrines (the new railways made a trip to Lourdes, Paray-le-Monial, or La Salette inexpensive enough for most). Here, not only the authority of the husband was ignored, but neither were the authority of the priest nor the masculine image of God given much attention. Between 1871 and 1876, more than 50,000 people visited Lourdes to honor and petition the Virgin Mary.[19] The reactionary stance of the church seems to have sent a great many people running from its rooms and from its clergy—even its most ardent believers and even those most dependent on the church as a social organization. To the skeptic, it seemed as if the newly secular society was bubbling with myriad new forms of superstition. In Owen Chadwick's words, the shrine pilgrims "were warm, cheerful, expectant of miracle, emotional, brash; and they could be offensive to reasonable or un-Catholic passers-by"

(124). Emile Zola visited Lourdes in 1892 and described it as a distasteful spectacle of humanity "hankering after the lie."[20]

As for the "un-Catholic," the substantial Protestant minority in France was made up of a religiously conservative majority, accompanied by an outspoken intellectual elite that saw itself as liberal, modern, and free of religious dogma. This elite had considerable influence on the secularizing campaign of the early Third Republic, particularly for Ferry's secularization of primary education. Ferry himself was raised Catholic by his devout elder sister (she was physically handicapped and was carried to mass every morning), but as John McManners has put it, "the males of the family were Voltairean free-thinkers."[21] He married into the Protestant world, and when he called for a revision of French education, of the ten in his cabinet, there were five Protestants. For these men the aim was to replace religion with devotion to science, progress, and country, but many Catholics and Protestants alike thought that Protestantism might be the answer for France, since it represented religion without such a historically problematic clergy. The smaller Jewish minority tended to be attached to the secular republic in a tradition of sympathy with the Revolution, through which they had been emancipated and granted citizenship. Through all these vicissitudes and variations in religion, class, and gender, the conviction persisted that religion and the political right were of a piece, opposed energetically by secularism and the political left. The history this book engages begins on the far left of that deep antagonism.

There existed in France, in the 1860s and 1870s, a population of materialist atheists who had lost their faith as part of a struggle against the authoritarianism with which that faith had been associated. Positivism was a well-entrenched ideology of republicanism in this period, but by the late 1860s a formidable rival doctrine, materialism, was gaining currency. Its differences from positivism were subtle but important. Positivism, in shunning all that was impossible to prove by empirical science, distanced itself from speculation on anything that was considered unknowable, such as the origins of life or the existence of God. In fact, as long as you kept it to yourself, you could believe in God and still be a good positivist. Materialists took secularism a step further and insisted that anything that smacked of the mystical or metaphysical simply was not true. So for materialists, even more than for positivists, the crucial criterion for so-called scientific work was the absence of supernatural causes. In this way, a lack of more meaningful criteria, such as experimental repeatability, empirical evidence, or even consistent methodology was sometimes ignored, so long as God was not invoked as a causal agent. There were, of course, excellent materialist scientists, but unlike positivists, late-nineteenth-century materialists had a strong motivation to fill in the gaps in human knowledge in order to lock out philosophy and religion—even when little evidence was at hand. Positivists and materialists both embraced evolutionary theory in the decades after Darwin published, but having an answer to the question of creation was particularly significant to materialists. They also tended to be much more hostile to philosophy than were positivists. They did find philosophy somewhat less insidious than religion, because it was not seen as wedded to either authoritarianism or dogmatic principles. Nevertheless, philosophy was attacked as fundamentally erroneous because it was concerned with questions that could not be experimentally adjudicated; philosophy was seen as opening a path for spiritualism. If we think of M. Homais, it is clear that the freethinking anthropologists' vision of themselves was not the dull pharmacist, offering platitudes and palliatives, but the impassioned, investigative scientist, with a scalpel in one hand and a ruler in the other. By contrast, in comparison to any scientists who were not primarily motivated by hatred for the church, the scalpels and the rulers wielded by the freethinking anthropologists were remarkably unproductive. As Comte was the heroic figure behind positivism, the materialists lionized the *philosophe* Denis Diderot, one of the few Enlightenment figures to take the step past deism and present an atheistic picture of the world. The materialists also made great use of Diderot's irreverent, quippish style, mocking the church and reveling in the freedom and pride of unbelief. Note that they did not choose a great scientist, but a great jester.

The freethinking anthropologists defined themselves through a loss of faith but experienced atheism as a tumultuous intellectual and cultural crisis, and they embraced anthropology as a response to it. The key figures of this original group would live to see the secularization measures of the 1880s but not the law of the Separation of Church and State of 1905. Secular, mandatory elementary schools were established in 1882; in 1884 the public prayers that began each parliamentary session were deemed inappropriate and abolished by the republican majority. The laicization of teachers in 1886 came down against any state-school teacher even teaching a single course in a religious school, let alone belonging to a religious order. Anticlerical passion died down for some at this point, but there were still fierce defenders of a more assuredly secular state. In 1904 a law actually banned any member of a religious order from teaching—a move that marked the beginning of sanctioned discrimination against the body that had previously discriminated against all others. In 1905 the final break was made: the republic would no longer recognize, subsidize, or pay ministers' salaries for any religion at all. As René Rémond has written, "Without consulting the Holy See, France unilaterally annulled the treaty laboriously drawn up a century earlier between the papacy and the regime that had emerged from the Revolution."[22] Their students would carry their materialist concerns, in a modified fashion, into the next century. But anticlericalism was essentially over in 1905. By then, artists, writers, philosophers, and other social theorists had begun to reimagine a place for the spirit, mysticism, and ritual in non-Catholic and even secular lives, and the church was beginning to find ways to reconcile itself to modernity and to offer a version of Catholicism that could have a less conflicted appeal.

One more thing needs to be said about the mood of the times in the early Third Republic. In the two decades of the Second Empire, the considerable section of the French populace that was secular and republican harbored a notion of the republic that was mystical: the famous republican *mystique.* Contemplation of this mystical republic served spiritual and emotional needs. When the Second Empire ended, with the Franco-Prussian War, and the Third Republic began, republicans were delighted. The situation, however, was not the stuff of long-awaited republican dreams: the new democracy was in the power of various types of conservatives, monarchists, and Bonapartists who merely believed that a republic would "divide [them] the least." For years, representatives of these forces vied with republicans for dominance in France, and there were moments when a return to monarchy or empire seemed imminent. Thus, though the Third Republic was proclaimed by Léon Gambetta in 1870 and established in 1871, it was not clearly in the hands of republicans for about a decade: the "republic of dukes" gave way with the abdication of its monarchist president Marshal MacMahon in 1879, ceding the terrain to the "republic of republicans."

The two great republican leaders of this period, Léon Gambetta and Jules Ferry, were both secular and scientistic. Gambetta was much more charismatic, much more the bearer of the republican mystique, but when he won power in 1881 he tried to effect change in sweeping gestures and clashed with coalitions on both the left and the right. He was out of power in sixty-seven days, and he never got another chance: within the year his appendix burst, and he was dead at forty-four. His death shocked even his enemies and shook the nation: the funeral procession "wound through Paris like a mourner's sash."[23] For many, this was the end of the heroic age of the republic, but the group of politically like-minded men who had gathered around Gambetta and joined his brief government remained together and helped shape the nation in the coming decades. Gambetta's brain continued to serve his scientistic France, but now as an artifact—material evidence of republican convictions.

Ferry took over in 1883. His was a duller personality, but his prudence helped him stay in power long enough to pass a host of republican legislation. Under his leadership, the French educational system, especially at the ele-

mentary level, was drastically overhauled to reflect the secular values of the republicans: students were going to forget class differences and old prejudices as
they were inculcated with a "lay faith" in science and country.[24] As Ferry described it in the *Revue pédagogique*, laicization was, "the greatest and most serious of social reforms and the most lasting of political reforms. . . . When
the whole of French youth has developed, grown up under this triple aegis of
free, compulsory, secular education, we shall have nothing more to fear from
returns to the past, for we shall have the means of defending ourselves . . .
the spirit of all these new generations, of these countless young reserves of republican democracy, trained in the school of science and reason, who will
block retrograde attitudes with the insurmountable obstacle of free minds and
liberated consciences."[25] Teaching French youth science and reason was thus
explicitly understood as a republican "means of defending ourselves" through
the creation of "countless young reserves" of democracy. René Rémond has
written of this in the religious and military terms it deserves: asserting that the
"great army" of primary school teachers would henceforth oppose the cleric
of even the smallest mountain village, as an "apostle of the new religion, an of-
ficiate of the cult of reason and science," in short, "a militant of the anticlerical
ideology."[26] As French republics had done in the past, the government also
created civic festivals to support the new mood of science and reason, and
these were enthusiastically celebrated in Paris and all over the nation. It had
been illegal to sing the Marseillaise under the empire; the ban was lifted in
1870, and in 1879 the song became the national anthem. Later that year,
Bastille Day was officially recognized as a national holiday.

The victory of the "republic of republicans" seemed to be the triumphant
end of a long, exhausting, and heroic struggle. In a way, this was true. The
Third Republic lasted until the end of the Second World War, and even then it
was replaced by reconceived republics, not monarchies or empires. Yet in the
last two decades of the nineteenth century, just as republicanism became
mainstream in France, it was assailed from both the left and the right. This was
not simply a product of republicanism's new vulnerability as the status quo.
Rather, specific and sometimes viable rivals assaulted the young republic, from
the heroic nationalism of the Boulangist movement of the late 1880s, to the
significant electoral gains of socialists in 1893, to the burst of anarchist violence in the mid-1890s. All these movements essentially agreed with the ideals
professed by the republic, arguing only that the republic was not serving these
ideals. There were also strong forces of antiparliamentarism and antifunc-
tionarism among republicans by the turn of the century. A powerful resent-
ment of the administrative and legislative bureaucracy of the republic existed
among the working classes, the middle classes, and the academic intelligentsia.

Furthermore, the regime seemed unstable and indecisive because of an almost constant shifting of ministries, despite the relative coherence of policy and of government personnel in general. As Jean-Marie Mayeur has written, "Because the regime did, in fact, last, we are liable to underestimate the discontent caused by the 'ministry waltz' and the consequent fears for the future of a regime that seemed, only a few years after its inception, so fragile."[27] Because of the proliferation of functionaries, the republic's image of instability was awkwardly accompanied by an image of the regime as bureaucratic and uninspired.[28] Political theorists struggled with the meaning of running a quixotic democracy wherein elected officials took brief turns overseeing a huge and relatively stable administrative machine.

To make matters worse for secular republicans, in the 1890s some cultural representatives of the Third Republic began making peace with Catholic ex-monarchists. This was partially a result of growing conservatism and the waning fear of the Catholic Church as a support for monarchist forces, and it was reinforced by Pope Leo XIII's ralliement—a call for Catholic support of the Third Republic. In many cases French republicans were turning to an accord with Catholicism in the 1890s. Thus, for those dedicated to a republican mystique of perfect democracy and strict materialist anticlericalism, the period of the 1880s and 1890s was full of triumphs (the republic was consolidated; free, lay primary schools were established), but these were attended by distrust, fear, and even despair.

THE SCIENCE AND POLITICS OF ANTHROPOLOGY
PRIOR TO 1880

The relationship between anthropology and politics has long been recognized in modern scholarship: Nancy Stepan's study The Idea of Race in Science offered an early and important history of scientific inequality, and her seminal article "Race and Gender: The Role of Analogy in Science" helped to set the terms of the discussion of the relationship between scientific racism and scientific sexism.[29] Pierre-André Taguieff and George Mosse have illuminated the relationship between racism and nationalism.[30] George Stocking's many works on the history of anthropology have explored the scientific and political battles of the discipline over the past three centuries, especially in Great Britain and the United States.[31] Works on the history of anthropology in France have endeavored to show the conservative subtexts of many scientific theories and have thus concentrated on the Broca period and the prejudicial assumptions inherent in his work. Steven J. Gould's Mismeasure of Man, Cynthia Eagle

Russett's *Sexual Science*, Robert Nye's *Crime, Madness, and Politics in Modern France*, and Susanna Barrows's *Distorting Mirrors* have all centered on demonstrating the ways in which biological determinism has been used to reinforce cultural hierarchies and group stereotypes.[32] Their concentration on the Broca period, however, has led to two difficulties: a slightly skewed understanding of Broca and a relative lack of knowledge about the fascinating period that followed his death. As I have said, some studies of the post-Broca period do exist: in French, Claude Blanckaert has written many articles on the subject, and Nélia Dias's book on ethnography in the period also includes much about the late-century anthropologists; in English, two dissertations, one by Joy Harvey and one by Elizabeth Williams, are the major reports on the period. And the political nature of anthropology in this period in France was already the subject of inquiry in one of the earliest studies of the subject, Michael Hammond's insightful article "Anthropology as a Weapon of Social Combat in Late-Nineteenth-Century France." Specific investigations of the scientific debates in anthropology (particularly monogenism and polygenism) and their political components are to be found in the works of these five scholars.[33] My own dissertation considered the late-century anthropologists of Paris in terms of their utopianism and their atheist and anticlerical campaigns.[34] Most general histories of the period mention these anthropologists very briefly or not at all, though Philip Nord's recent *The Republican Moment: Struggles for Democracy in Nineteenth-Century France* includes four pages on "the human sciences" discussing Broca, Mortillet, Letourneau, Bertillon, Hovelacque, and Lefèvre, their republican convictions, and their materialist anthropology.[35]

The Broca period is better known partly because of a late-twentieth-century interest in the conservative messages of science. In this context, it is interesting to note that the peculiar truth status accorded to science, and specifically to biological determinism, has been used to justify and naturalize progressivist egalitarian principles as well as conservative ones. Were scientists who held racist and sexist beliefs more likely to forgo the tenets of "good" science than scientists dedicated to egalitarian politics? The relationship has gone largely unquestioned, in part because of the common assumption that the revelation of empirical truth would support present-day notions of equality. It follows from this assumption that there would be no reason for egalitarian scientists to manipulate their theories or data in order to have their results accord with their politics. But this is belied by two notions: First, the human sciences are always a human art based on the ideas of the moment and the ideology of those who articulate them.[36] Second, when egalitarian scientists inherited a scientific theory and methodology with conservative implications, they did not always jettison the entire system but rather chose to twist and turn it so that its results accorded with their convictions. One should thus assume that there have been scientists whose relatively egalitarian political beliefs led them to formulate particular theories or, indeed, to manipulate data in ways as real (and sometimes as flagrantly self-serving) as those performed by politically conservative scientists. Of course, not all egalitarian science was consciously informed by political imperatives. Nevertheless, ideology does inform science on both ends of the political spectrum.

Before founding the Société d'anthropologie de Paris, Paul Broca was a physician with a special interest in neurology.[37] His innovations in brain surgery, such as using trepanation to treat abscesses of the brain, won him considerable prestige at a young age. In 1861 he demonstrated through human postmortems that the left, third frontal circumvolution of the brain controlled speech functions. As I have noted, this was the first clear confirmation of a relationship between specific areas of the brain and specific abilities, and it revolutionized the way many people thought about the brain. The discovery, and the assumptions that had led to it in the first place, gave birth to a whole range of suppositions concerning the relationship between brain morphology and human faculties, characteristics, and intelligence. Broca's work pushed the boundaries of accepted medical practice. His relationship to authority was rebellious in other ways as well. In 1848, while still in medical school, he and his fellow students formed one of the first freethinkers' societies. As Broca reported in an enthusiastic letter to his parents, "about a hundred people joined right away."[38] This society does not seem to have lasted very long, but it does demonstrate the young man's mindset. Furthermore, Broca was, under the authoritarian eyes of the Second Empire, an avid republican. He made explicit claims to scientific objectivity, so his political beliefs were not supposed to have anything to do with his scientific thinking. In an odd way, however, this republican objectivity was self-conflicting, because the powerful connection between scientific positivism and republicanism made Broca's desire to keep science "pure" a republican act. But Broca was also being careful. There had been several false starts for French anthropology. All of them had distinct political leanings, and most had fallen prey to political censorship.

Anthropological societies in France had had their start in the Société des observateurs de l'homme, an anticlerical and politically left-wing association that was born in the French Revolution and died when Napoleon became emperor. The success of specific scientific theories also followed political fortunes: Jean-Baptiste Lamarck's progress-oriented, egalitarian vision of transformism brought him favor during the First Republic; under Napoleon and the Restoration, Georges Cuvier's belief in the fixity of species and the fixity of the social strata reigned. Elizabeth Williams's study of the French medical and anthropological "science of man" well demonstrates the deeply ideological content of the field in the century from 1750 to 1850.[39] After 1851 anthropology's left-wing politics were attacked under the fledgling Second Empire when discussions on the origins of humankind, held in the Société d'ethnographie, were deemed politically dangerous, and the society was suppressed. Many of its members then joined the Société de biologie, this time studiously avoiding politically questionable topics.[40] Indeed, the president of this society, Pierre Rayer, had been a member of the now-defunct Société d'ethnographie and was particularly concerned when the young Paul Broca began a presentation on animal hybridity. Rayer stopped Broca midspeech and requested the withdrawal of the paper.[41] Broca was probably not as surprised as his consequent indignation implied. Through his discussion of hybridity, he was clearly supporting polygenism—the idea that the human race had multiple origins (multiple "Adam and Eve" pairs)—a doctrine that ran counter to the Second Empire's fervent Catholicism but seemed to many to be the only reasonable explanation for the variety of racial types.[42] Broca was committed to rational explanation and the questioning of biblical truth and frustrated with the limitations of the Société de biologie. He founded the Société d'anthropologie de Paris on the night his presentation was cut short.

Broca's new society was thus controversial and defiant from its inception. The conservative Second Empire did not give it official sanction for a full year, during which time it kept its membership to nineteen because there were no laws preventing associations of less than twenty people. The cause was probably not aided by the fact that sixteen of the members were connected to the Paris Faculty of Medicine, for as prestigious as that institution was, it was well known for its materialist and even atheist philosophy.[43] In any case, some of the original nineteen members were independently known for their materialism and free thought, such as Broca's close friend, the doctor and demographer Louis-Adolphe Bertillon. Even after the government gave in and allowed the Société d'anthropologie to expand its membership, the empire ordained that a police officer be present at every meeting to ensure that nothing political was discussed.[44] In 1864 the society was declared to be "of public utility," and the police surveillance was removed, but even before this there were subtle political implications to almost all the anthropological questions the group addressed.

Over the years, Broca led his society in supporting a number of scientific positions that had distinctly left-wing or anticlerical political implications. His belief in polygenism was one example. Another was his countering of the claims, most energetically made by Count Arthur de Gobineau, that any mixing of human racial groups would lead to degeneration. Gobineau's stance was particularly antiegalitarian because his theory implied that social classes were actually distinct biological races. Broca argued that some racial "hybridization" could result in strong mixes, which suggested less biological difference between the classes at home and also served to temper fears of miscegenation in the colonies. People had worried that French men and women who entered into unions with colonial peoples would unleash a great destructive force on their own race; Broca argued that mixes could result in strong offspring who had tendencies toward the higher portion of their pedigree. The idea served to support the republican government's new imperialism by dissipating one of the anxieties that had accompanied it.

Broca also argued that human beings became less religious as they became more educated, stating that in most cases religiosity was "nothing more than a type of submission to authority."[45] In 1868 the French senate expressed extreme concern over the "atheism and materialism" being taught at the Paris Faculty of Medicine, and Paul Broca was singled out as one of the worst offenders.[46] Not only were there accusations of atheism and materialism and verbal hostility to church authority, but Broca actually pitched a fight over the issue that came to be known as the Broca Affair. He came from a Protestant family, and though we do not know much about them, we know that he was acceding to their wishes when he went looking for a Protestant wife. As one biographer tells it, "They would not let him marry a Catholic girl and pious Protestants would not give him their daughters on account of his undisguised lack of religious fervor." He found a match with Augustine Lugol, a Protestant young woman with a scientist father (the inventor of the iodine treatment still known, and sold, as Lugol's Solution).[47] That obedient Broca may have been, but in the 1860s he protested the fact that you had to come to the church in order to register as a voter on the Presbyterian Church's electoral list. The church council required attendance. The church council wanted both to draw in people who otherwise would have stayed away and simply to lower the number of liberal voters. Broca wrote a pamphlet of over six thousand words as an open letter to the minister of justice and religious affairs, accusing his church council of intentional deceit and calling for registration by mail.[48]

Broca was not the first to rally around the idea of Darwinian evolution, but he welcomed discussion on the subject and accepted it long before most of his French colleagues.[49] As early as the 1860s, he was certain that the variety of life on earth was to be explained by evolution, though he was not sure Darwin's particular understanding of the process would win the day.[50] Reiterating the theme of Thomas Huxley's famous speech at Oxford, Broca added the French emphasis on exchanging scientific visions of progress for religious visions of decline: "I would much rather be a perfected ape than a degraded Adam. Yes, if it is shown to me that my humble ancestors were quadrupedal animals . . . far from blushing in shame for my species because of its genealogy and parentage, I will be proud of all that evolution has accomplished, of the continuous improvement that takes us up to the highest order, of the successive triumphs that have made us superior to all the other species." Broca lauded "the splendid work of progress" and made it clear that without much faith in Darwin's natural selection, one could still come to a position that equally rejected the traditional model of species as created and fixed. Wrote Broca, "I will conclude in saying that the fixity of the species is almost impossible, it contradicts the pattern of succession and the distribution of species in the sequence of extant and extinct creatures. It is therefore extremely likely that species are variable and are subject to evolution. But the causes, the mechanisms of this evolution are still unknown."[51] This was radical in France at the time.

Also, despite his conviction that women were slightly less intelligent than men, he considered this to be a result of inherited effects of cultural inequality and believed that improved education could do much to redress the discrepancy. When Darwin's first French translator, Clémence Royer, lamented that women were not allowed to join the Société d'anthropologie, Broca replied that he knew of no reason for such an exclusion and welcomed her to join. This was, by any standard, an extremely bold gesture, for women were largely barred from learned societies—even when she wanted to join the Freemasons, Royer had to find an "independent" Masonic lodge not directly affiliated with the Grand Orient, as did the well-known feminist Maria Deraismes.[52] But Royer was a bold addition to the Société d'anthropologie for more reasons than gender: this particular woman had written one of the most radical and controversial anthropological texts that had ever been penned in France and would become one of the creators of freethinking anthropology. One more fact will add to an understanding of Broca's worldview: among the active members of the women's committee of the Paris Ligue de l'enseignement we find Madame Paul Broca—his wife, Augustine, was a modern woman engaged in republican campaigns (106).

While Royer was an extreme case, the inclusive nature of Broca's society extended beyond any individual person or position: the Société d'anthropologie allowed antiestablishment scientists, along with not-yet-established scientists, to gather and debate in a relatively free environment. While the society's welcoming attitude toward new ideas was always more explicitly extended to the left, during the Broca years conservatives were also welcome. So while the most celebrated living positivists, Emile Littré and Ernest Renan, joined the society in the early 1860s, adding to the society's already explicitly positivist stance, other scholars were invited despite their opposition to several of Broca's more unorthodox positions. Armand de Quatrefages, Isidore Geoffroy Saint-Hilaire, and Henri Milne-Edwards, three of the most illustrious French naturalists, all of whom held chairs at the Museum of Natural History and all of whom were strict monogenists, were invited to join the society as honorary members. Quatrefages remained active in the society over the coming years. He held the chair in anthropology (renamed for him) at the museum, which was the only establishment position devoted to anthropology, and came to represent the rearguard voice of the society.[53] The regular debates of the society were published with great detail in its two journals, the *Bulletin de la Société d'anthropologie* and the *Mémoires de la Société d'anthropologie* and involved the general public through annual anthropological essay competitions with significant monetary prizes.[54] Until the mid-1890s, the society's membership rose steadily, reaching a high of nearly four hundred members. In short, Broca's Société d'anthropologie became the most important center for anthropology in France while remaining outside of establishment scholarship, unattached to any university or state institution. Broca, who seems from his photographs to have been a smallish man, with round features and pronounced muttonchops, presided over it all with a good mixture of grace, scientific pomp, and light-hearted bonhomie.

The other Paris center for anthropological discussion was the Ethnological Museum at the Trocadéro, over which the Eiffel Tower would soon loom. It is worth noting that much less quantification went into the "science of man" there. This institution was dedicated to classifying the massive quantity of artifacts from Mexico, South America, and Africa that were being brought into France in the late nineteenth century.[55] Dominated by the value-laden classificatory systems of Edme-François Jomard and, later, Ernest-Théodore Hamy, these collections came to shape images of the colonies and of "new worlds" in the French imagination. A host of scholars in this period were fascinated with the cultures, bodies, and landscapes of "new" territories, but from those interested in racial anatomy to those interested in acclimatization (the variable ability of Europeans to survive in different climates and settings), these scientists were overwhelmingly concerned with the anthropological other. This was not true of Broca's society, school, and museum of anthropology. Since its origin in 1859, this center—casually called "the institute" by those involved with it—had been dedicated to an anthropology that more frequently directed its gaze toward the French people, inquiring into their development over time and dividing them into any number of meaning-laden subgroups.

Though we do not know exactly what the phrase meant to him, we do know that Broca considered himself to be on "the extreme left" politically. In a private note to Carl Vogt, the Swiss materialist anthropologist, Broca wrote that "one cannot hide the fact that if, in the history of progress, movement is provoked by the extreme left, it is actualized gradually by men less advanced and less logical but more in accord with the masses."[56] The quotation demonstrates Broca's equation of the political left with scientific ability and rationality in general. As he hinted, Broca was not exactly "in accord with the masses." In 1868 conservative members of the French senate called for a petition against the materialist interpretations of the mind that were being taught in the Paris Faculty of Medicine.[57] They were particularly incensed by Broca's materialist approach to the mind-body question, but the senatorial outcry, led by Cardinals Donnet and Bonnechose, generally warned that at the "lectures of MM. Vulpian, Sée, Broca, Axenfeld, [Charles] Robin and others," one can "look at the jammed hall" and see "1,500 young men eagerly listening . . . all determined adepts and defenders of science, i.e., materialism."[58] In response, a petition against such lectures was circulated and eventually signed by two thousand religious and political leaders across France.[59] The conservatives were right to be concerned. Robin alone directed the theses of such soon-to-be-eminent republican leaders as Paul Bert, who would help perform the autopsy on Gambetta's brain, Emile Combes, and Georges Clemenceau—this last having included in his thesis a proof of the impossibility of the existence of the soul.[60] Broca died just after the government had finally come into the hands of committed republicans and was honored posthumously with the title of Unremovable Senator. The archives reveal that he knew it was in the works. Just before his death, Broca wrote the following note to Victor Hugo: "The unexpected honor of being considered for the Republican Union of the senate filled me with great pride, but this pride grew to arrogance when I learned in what terms you had promulgated your adhesion to my candidacy. Let me then tell you that your vote is for me the most precious of all. I may be beaten in the second vote, but I will have had Victor Hugo on my side."[61] Thus Broca was unquestionably republican if taken at his word, he was recognized as an important republican by his peers, and his values were equally republican if understood within the context of his historical moment.

There was, however, room to his left: Scientifically, Broca was responsible for guiding French anthropology into the quantitative study of human bodies and away from ethnology and linguistics. His studies led French anthropology into a deep preoccupation with numbers and measuring devices. In the journals and archives of the Société d'anthropologie, stacks and stacks of paper list the measurements of thousands upon thousands of people's bodies. The sheer quantity of recorded numbers can hardly be overestimated; speaking of statistics in the nineteenth century, rather than particular measurements, Ian Hacking has referred to "the avalanche of printed numbers," and it is an apt description of anthropology as well.[62] Broca, of course, did not take the great cascade of measurements himself, though his output was quite impressive: he trained hundreds of "amateurs" to perform the calibrations. The preoccupation filled thousands of books and articles and thousands of hours. This immense collection of now meaningless numbers had tremendous cultural consequence, buoying up an array of sociobiological assumptions. Thus, despite his left-wing ideals, the physical anthropology that Broca championed blossomed, both during and after his life, into a frightening array of racist and sexist theories. An interesting balance emerges. Broca created a doctrine with deeply conservative tendencies but was himself a supporter of the republic and of mild social reform. Further, though he did not run his life and his science as a great campaign against theism, his moderate free thought led him to create a welcoming place for such a campaign—and for those who would devote their lives to an impassioned cult of radical unbelief.

THE FLOWERING OF THE FREETHINKERS

In the 1880s the Société d'anthropologie de Paris underwent what amounted to a pirate takeover by a group of leftist freethinkers dedicated to materialist atheism. The balance of power in the society had been shifting from positivist (and ostensibly objective) to materialist (and actively political) for some time, but with Broca at the helm the change had been modest. Now the shift accelerated, and the freethinkers took over. In fact, Broca's Société d'anthropologie came to embody the most cohesive freethinking movement in nineteenth-century France. Considering the importance of the freethinking movement—booklength bibliographies have been compiled on the subject vis-à-vis England and the United States—it is surprising that so few historical studies have taken as their subject the French contribution. Jacqueline Lalouette's recent *La libre pensée en France, 1848–1940* is the first comprehensive study, admirably covering an extraordinary number of ways in which freethinking manifested itself over almost a hundred years.[63] Lalouette isolates three "first" freethinking societies in the Second Republic—the one formed by Broca and his medical student colleagues is among them—and reports that all were very short lived and that "none of them were preoccupied in the least with civil burials, those ceremonies that constituted the first and principal raison d'être of other freethinking societies."[64] Though Lalouette finds a tremendous amount of activity in this regard over the next few decades, the movement seems generally limited to concern with burial laws and never appears to be as sustained, organized, and substantial as the movement in other countries in the same period. I would argue that a cohesive movement existed, but in a covert form. One of the first and most active freethinking groups in France operated as Broca's society of anthropologists. Between the early 1860s and the First World War, these French freethinkers infiltrated and transformed Broca's society, by his invitation. There, they created an anthropology that spoke to their own preoccupations and, with it, drew other freethinkers into the new science. They also managed to further the practice and prestige of anthropology and to create some of the classificatory and thematic procedures that would define the discipline in the early twentieth century.

A brief perusal of any copy of the *Bulletin de la Société d'anthropologie de Paris* from the 1880s or 1890s demonstrates that several figures overshadowed all others in terms of the number of articles published and positions held. The core group comprised André Lefèvre, Eugène Véron, Gabriel de Mortillet, Charles Letourneau, Abel Hovelacque, and Henri Thulié. While most of the society's over three hundred members merely attended meetings, this handful of men published numerous anthropological books and articles, taught anthropology at the School of Anthropology, edited the journals, filled posts at the society, laboratory, and museum, and, except for Lefèvre, served at least once as the president of the Société d'anthropologie. In 1860, however, most of them gave no sign of knowing what anthropology was. Letourneau was a doctor working among the poor of Paris, Hovelacque was a lawyer, Lefèvre was a poet and translator, and Thulié was part of the "realist coterie," hanging around with the artists Courbet, Bonvin, and Champfleury, and founding the journal *Le réalisme* (soon suppressed by the empire).[65] Véron was an author, publishing passionate prosocialist works such as *Les associations ouvrières*.[66] Mortillet, who was already engaged in archaeological research, was in exile for having written socialist pamphlets in violent opposition to Louis Napoleon's empire.[67]

What brought these politically like-minded men together was their friendship and intellectual sympathy with the freethinkers Louis Asseline, a lawyer and writer, and Auguste Coudereau, a medical doctor. Their intellectual roots were solidly in Enlightenment rationalism, especially that of the Encyclopedists. Asseline would later come to be known for his highly praised history of Austria and for the publication of a two-volume collection of Diderot's works.[68] As Lefèvre later described, in the early days of the group, these were angry young men. They were furious at their lack of opportunity under the empire and felt that they had been born at the wrong time. Their anger was primarily directed against the people their own age who were "ralliés," young men and women who would have preferred a republic but opted for reform rather than revolution. As Lefèvre reported, they asked, "Why not perfect *our* institutions and thus hasten the coming of the liberal Empire?" These "detestable ralliers" laughed at the "absolutist dreamers," who believed in "all or nothing." Overall, Lefèvre's sense of his early years and those of his colleagues was deeply mournful. "These intransigents," he wrote of himself and his friends, "these irreconcilables, entered into life at the moment when the coup d'état cut off the road under their steps." As outcasts they found communities in cafés and bars where people, brought together by a common "hate for the empire and love for liberty," would speak of art and science between games of billiards, dominoes, or chess. In Lefèvre's memory, such gatherings were the "happy moments of sad years."[69]

These were not the frustrated accusations of underachievers. Many young republicans refused positions in academia because of its connections with the imperial state. The future anthropologist Eugène Véron, for example, like many republican-minded graduates of the illustrious Ecole normale supérieure, resigned his post because of his frustration with the empire.[70] In a series of articles that ran in the *Revue des cours littéraires* (a temporary name of the *Revue bleue*) in 1865, Véron explained that he left academia because of the empire's educational policies.[71] He claimed that in the contemporary university one could work only within the "official science, which had the state, the ministers, the budget, and sometimes even police and tribunals behind it" (436). Véron wrote that despite general claims to French educational freedom, any student who expressed minority opinions at his university exams would fail. In a long and impassioned list of such possible opinions, he asked his readers to imagine that a student "dared to regret that in the long struggle against feudalism, monarchy has triumphed" or to imagine that a student would risk "the assertion that the eighteenth century, taken as a whole, seemed to him superior to the seventeenth and that he found more genius, more grandeur, and a more noble use of human faculties . . . in the *Dictionnaire philosophique* or in the *Esprit des lois* and in the *Histoire naturelle* than in the compilation of the maxims of theocratic politics." Véron claimed that any one of these heresies would suffice to enrage the university orthodoxy (449–450).

The disappointments of the group's earlier years help clarify their behavior in the 1880s and 1890s: their wild dedication to a cause—the establishment of a secular republic—which was already essentially won, and the radicalization of that cause. To be sure, in the 1880s and 1890s there were many signs that the Catholic Church was regaining some of its social and cultural prestige, and the freethinkers were reacting against this. But the passion of their commitment originates in their frustration with the empire years earlier.

As a group, they felt that the empire had robbed them of their youth, their vigor, and their glory. Under the republic, having finally established prominent, respectable positions for themselves, they tried to reclaim that vigor and that glory, though youth was gone. As Lefèvre wrote of those who would not rally to the empire: "They arrived at maturity without having exercised their strength, without having lived; and life will not begin again. History may well honor them, as a group, with a benevolent glance; but 'time, which always marches on' will not bring to them, or at least not to most of them, the compensation that is due to them for their long sacrifice. Their hour has passed." Despite this lament, he claimed they had done a great deal for the cause, even by merely avoiding "the shame of manifest complicity" and "the moral diminution of profitable collaborations." They were partly responsible for the "slow return of universal suffrage," and "they [were] the ones who would have made the republic if the terrible year had not done it."[72]

Napoleon III's empire entered a liberal phase in its final years. According to the future anthropologists, it was the war in Italy that first began an open revival of public discussion of politics. "Under the pretext of temporal power and the French occupation," it reopened public debate on "the grand discussion of clericalism and religion." Lefèvre saw the coming of the liberal empire not as a new period of freedom but rather as a Machiavellian ploy: the government, he argued, had decided to let some freethinkers talk and publish in order to control the church. "To let a few unbelievers speak was to remind the church of its need for official protection." Most prominent among the journals published in this new climate was *Philosophie positive*, run by two of Comte's most prominent disciples, Emile Littré and Grégoire Wyrouboff. It was dedicated to questions of secular morality, psychology, and physiology, but Lefèvre and his group rejected the Comtean method, which they saw as "not without a certain systematic and grim mechanism that alienated outsiders." The new generation was not at home among the rationalists of the old school or among the "excessive admirers" of Germany (Lefèvre liked to stress that French materialism was of French origin), or among the positivists, who "despite being heretical" were "faithful to a philosophy that had already been surpassed on the path it had opened." The new generation "wanted certainties, not systems." They wanted "the secret of the universe and of organisms," and it was just at this time that "translators popularized in France the views of Lyell on the slow formations of the earth and of Darwin on the origins of species" (120).

If Asseline and Coudereau's freethinking had brought the group together, it was Clémence Royer's writings on evolution that brought them to anthropology and gave them a doctrine. In Lefèvre's view, those French thinkers who celebrated Darwin in the first hour were already profoundly materialist, and they immediately set out to use evolutionary theory as a political weapon. It is useful to note that until they found a scientific credo, most of what the free-thinkers stood for was renunciatory. They did not have a full doctrine of their own until they discovered anthropology, and when they did, they turned the new science into a profoundly cultish endeavor. It was Royer's translation and preface of Darwin's *Origin of Species* that drew them in. In Lefèvre's words: "All those whose hate of the imperial regime had thrown them easily against all re-ceived ideas sanctioned by the government and the official bodies used all their wits to take from evolutionary theory the meanings that were the most hostile to religion and to metaphysics." Still, they considered Darwin to have been "surpassed before he was even fully understood," for while Darwin was essen-tially silent on religious questions, the preface to the French translation of his work insisted on an antireligious interpretation. Indeed, they scolded Darwin for having disavowed "the remarkable preface into which Madame Clémence Royer condensed all the significant substance of the *Origin of Species*. The trans-lator had seen more clearly and farther than its author" (125).[73]

Lefèvre's description of Darwin's French translator, Clémence Royer, was astute: while Darwin's evolutionary mechanism impressed her, she credited Lamarck with the discovery of evolution. She was already an evolutionist be-fore reading Darwin, and she was also much more of a materialist than he would ever claim to be. Again, confusion about the relationship between reli-gion and scientific knowledge has led historians to caution that the blunt truths of evolutionary anthropology did not shock people out of their faith in God; many factors of modernization changed people's relationship to traditional ways of behaving and believing. But the intellectual issues did have some mean-ing. "The argument from design"—the idea that the world's intricate wonder proved God's existence—was dealt a considerable blow by Darwinian evolu-tion's alternative explanation. Yet Darwin's text respectfully and admiringly mentions the work of "the Creator." As Peter Bowler has argued regarding the English case, for religious people the notion of divine creation was quite com-fortably replaced by a notion of divinely ordained, purposeful evolution, and Darwin's evolutionary theory was read in this light.[74] By contrast, agnostic or atheist republicans were eager to find an alternative cosmology when evolu-tionary theory turned up with some answers. For them, the new respectabil-ity of a mechanism for natural development was a real windfall.

Royer's famous, lengthy preface to her translation of the *Origin of Species* (and the translation of Darwin's text itself) gave the work a vigorously antire-ligious character that was not present in the original.[75] Royer wrote with brash iconoclastic fervor, claiming that the power of the book lay in its support for a materialist worldview. Her preface began dramatically, as if she had been asked *Do you believe?* "Yes," wrote Royer, "I believe in revelation, but a permanent revelation of man to himself and by himself, a rational revelation that is nothing but the result of the progress of science and of the contemporary conscience, a revelation that is always only partial and relative and that is effectuated by the acquisition of new truths and even more by the elimination of ancient errors. We must also attest that the progress of truth gives us as much to forget as to learn, and we learn to negate and to doubt as often as to affirm."[76] The thirty-seven pages of preface dealt almost entirely with the relationship between religion and science and mentioned the specifics of evolution and the method of translation only in passing. Quoting Diderot more than Darwin, she specifically wrote of Jesus as an "incomparable man," the "rabbi of Nazareth" who "is more of a God today than he was in his century." She also indicted mysticism in general, calling it a "sickness of exhaustion" and writing that wherever it appears it "brings weakening and moral torpor" (128–129). Perhaps most important, she wrote that Darwin's theory "despite its eminently pacific character" will be "exposed to attacks from the great and immobile Christian party," but she promised that it would also be "a powerful weapon in the hands of the opposition, that is, the liberal and progressive party" (136). The freethinkers would take up this "powerful weapon" with great enthusiasm.

Royer also used her preface to discuss eugenics—quite remarkable considering that Darwin's text only mentioned the development of humanity in a single enigmatic sentence in the final passages: "Light will be thrown on the origin of man and his history."[77] Her particular interest, moreover, was the development of women. Evolution had weakened them, made them beautiful and docile. "In order to hasten the rapid progress of the race in all senses, we must ask of women a part of what up until now we have only asked of men, that is to say, strength united with beauty, intelligence with gentleness." She believed that intellectual women like herself were biological anomalies who had "men's brains."[78] Her belief in biological hierarchy—despite a profound desire for general equality—was consistent: she also wrote about the natural inequality of individuals and races and insisted that an egalitarian regime would breed out these differences. Indeed, she claimed that the theory of evolution proved this and, in general, that the theory had in it "an entire philosophy of humanity. . . . One could say it is a universal synthesis of the laws of economics, the quintessential natural social science, and the code for living beings of all races and of all epochs." The preface also called Darwinism a "scientific revelation" and asserted that it teaches us "more about ourselves than any sacerdotal philosophy about original sin by showing us, in our brutal origins, the source of all our bad tendencies." It also shows us "our continual aspirations toward the good or the better," as a function of "the law of perpetual perfectibility, which rules us." These ideas—that evolutionarily informed anthropology was now an entire philosophy of humanity, a universal code of living, and a replacement for ethical monotheism—would have tremendous impact on French freethinkers, particularly those who turned to anthropology as their lifework. In the closing of her preface, Royer set the terms of French anthropology for the next several decades: "The doctrine of M. Darwin, which is the rational revelation of progress, is set in a logical antagonism to the irrational revelation of the fall. These are two principles, two religions at war. . . . For me, my choice is made: I believe in progress."[79] The combination of the freethinkers' passion with Royer's political anthropology was especially potent because Royer had so clearly articulated the religiosity of her opposition to religion. Darwin, as Lefèvre hinted above, seems to have been amused by Royer's work at first, but a few years later he had a new translation done by someone else—without the combat mood.

Royer had been raised as a Catholic and a monarchist; her schooling concentrated on the religious and the domestic and afforded her little contact with science. A brief period spent in convent school had terrified her with notions of original sin and eternal damnation. Years later she came upon a library of books that had been forbidden to her—Voltaire, Rousseau, and Diderot—and underwent a profound and angry transformation away from all religious dogma and to a simple belief in God—deism.[80] She grew fascinated with science and later wrote that it was after ten years of studying natural history that she embraced an "absolute negation" of God (39). Along the way, Royer earned the rigorous qualifications to become a secondary school teacher and worked as such for a few years. The Second Empire, however, abolished secondary school for girls, as well as teaching by lay teachers. As Joy Harvey demonstrates in her intellectual biography of Royer, the young scholar left France for Switzerland feeling betrayed by the church and the empire (40–41). In Switzerland and later in Italy, she augmented her small inheritance by giving lectures on natural history, some even touching on Lamarckian evolution. It was while abroad that she translated Darwin's book, and when she returned to France it was with some small renown.

Royer soon became a very active participant in the late-nineteenth-century French scientific community. A photograph of her from this period shows her longish brown hair pulled back over her head in rolls that ended in two long, tidy curls that she drew in front of each shoulder; her face is round, with small features, and she wears tiny earrings and a round brooch at the neck of her black shirt and fur collar. As well as being a member of the Société d'anthropologie, Royer was a member of the Association française pour l'avancement des sciences, and she contributed regularly to the science column of Gambetta's newspaper La république française, edited by Paul Bert.[81] She took part in a number of conferences on women's rights, as well as on science and free thought, published many feminist articles, and in the last years of her life wrote a regular science column for Marguerite Durand's feminist (and adamantly Dreyfusard) daily newspaper *La fronde*. As a woman and as a scientist of strong opinions, her situation was difficult. To some extent, the relative egalitarianism of the freethinking movement did open doors for her, but it would probably be more accurate to speak of her as having helped to create the respect for female intellectuals that existed in the movement. The Second Congress on Free Thought, for example, was dedicated to "The Rights of Woman," but that was in 1893; Royer was already sixty-three years of age and as well established as she would ever be.[82]

It should be mentioned that at midcentury a materialist wind blew into France from the general direction of Germany. It was most associated with the works of Karl Vogt, Jacob Moleschott, and Ludwig Büchner. Vogt was a professor of geology who had been exiled from Germany to Switzerland after his part in the Revolution of 1848; Moleschott was a lecturer in physiology in Heidelberg; and Büchner lectured in medicine at the University of Tübingen. All freethinkers, they wrote books to this effect and toured Europe giving rabble-rousing lectures—especially after Darwin had provided such wonderful ammunition in 1859. The world they described was a hard-edged, meaningless accident, and, worse, their materialist determinism left no room for free will or moral feeling. The reasons for their enthusiastic, stark materialism were surely as complex and context-bound as those of the French freethinking anthropologists and lie outside my project here. What needs to be noted is that this profound materialism was not original to the Paris group, but it seems they came to it on their own, through their own development. They certainly took it in new directions, as I will demonstrate. Here, it is enough to note that they consistently credited their science and philosophy to the French woman rather than the Germanic men. They may have done so even if both sources of materialism had been an influence: many French intellectuals considered the extremes of positivist determinism to be essentially French. They had understood Germany, by contrast, to be the country of mystical philosophy: Kant was not much read in France until the late nineteenth century, but the French were aware of his critique, and they saw it as strange, mystical stuff. In Owen Chadwick's words:

> Frenchmen were surprised to see Büchner and Vogt. They thought atheism particularly French and Germany the home of idealism and mysticism. Accustomed to deride Germans for imagining matter not to exist, they were astonished to find Germans who maintained that the mind did not exist. . . .

French positivism was far more agnostic, and far more reverent, than the ma- terialists. Positivist utterances could be as interesting as when Littré and [Charles] Robin used a dictionary to define soul as: "anatomically, the sum of functions of the neck and spinal column; physiologically, the sum of functions of the power of perception in the brain." Still, the French were surprised at German materialism, and the surprise is a little piece of historical evidence about the nature of influence.[83]

The French freethinking anthropologists did not much refer to these Ger- manic materialists; they had their own system of thought and were concerned to fix their own doings to the Enlightenment in one seamless narrative of French antireligious genius. Their tolerance, even celebration, of a woman scientist was rooted in this nationalist pride and in their respect for Royer in particular.

Royer and Mortillet did have some contact with Vogt. At roughly the same time that Royer was abroad, Gabriel de Mortillet was living in exile for hav- ing published socialist pamphlets opposing Louis Napoleon's coup. He was from a noble family, supporters of monarchy and Catholicism, that had sent him off to be educated by Jesuits. A pamphlet he wrote in 1849, Les jésuites, de- scribes the humiliating physical and emotional punishments he suffered at their hands.[84] By the time he was nineteen he was a wanted man, in exile, working as an engineer, building railroads in Switzerland. Already interested in prehistory, Mortillet collected a wealth of artifacts turned up by this rail- road work and used them to argue against the biblical explanation of life on earth. It was here that he met Vogt and began a professional friendship that was to last many years. (Royer may have met Vogt here as well.) After a while, Mor- tillet was deemed politically undesirable in Switzerland, too, and was again ex- iled. He continued to assert his antibiblical position in Italy, where he found more work engineering the construction of railroads. In 1864 Mortillet began publishing the first journal of prehistoric archaeology, Matériaux pour l'histoire positive de l'homme, in which he advertised Royer's Italian lectures. He contin- ued to produce this journal after returning to Paris later that year.

By now, Lefèvre and his cohorts had begun publishing their own journals. The first was a radically anticlerical journal called Revue encyclopédique. In 1866 they started a new one, boldly titled Libre pensée, which was dedicated "to free the human spirit from all hypotheses, all superstitions, and all irrational doc- trines."[85] In its seventh issue, Libre pensée ran an article on anthropology, which it celebrated as the science of the perfectibility of human beings.[86] When Mor- tillet returned to Paris, he began submitting essays to these freethinking jour- nals, displaying his own brand of scientific materialism.[87] Libre pensée soon became intolerable to the Second Empire, despite Napoleon III's new liberality, and the journal was suppressed in 1867. Undaunted, the freethinkers began publishing *Pensée nouvelle* in 1868. According to Lefèvre, these journals gave the freethinking movement its name, and they do appear to have been the first in a long line of journals to use those titles. In its origins, the group of free-thinkers included Asseline, Lefèvre, Letourneau, and Coudereau. It grew to include Royer, Mortillet, and Thulié, who had spent most of the 1860s in med-ical school because his realist "literary theories [had] led him to natural sci-ence."[88] The future government minister Yves Guyot was also very much in-volved, as was Broca's freethinking friend and colleague, the demographer Louis-Adolphe Bertillon. These journals published a few pieces by Büchner and Vogt and a number of articles by future sociologists and philosophers of the Third Republic, but the core group of freethinkers wrote the vast major-ity of the articles and reviews themselves.[89] The content was strictly limited to the defense of atheism, attacks on the church, and, in this spirit, discussions of the Enlightenment, natural history, the nature of morality, and philosophy in general. When *Pensée nouvelle* was suppressed in its turn, many of these writ-ers began contributing anthropological essays to Mortillet's *Matériaux* and to the positivist *La philosophie positive*.[90] They were becoming a cohesive group of anthropologically minded, outspoken atheists.

THE CONVERSION

This group joined the Société d'anthropologie to use it as a base for their evan-gelistic atheism. As Lefèvre explained it: "Mortillet, I think, by a masterstroke, led us to Broca; and we entered—without ourselves dissolving—into the So-ciété d'anthropologie de Paris, which furnished us with precious recruits, sci-entists rightfully attached to their specialized studies but who adhered, with-out equivocation or bashfulness, to these grand principles: that incredulity is the first step toward science, that the experimental method is the mother of all philosophy, and that absolute secularism is the sine qua non of all teach-ing."[91] Letourneau and Mortillet became members first and were soon joined by Coudereau, Thulié, Asseline, Lefèvre, Hovelacque, and Royer. These in-ductions were not always without some resistance from the society. Royer's entrance was certainly the most dramatic. As Letourneau would later write, her entrance into the society was "doubly revolutionary as a woman and as Darwin's translator."[92] True enough. Yet to the more traditional members of the society, many of the freethinkers seemed a bit revolutionary, and they ac-tively encouraged this image of themselves. When Mortillet founded a new anthropological journal, *L'homme: Journal illustré des sciences anthropologiques*, in 1884, he bluntly described it in its first issue as "an intermediary between science and politics."[93] This claim should not be misread: the journal hardly ever weighed in on any specific political question of the day.[94] What its authors understood by "politics" was fierce atheism combined with general attestations of feminism, socialism, and egalitarian democracy. Indeed, *L'homme* was so explicitly antireligious that defenders of the church soon came to call it *L'homme mal élevé*.[95]

Yet by 1886 the freethinkers had so long cloaked their atheism in science that they began to worry about their legacy. They were sensing that the subterfuge had succeeded a bit too well: now they were known as anthropologists but not as originators of the freethinking movement. "Now," in republican France, wrote Lefèvre, "everywhere the avant-gardes carry the flag of free thought. This flag is ours." He claimed that the journals *Libre pensée* and *Pensée nouvelle* "were and remain one of the philosophical monuments of our age," and he collected his own articles within his book *La renaissance du matérialisme*, under the subtitle "Militant matérialism."[96] Letourneau likewise collected and published his essays from this period as *Science et matérialisme*, referring to them as his "combat articles."[97] Still, the freethinkers complained that they were successfully using anthropology to argue atheism "and meanwhile, all around us, freethinking societies have been growing up from the ground in crowds in Paris and in the provinces, ignorant, perhaps, of their origins, heedless of their older sister."[98] Because, of course, she was in disguise. As a result of this rather odd confluence of events, it becomes clear that Broca's creation of one of the first freethinking societies had considerable impact. It was a youthful public gesture expressing what seems to have been a more private belief in the mature man. Yet he went on to create a scientific society that was attractive and comfortable for the brazen group of freethinkers who would take it over. As such, Broca's freethinking society may be seen as the beginning of a coherent scientific-atheist movement that lasted into the following century.

The doctors and other professionals who had joined the Société d'anthropologie under Broca's stewardship were by no means unanimously willing to agree on atheism. Though the freethinkers would be even more dramatic in their rejection of positivist objectivity after Broca's death, they campaigned against religion from the moment they became members. Broca may have even encouraged this; at the least he winked at it, for Mortillet was surely among the most outspoken, and Broca always spoke of the School of Anthropology as cofounded by himself and Mortillet. It was Broca's society, and he had tremendous prestige. He need not have collaborated with anyone whose politics or behavior gave him pause, and while he may have been an intellectually generous man, this project was the central teaching institution of the science he had created.[99] It is a good indication that Broca enjoyed the free-thinkers' antics. As early as 1867, one can find brazen attacks on the very notion of religion, and there are important signals that the freethinkers had considerable opposition within the Société d'anthropologie. Wrote Coudereau of some of his colleagues, "They want religion, which is a banality for some of them and a sacred institution for others of them, to be always placed outside of the subject of debate. I afford it neither that much respect nor that much suspicion."[100] Quite to the contrary, Coudereau insisted that the Société d'anthropologie ought to study religion because it had exercised an immense negative influence on the progress of civilization. "Since time immemorial, society has tossed about between two worlds: 'science' and 'religion.' Science embraces all . . . religion . . . is a synthetic system exploiting its monopoly on the circle of the unknown" (591).

There were many direct attacks on anthropologists who were not freethinkers. In 1884 Lefèvre included such an indictment after asserting that "it is anthropology that will give the final assault and will bring the supreme blow to metaphysics, which is already on its last legs." As Coudereau had done before him, Lefèvre made it clear that some of his critics were religious believers but that a good many were positivists, that is, they thought that such questions should be left to religion and philosophy. "Many anthropologists," conceded Lefèvre, "whether themselves inclined to some of the doctrines or superstitions being menaced, or whether they have mistaken the character and the utility of philosophy, find this militant role repugnant." These people were not openly hostile, they simply "tried to stay outside the debate."[101] For the freethinkers, however, silence was a political act, and they expressed real surprise that people possessed of this powerful information could even imagine keeping it to themselves. For them, materialist science had finally been born and was poised to deliver humanity from millennia of superstition. In her preface to Origin of Species, Royer made it clear that she saw the advance of science as dependent on a direct confrontation with religion. Marveling at the very idea of creationism, Royer mused that "one might well ask oneself how a doctrine that necessarily involves supernatural intervention could have remained so long established in science, to the point where it reigned without rival." The answer she offered was that "in science, the supernatural retreats to whatever degree that the natural gains ground, and that the amount of direct action attributed to God has always been the same as that of our ignorance of the real laws of the universe."[102] The freethinkers took this as their credo and were shocked that all scientists did not see it as their duty to bring the discovery to the wider society. Some scientists, they marveled, did not even agree that the fight against religious dogma was inseparable from the rest of science. "They sincerely believe," wrote Lefèvre, "in distancing themselves from a struggle that disturbs their work and could compromise the serene dignity of science." The great positivist error, according to materialists, was that they desired to limit the influence that science had on questions of spirit, origins, and other unknowns. According to the freethinking anthropologists, "this influence itself is the mark and the measure of progress."[103]

Jean-Louis Fauvelle's attack on philosophy in *L'homme* in 1885 was just as scathing. As far as the freethinking anthropologists were concerned, there were no philosophical questions, just as there were no religious questions; there were only materialist answers. "Philosophy," wrote Fauvelle, "is dead and well dead." The anger apparent in that "well dead" comes from the belief that philosophy, by accepting and promoting a division of the world between the material and the ethical and metaphysical, creates the space in which religion functions. Asserting that there are some things whose truth or falsehood are not amenable to proof, positivists also protected religion from scientific scrutiny and dissection. "The partisans of philosophy," he asserted, "in the effort to save the principles of religions, would do well to make an alliance with the positivists, their worst enemies: the method of Francis Bacon has made its entry into the domain of the natural sciences, and even were our adversaries to have recourse to a *manu militari*, as in the time of Galileo, this conquest by free thought is most definitive."[104] In another article, entitled "Il faut en finir avec la philosophie" (We must finish with philosophy), Fauvelle reiterated these points, confidently adding that the philosophy of psychology must be replaced by physiology and the study of logic replaced by the scientific method.[105] On the subject of materialist morality, Fauvelle was much less self-assured, managing to do little more than suggest an approach to individual rights and responsibilities.

The war between the freethinking anthropologists and the church was neither discreet nor one-sided. On the first page of *L'homme* of October 1887, Mortillet reviewed a Catholic conference on evolution, writing that "while free thought, in the name of the liberty of conscience, has sought to separate the church and the state, the Catholic Church has made vigorous efforts to monopolize science."[106] Indeed, the Catholic conference had specifically derided Clémence Royer, Broca, Vogt, Virchow, Darwin, and, among others, Mortillet himself (611). It does seem that Broca felt the heat. On at least one occasion he actively suppressed a scientific essay that he considered to be dangerously political. In 1875 he refused to publish an article by Clémence Royer on the grounds that it could lead to political trouble outside the society. The article held that all assumptions regarding the causes of the falling birthrate in

France were wrong: this was no decline in national vigor or anything of the sort. French women were limiting the sizes of their families on purpose, for pragmatic social and economic reasons. Broca wrote to Royer saying that the society would be unable to publish her article because they were already under attack by the Catholic press as a "school of freethought."[107] For the annals of slow but dogged justice, one of Royer's historians, Claude Blanckaert, published Royer's "On Natality" in the still-running journal of the Société d'anthropologie in 1989, a hundred and fourteen years after Broca had turned it away from the same journal.[108]

PURGING THE INFIDEL

In 1889 the freethinkers solidified their control of the Société d'anthropologie by successfully expelling Paul Topinard from his chair at the Ecole. They locked him out of his office, rerouted his students, and publicly accused him of mischief. After his expulsion, Topinard published a pamphlet explaining his situation. His estimation of the events that led to his dismissal was, no doubt, skewed in his favor, but in light of analogous statements from within the ranks of the freethinkers, the basic narrative appears to be reliable.

In his pamphlet, Topinard argued that the Société d'anthropologie and the School of Anthropology were created as two "absolutely distinct institutions . . . that nevertheless presently have in common that they are governed by the same majority." He went on to say that this majority was "materialist and in all points of view intransigent. They do not deny it, they have made overt professions of this and call themselves, alternately: Dinner of Free-Thought, or Dinner of the New Thinking; Society of Mutual Autopsy, Group of Scientific Materialism."[109] For Topinard, "the first cloud" appeared while Broca was still alive. Hovelacque had invited Topinard to a dinner for "friends of the Société d'anthropologie." Topinard accepted the invitation and was surprised to find a considerable number of people there who were not members of the society. At a later dinner, the group proposed to publish a series of books under the title "Bibliothèque des sciences contemporaines." They requested that Topinard write the text on the subject of anthropology, while other volumes would be prepared on biology, linguistics, archaeology, philosophy, and so forth. Topinard was pleased with the commission and signed a contract with the publisher Reinwald along with Asseline, Mortillet, and Lefèvre. When Topinard submitted his completed manuscript, "the committee gave it to one among them to correct any ideas that might be found therein that were not in accord with the ideas of the group" (15). Still, he continued to believe that the dinners he was attending were for "friends of anthropology," until he received a written invitation that identified the event as a "Dinner of Free-Thought." At that point, asserted Topinard, it became clear to him that anthropology was being taken over by a fiercely polemical group with only secondary interest in anthropology as such. "Once my eyes were open," he wrote, "it became evident to me that anthropology was nothing but a screen and that a sort of systematically materialist confederacy, at once political and social, was disguising itself behind that screen—a confederacy that had other concerns than those of science" (16).

The freethinkers, he reported, then began to found new societies, like the Society of Mutual Autopsy. They also began to create new events such as the Lamarck Dinner, which regularly gathered to raise money for a statue of the famous French evolutionist, and the Voltaire and Diderot Dinners, which gathered to raise money for the republication of the work of these authors. The freethinkers also began to take over the Société d'anthropologie's essay contests and to organize the memorials for deceased members. According to Topinard, "the moment one dies, they organize a subscription to raise a monument, a medallion. All civil burials were a triumph. Medallion, procession, autopsy—they were all connected" (16). Further, in the late 1870s the freethinkers proclaimed their intention to create a dictionary of anthropology. It has served as a reference guide ever since (it still graced the shelf of scientific dictionaries at the Bibliothèque nationale when last I checked). In Topinard's account, this work had its beginnings at a freethinkers' meeting when one of the members said "that anthropology has not yet given anything; that it was necessary that we affirm our ideas." It was time, they agreed, to start an "anthropology of combat" by creating a dictionary and asking Broca to write a preface. Topinard wrote that he resigned from the group when he heard this, "but Broca, ignorant of the facts that I could not reveal to him, let them put his name on the cover, as did Quatrefages, so I let them do the same with mine" (16).[110] A photograph of these men from this period—to my knowledge the only photograph of them together—shows Broca looking relaxed in a bow tie and tuxedo, a comb-over, a small smile, and white muttonchops. Quatrefages sits more erect and posed, also very soberly dressed, with a little scowl on his face, white hair, and a white beard running from ear to ear under his chin. Mortillet wears light pants with his black tuxedo jacket. He is the only one in a long tie and sports a full white beard. Topinard is a larger man than all of them, tall and a little heavy, still dark of hair, wearing the master's muttonchops and a pair of wire glasses. The sides of his mouth turn down, as if he were pouting.[111]

Lefèvre's statement that the freethinkers joined the Société d'anthropologie "without ourselves dissolving" is complemented by Topinard's assertion that freethinkers, "so totally ignorant of anthropology that they hardly knew what the word meant," were often brought into the Société d'anthropologie and then shuffled "one year later, after a strictly regulated delay," into the central committee. Charles Issaurat fits the description: in the 1860s he wrote for *Libre pensée* and *Pensée nouvelle*; he joined the Société d'anthropologie in 1874, and though the most anthropological writing he ever did was to popularize his colleagues' work in other journals (he was still calling himself a "man of letters" in 1893), he served as vice president of the society and was a member of the central committee until the end of his life.[112] His son became a member of the society in 1888. This last bit of information both highlights the significance of the society for Issaurat and explains Topinard's sense that a tide was rising against him. Topinard claimed that he had on his side "the older members; against him, the entire group"—and the older members were rapidly dying off.[113] As this happened, the freethinkers gained more and more control. In Topinard's estimation, they did not consider themselves the new stewards of anthropology; rather, they saw anthropology as ground for plunder. "In brief," wrote Topinard, "the group had but one objective that became more and more evident: to become the complete masters of the society and to dispose of it, at their will, for their own particular designs" (22).

This interpretation of the events was shared by Eugène Dally, a medical doctor and anthropologist who had been a close friend of Broca and had penned the first French translation of Huxley's writings in 1868. Huxley's popularization of Darwin's work and his explorations of evolution's religious and social implications were radical in France under the empire, so Dally was not an archconservative. The freethinkers recognized this, and his work had been several times celebrated in *Libre pensée* and *Pensée nouvelle* back in the 1860s.[114] But in the 1880s Dally found his extremism both outdone in temper and betrayed in theory. "Some of our colleagues," wrote Dally in 1882, "would very much like to transform us into a sort of church council and have us promulgating the truth, which they, as much as their adversaries, believe themselves to have accessed completely. But the absurdity of such a role could escape no one: we have not gotten out of creeds only to reenter them from the other direction."[115] Frustrated, Dally broke with the society soon afterward and was accused of positivism by the freethinkers.

Struggles between the freethinkers and the rest of the society decreased rapidly as the rest of the society became smaller and smaller throughout the 1880s. Still, tension was obvious. In 1888, for example, the marquis de Nadaillac, one of the dwindling group of more conservative members, attacked Mortillet for writing, in a study of cannibalism, that it was religion that led human beings to eat human flesh.[116] De Nadaillac protested, saying that, "if perverted religious sentiment" led man to cannibalism, it was a rare exception. He explained cannibalism by "hunger, cruel hunger, and the madness that it engenders" and "a depraved taste for it, the bestial sentiment that is in us and neither education nor the progress of civilization manages to destroy completely" (27). Mortillet counterattacked with the insistence that aside from a few cases "occasioned by accidental or habitual lack of food" cannibalism was the result of "religious perversion." In case anyone should think he was only speaking of "savage" religions, Mortillet pointedly cited the Christian Eucharist as proof of the intimate connection between eating human flesh and belief in God (43).

It is worth noting that modern opinion gives this one to Mortillet. Mortillet took several positions for antireligious reasons that seemed extreme at the time but are now in accord with dominant scientific theories. To take one important example, before Louis Pasteur, fermentation and mold were frequently assumed to be the result of some form of spontaneous generation. Materialists' explanations of life happily likened its origins to these effervescent physical phenomena. When Pasteur boiled a beaker of liquid and then melted the top closed, everyone could see that nothing grew, and spontaneous generation was dealt a serious blow. Because that seemed to mean that life required God to get it started (Pasteur was a member of the Catholic Scientists Association and would not mind this interpretation), Mortillet railed against it, going so far as to argue that it had been faked.[117] Again, present consensus leans toward Mortillet: current theories seek to isolate how chemical systems that show less self-replication than a virus develop into chemical systems with as much self-replication as a virus or more—perhaps under conditions such as the extreme pressure and heat of our younger planet or, continuously, near volcanic spigots at the bottom of the ocean.

Also in 1888 the freethinking anthropologists came into conflict with one of France's best-known popularizing social scientists, Gustave Le Bon. Le Bon has figured prominently in several studies of late-nineteenth-century French social sciences because of his extremely low estimation of the innate intelligence of women and of various races.[118] But though he had worked with Broca and had been a member of the Société d'anthropologie since the 1870s, Le Bon was never representative of contemporary French anthropology. In 1879 Broca had approved Le Bon as the winner of the society's Godard Award for his essay on brain size and intelligence, but it was recorded in the minutes of the society's central committee that Broca "did not approve of all its views."[119] Indeed, Broca used Le Bon's data on the inferior brain size of women to argue that girls' education needed be more rigorous to correct for this socially created deficiency—which was not at all what Le Bon had in mind.[120] Letourneau's reaction to Le Bon's prize-winning essay in the *Bulletin*'s notes on the prize was even more strongly negative, and toleration of Le Bon did not long outlast Broca's lifetime.[121] By 1882 Le Bon's work was being boldly attacked in the *Revue anthropologique*, and the *Bulletins* recorded a sharp polemic between the anthropologist Léonce Manouvrier, who argued that women's brains were proportionately equal in size, if not larger than the brains of men, and Le Bon, who insisted they were comparatively puny. By 1888 Le Bon recognized that the Société d'anthropologie was completely opposed to his views, and he resigned his membership. As he wrote in one of his several letters of resignation, "I am much too independent for this society, which now, having its completely settled doctrines, its official science, and its certainties, has need of benevolent auditors and has no need at all of scientists or of researchers. It is in the conviction that I have as little need of it as it has of me that I have decided to resign from it."[122]

While such squabbles continued to ruffle feathers, it was Topinard's expulsion from the Ecole that confirmed the freethinkers' dominance. Topinard's supporters within the Société d'anthropologie were few—his only active support came from de Nadaillac and Dally—and when he brought a legal suit against his accusers he was quickly defeated.[123] Topinard's best argument was simply that he had been Broca's right-hand man and that this entitled him to succeed Broca as the conceptual, if not titular, leader of the society. The freethinkers generally ignored this notion, but in any case they did not see Broca's attachment to Topinard as a reflection of the latter's worth. In Fauvelle's words, "Our illustrious founder, with the authoritarian character that seeks to accomplish a determined end by any means, chose to surround himself only with persons on whose docility he could depend absolutely."[124]

One gets the sense that the freethinkers did not like Topinard much but also had some real and substantive problems with his work. For one thing, it seems that his works of synthesis did not adequately credit the original scholars for their contribution.[125] A larger problem had to do with his adamantly narrow definition of anthropology. Whenever the question of anthropology's role arose, Topinard took the position that only the most strictly physical—that is, biological—interpretation of anthropology would be admissible. One of the many showdowns on this issue took place at the 1889 Congress of Criminal Anthropology, where Topinard insisted that anthropology define itself as a physical science of the human races, completely devoid of a political agenda or, indeed, any application to the workings of society.[126] In Topinard's estimation, anthropology was, "the zoology of man." He claimed that Broca and Quatrefages allowed no infiltration of ethnography, sociology, or psychology into the science of anthropology (491). "In anthropology," wrote Topinard, "one must separate pure truth from its applications to medicine, social economy, politics, and religion. . . . The zoology of man must be able to work without the slightest worry over the consequences drawn from the truths it discovers; it must work above the passions that it awakens" (492). This definition was in complete conflict with the freethinkers' idea of anthropology, but it was also at variance with the vision of the more sedate Léonce Manouvrier. For Manouvrier, anthropology was the science of human beings in all their aspects. As he discussed in "L'atavisme et le crime" (1891) and "L'anthropologie et le droit" (1894), Manouvrier refused to have artificial limits set on the study of humans.[127] In defense of this position, he quoted Broca at length, showing him to have been significantly concerned with the study of society and not shy about it, writing, for instance, that "the condition of women in society must be studied with the greatest of care by anthropologists."[128] In any case, suggested Manouvrier, anthropologists were not honor bound to uphold Broca's conception of the science.

Conservatives outside the society understood Topinard's predicament as decidedly political. The newspaper of the Ligue de la patrie française, La patrie, which identified itself as an "organ of national defense," ran an article entitled "Un coup d'état à l'école d'anthropologie." In it, the Ecole was described as being "governed by a materialist and intransigent majority among whom we find the names Mortillet (you know, the Mortillet who fought against the cemetery cross at Saint-Germain), Mathias-Duval [sic], Fauvelle, Hovelacque, etc." (Mortillet's attack on Christian symbols at the local cemetery will be discussed below.) The article went on to say that Topinard had been evicted because the materialist leadership of the Ecole could not bear to associate with someone who "refused to make anthropology into a weapon of social and political combat." Indeed, the article expressed the sense that the materialists were attempting to take control of French culture in general: "The materialists want to be our masters these days. It is already more ferocious than in the Rabagas cafés where whoever speaks the name of God has to pay a fine. Still, one must pronounce the name of God. Nowadays this costs more than fifty centimes, and one has to pay a fine if one is even suspected by the sectarians of entertaining, within oneself, an idealist or deist sentiment. When will it be: atheism or death?"[129] The struggle that was taking place in the Société d'anthropologie clearly reflected concerns and struggles in the society at large.

The role that anthropology had assumed was by no means lost on the defenders of the Catholic Church. Indeed, the "anthropology question" became increasingly important as the century drew to a close. For example, at the Congrès scientifique international des Catholiques (of Belgian origin, held in Paris in 1888), Canon Duilhé spoke on "le problème anthropologique et les théories évolutionnistes." This was no mere explication of the contradictions between evolutionary theory and biblical narrative. Rather, it was a specific indictment of the attitude of French anthropological institutions, with specific attention to the Ecole. "The Ecole d'anthropologie," asserted Duilhé, "seems to have only one goal: to efface the irreducible characters that make the human soul a special creation of God in nature."[130]

THE CULT TAKES SHAPE:
ANTHROPOLOGICAL RITUAL AND LITURGY

That the freethinking anthropologists were functioning according to a dogma as narrow as that of any religion was put forward not only as a critique by the Catholic Church but also with appreciation by Littré's positivists at the *Revue occidentale*. In an 1885 article on the positivist delegation to the Congress of Freethinkers (1884), Charles Jeannolle made clear the positivist understanding, and acceptance, of the limited freedom involved in freethinking.[131] He wrote jubilantly that "freethinking, too, is a religion, because it reproaches, because it inspires devotion, and because it proclaims rules of conduct to which its adherents must submit or else risk exclusion, such as to have a civil marriage, not to confess, etc. . . . [Freethinking] has a dogma, as it is not content to banish old chimeras from the human mind but rather intends to substitute them with science" (241).

The freethinking anthropologists did not see their doctrine as dogma, but they recognized that theirs was a passionate position and were quite clear in their hostility toward the unenlightened. They believed themselves to be the representatives of the future, and, to a significant degree, they were—despite the fact that they were too passionate about their subject to be accepted as champions of reason once the specific issues of their day had faded. Their own estimation was almost ecstatic: "More than ever," they proclaimed, "the fetishists . . . these mastodons of thought, seek to conspire against the progress of science and reason. We walk among future fossils. Was it not the same in ancient times for those who live on today? Didn't they have representatives of imperceptibly disappearing races as companions and as adversaries? In the same way that this comparison comes to the aid of hypotheses in paleontology, the successive discoveries of our archaeological digs strengthen the case of free thought."[132] It is hard not to smile at the hubris of such a claim. Ethnological studies of religion centered on fetishism and totemism in the second half of the century, so this jibe at the religious in France specifically compared believers to savages, before advancing the metaphor so that believers were mastodons. It should further be noted that, in the above passage, the important notion for the anthropologists was that science helped free thought, not that free thought strengthened science. The purpose to which they would put anthropology was equally clear. Hovelacque's *Les débuts de l'humanité* claimed that anthropology "teaches us . . . that all religious disciplines being essentially intolerant and cruel, it is useful and moral to peacefully deliver humanity from them . . . and it teaches us how we can and how we must achieve the goal of civilization: social equality."[133] In a book marvelously entitled *Plus les laïques sont éclairés, moins les prêtres pourront faire du mal* (The more the laity is enlightened, the less priests will be able to do harm), Hovelacque wrote that "we live in an epoch where the battle between the revolution and the counterrevolution is engaged on all sides at all times. . . . Clericalism is certainly not the only enemy of modern society, but it is the common link of all the elements hostile to the republic and to social progress."[134]

Church dogma and ritual were clearly the enemy, but dogma and ritual themselves were by no means anathema to the anthropologists, and the dogma and ritual they chose was concentrated on the macabre. One point of religion had always been to focus the congregation's attention, in a carefully controlled way, on the terrors of the abyss, the meaning of it all, and the nature of our most authentic selves. Anthropological mystery was just as earthshakingly strange as church mystery and just as effective at drawing attention to difficult meditations: death and cosmic abandonment, the insignificance of individuals, and the burden of carrying a dizzying new truth to the unenlightened. In this sense, the freethinking anthropologists, while rejecting Catholicism for ethical, political, and intellectual reasons, were religious personalities. They thought about life-and-death issues a great deal. Skulls sat on their desks and filled cabinets. One cannot help thinking of the medieval monastery's *memento mori* (reminder of death): a human skull placed in view to keep the monk's attention where it should be. At the Laboratory of Anthropology, the freethinkers were surrounded by tokens and representations of death, of the material stuff of which humanity is composed, and myriad reminders of our animal nature. Thousands of human and anthropoid skulls lined the walls and tables. In the archives, irregular, handwritten meeting notes speak of strange arrivals in the mail: a box came from a Madam Masmenier who was presenting the society with "twelve Negro heads, massacred after a revolt. They are in the same box in which they had arrived [when she acquired them], and which contains several indications of their origins. Also, there is an inscription on each head."[135] A Constantin Snow sent a large collection of Russian men's and women's hair.[136] The municipal councillor Charles Gras sent pottery shards and bones from the Grotto of Salpêtrière and asked what the anthropologists would like him to bring them from his next trip.[137] Assorted bones and skulls were delivered to the anthropologists frequently and usually without much background: a full skeleton arrived on February 4, 1895, with the singular indication that it was "from the Canary Islands."[138] Of course, some of the skulls, brains, and other body parts were the remains of friends, colleagues, and lay members of the autopsy society.

Jacques Bertillon described the Museum of Anthropology adjoining the lab to his readers in La nature, saying the place had "around four thousand skulls of diverse races, a considerable number of bones and other anatomical pieces, a series of forty skeletons, in brief, several thousand objects of ethnography and prehistoric archaeology."[139] There was a tone of pride and wonder as he described it: "The museum contains a great number of skeletons of large animals. The most precious are three complete skeletons of gorillas and some fifteen gorilla skulls, as well as a mannequin very exactly representing the muscles and other organs of this remarkable animal." There were also "collections of Parisian skulls dating from more or less ancient periods" that Broca had dug up and measured and upon which he had based many of his anthropological conclusions, as well as "thirty or so microcephalic and partially microcephalic skulls accompanied by a complete skeleton of an adult microcephal." The museum also possessed "a beautiful series of artificially deformed skulls," which Bertillon described with considerable relish, adding that "what makes the study of the skull so important is that it contains the brain, the organ of thought. But how much more interesting still is the study of the brain itself!" (40–41). He then proceeded to delight his readers with a description of the museum's "rich" collection thereof, which included "desiccated brains as well as plaster models" (141). They also had the skeleton of a giant, from Burgundy, two meters fourteen centimeters tall (around seven feet), who had lately worked the fairs for his living.

That essay was written by Jacques Bertillon in 1878; after 1883 his own father's full skeleton and preserved brain would make up part of the exhibit. The composite effect was that when the freethinkers went to do anthropological work they walked out of modern Paris, with its celebrations of the civilized, the mechanical, the literate, and the orderly, and into a rather macabre environment. They created the environment themselves, without the institutional backing and rigorous, methodical training that would today separate a museum of natural history from, say, a private citizen's enthusiastic collection of human remains. Add the biological anomalies, and the dyad is medical museum versus freak show. Add the passionate atheism, and we are reminded that such strange collections would have once been understood in terms of miracles and demonic monsters. Jacques Bertillon's four-part article "Monstrosities" for La nature began by arguing that Etienne Geoffroy Saint-Hilaire and his son, Isidore, had elevated teratology, the natural history of monsters, to the rank of a real science by "showing that one could press monsters into a Linnaean classification and submit them to a scientific nomenclature, demonstrating that one could formulate laws about their configuration, and, above all, by establishing the importance of their study in the eyes of embryologists and in the eyes of philosophers."[140]

The phrase "eyes of philosophers" is a giveaway that Bertillon and his colleagues were interested in odd bodies for metaphysical reasons. Up to the midseventeenth century, Bertillon explained, monsters were understood as either sent by God or "made in the image of the devil." He characterized the period as representing "vague, incomplete observations, recorded by chance; the grossest errors, the most absurd prejudices admitted without hesitation and obstinately upheld; and explanations made childish by superstition and dignified only of a childish origin—these are the sad characteristics of this clumsy period of teratological science" (209). Since he mentioned his disdain for absurd prejudice, it is fair to note his claim that most giants are indolent, that dwarfs are not usually very intelligent and that they are so often sterile that the case of a dwarf named Borwilaski, who married a normal woman and had two normal-sized children, "raises real doubts about the true origin of these children" (244–245). Casual prejudice and unkind assumption were a matter of course, but here the point is that Bertillon and his colleagues were deconsecrating "monsters" and using the opportunity to think about what they meant about humanity. The article brimmed with anecdotal detail, arbitrary classification spoken of as law, and careful descriptions of a very wide range of anomalies.[141]

There were lurid engravings of a foot with eight toes, a headless baby (*monstre acéphale*), a Cyclops baby with a horn, a harelip, a baby with fused legs, a gruesomely monstrous portrait of a normal embryo face at thirty-five days, and several conjoined twins. Some of them shared limbs, some were connected together by a tube of flesh, or sternopage, and Bertillon mentioned that a "remarkable sternopage is conserved in alcohol at the museum." There was also a woman with a full-sized head growing out of the back of her own. It was upside-down, chin pointing up toward the sky. The anthropologists did not have much more to say about these abnormalities than did anyone else, but they spent a good deal of effort discussing and cataloging such wonders. Preserved, tattooed skin was also a favorite in their collection. A note in the archives mentioned the preserved, tattooed head of a Maori from New Zealand.[142] From all over France and from the wider world, people sent the Société d'anthropologie sketches, descriptions, and specimens of physical oddities, suggesting that many people felt the need to register the amazing, stimulating productions of the natural world.

By collecting specimens and publishing analyses of these remains, the anthropologists were transforming these items and ideas from objects of awe to objects of knowledge. This is one function of measuring and classifying. Ritual and orthopraxy sanctify mundane things; measuring makes the sacred profane. The objects were being used in a kind of religious way: they were not relics nor yet mundane things but highly charged not-relics. Measurement, then, transformed objects from sacred to profane on their way to becoming mundane. The anthropologists exemplified this in myriad ways. To take an example from outside their group: in 1878 the freethinker (and future deputy to the chamber) Jean-Marie De Lanessan expressed fury that a Catholic medical school had been founded in France. Yet Lanessan was not too worried, because as he saw it, "observation kills faith."[143] Observation does not, on its own, kill faith, but measuring and reordering can remove objects from the altar (or the imagined "black mass") and place them, labeled, in the medical museum and textbook. An object that inspires fear or deferential care changes meaning when it is grabbed in a matter-of-fact manner and held up to a ruler. That is what the expression "matter of fact" is all about: it refers to a scientific, objective attitude and claims its own truth status.

In this context, the "rotting garbage" jettisoned by the Society of Mutual Autopsy takes on further meaning, for the gesture turns out to be part of a very large project. The freethinking anthropologists deconsecrated a lot of things. As I will discuss, many of the anthropologists held political positions. When Mortillet was deputy and mayor, his radical anticlericalism was minutely cataloged in a running report and article repository in the Paris police headquarters.[144] According to these reports, Mortillet mounted a campaign to change the monarchical or theistic street names in Paris. In one instance, he wanted to change rue Saint Louis to rue Diderot, but when this met with too much resistance he settled for rue Louis IX. He also insisted that all government stationery bear the words "République française." More seriously, he fired a government employee because the man had sent his child to be educated in a Catholic school, and, in an act that made him both infamous and, in many circles, a bit ridiculous, he had the main cross pulled down in the local cemetery. Many of his decisions were appealed and overturned. A sense of the notoriety of Mortillet's doings may be gleaned from a police department list of the journals and newspapers that reported them. They include *Le figaro, Le monde, Le dix-neuvieme siècle, Le petit corporal, Le matin, L'autorité, La justice, L'univers, Le français, Le moniteur universel, La patrie, La défense, La liberté, Le gaulois, Le temps, Le soleil, Le petit moniteur, Le national, L'industriel,* and *L'instransigeant,* among others.[145]

Abel Hovelacque was also a deputy and president of the Paris municipal council. Like Mortillet, he, too, went looking for battles with the sacred. It may be remembered that Hovelacque started out as a lawyer. He was taking a variety of courses while studying for the law exam and thus came under the influence of two well-known scholars: the pioneering linguist Honore Joseph Chavée, with whom he founded the *Revue de linguistique*, and Broca, whose anthropological society he joined in 1867, at the age of twenty-four. He spent the next decade measuring skulls in the lab, writing essays, and teaching linguistic ethnography at the Ecole. In 1878 he was elected municipal councillor and from this post he organized a petition to claim two Paris convents for the city.[146] The buildings were communal property but Napoleon's Empire and the Restoration had allowed the Sisters of Saint-Vincent and the Sisters of Ignorantins to use the buildings, and they had been there ever since. Hovelacque argued that the arrondissement needed buildings for secular schools, and he collected enough signatures to kick the poor nuns out. One convent was transformed into a vocational school for boys, the other into an upper-level primary school for young girls.

Thulié, the freethinker who had begun life in Paris as a founding member of the realist art movement, had gone on to earn a medical degree specializing in mental illness. Just after serving as a surgeon in the Franco-Prussian war, he was elected to the municipal council of Paris and the general council of the Seine. He was secretary of this latter when it voted for free, obligatory secular education, so his report was published "by the entire republican press." Thulié then began an energetic anticlerical campaign in the manner of his colleagues. In the words of an outside observer, "Considering clericalism as the union of all the enemies of the Republic, [Thulié] began his anticlerical conferences, which became his specialty." We do not know exactly what these anticlerical conferences were, but we have got a sense of the thing. He had ample opportunity to put on such events, for in the following decade he was four times elected president of the Paris municipal council. He was chosen by this body to create a conference on Voltaire's centenary anniversary—quite likely on his own suggestion, as this was a major preoccupation of the freethinking group. From this post he also wrote "combat brochures," such as his 1875 *La coalition cléricale*, which apparently sold out at 200,000 copies.[147]

The freethinking anthropologists thus worked to deconsecrate human remains, wonders of nature, government stationery, government personnel, several buildings, a city cemetery, and burial rites in general. They also worked to create popular secular education and festivities, such as antireligious lectures, conferences, and celebrations of Enlightenment heroes. This work was about revising the epistemic setting, the worldview, for the new secular democratic era they were planning. Government and populace alike were willing to support the project—financially and otherwise. The Société d'anthropologie had invented and ritualized so many activities that it was beginning to parallel the heavily marked calendar of Catholic worship. The group had long been meeting once a month for a Freethinker's Dinner, which was retitled the Dinner of Scientific Materialism to suit their new affiliation. In addition, they met for the Lamarck dinners, and the dinners for Voltaire and Diderot. There were elaborate parties held to welcome new members, to celebrate the accomplishments of old members, and to raise money for the group's many related endeavors. To examine these festivities is to see a group actively forming its members' identities, reinforcing its perception of the social and political world, and cajoling group uniformity and loyalty.

For instance, on January 7, 1886, the freethinking anthropologists held a dinner for Mathias Duval on the occasion of his ascendance to the chair of histology at the Faculty of Medicine. He was replacing the famed Charles Robin, who had essentially founded French histology, creating a lab to study it "at a time when the microscope was an object of derision for most doctors."[148] Robin was also famed as a freethinker. As I mentioned earlier, he was singled out along with Broca as one of the most dangerous materialists at the Paris Faculty of Medicine. So, at least in terms of his freethinking, it was not extraordinary that the respected Dr. Mathias Duval would replace Robin. Still, for a member of the Paris anthropological coterie, this was an extremely prestigious post at an illustrious institution. The anthropologists wanted to celebrate this step into legitimacy, but they also wanted to claim it for themselves and use it as an occasion for materialist ritual. Consider their own description of the feast and the decor at Duval's dinner: "The menu, which was most fantastic, was presented by a human skeleton and a skeleton of a gorilla. Below that could be found the bust of Duval surrounded by symbols of science and of the arts. The hors-d'oeuvres were served in prehistoric pottery. Some of them were even to be found in the cavities of skulls. The grand piece mounted at the center of the table was an immense nougat representing a group of human skulls."[149]

With celebrations like this, the society was bringing its deconsecrating project into the realm of religious ritual. The food was served inside the cavities of human skulls! We may assume they had a professional confectioner shape the "immense nougat" group of skulls, at a cost of much time, money, and effort, and the simple fact of having thought of it and deciding to act on the thought is remarkable. Consider that they then ate the nougat human skulls. These are, we should remember, the same people who argued that cannibalism is always religious. That they were also handling each other's brains and skulls, year after year, makes this display more significant still. How might it have felt to look across a dinner table laden with skulls and raise a toast with someone who would likely, someday soon, lift your own brain out of its shell of bone and heft it to the light? What expressions were there exchanged? For that matter, there must have been some widened eyes and purposeful glances traded over the autopsy table as well. Sitting down to a feast table full of skulls charged the atmosphere. These men and women, intimate friends, were symbolically enacting an antireligious rite together.

The human and gorilla skulls displaying the menu for the feast were potentially disturbing on two levels: they reminded one of death, of course, but, because of their juxtaposition and given the context, they were also a reference to the animal nature of humanity. The human skeleton was dramatically being treated as just another object, as if there were nothing special in it or in its living counterparts. But remember that Jacques Bertillon said the gorilla skeletons were the most precious things in the Museum of Anthropology. The anthropologists could still get a rise out of dining with a gorilla skeleton, but the point was to treat such wonderful objects as if they were no longer wonders. Even the casual respect that one might pay this remarkable item was actively undermined by the silliness of having it display a menu card. The effect was disturbing, and that was the point. The freethinking anthropologists wanted their celebration to be upsetting to the uninitiated, and at least a bit unsettling to themselves. As they reported, "Some people found this decor to be depressing, but that did not hinder the most frank gaiety from reigning throughout the meal" (25). In their eyes, all this was cheerful because they had exchanged the comfort of authority (of the priest and of God) with the pleasure of rebellion, the pride of independence, and the delight of existential courage in the service of truth. But we must also see that they themselves could not avoid mentioning that the decor was depressing. They had purposefully gone rather far in re-creating the spooky atmosphere of religious ritual, communion, and feast. This points to a curious paradox of the freethinking anthropologists: what they said was that they wanted the power of science and that they did not need the comfort of religion, but they created a science that gave them no power (capable of curing nothing, predicting nothing, moving nothing) but was able to fulfill the social roles of religion, to move them and comfort them.

After the meal, there were toasts and testimonials. Lefèvre's speech was described as containing a "very faithful" history of the freethinking dinners. The speech was preserved and is worth an extended look (26–28).

> At the moment when one of us receives the much merited recompense for his work, and the title that consecrates his talent, at the moment when our learned friend Mathias Duval managed to bring into higher education, like a fresh breeze, the free spirit—the freethinking that animates us all—isn't it

natural that one of the oldest members of our group give a little history and show the line that attaches the foundation of this modest dinner to the happy success that we celebrate tonight?

Assuredly, if the doctor Mathias Duval has a place at the Ecole de médecine, it is not at all because he is one of ours; it is not the fraternity with which he honors us that could recommend him to the choice of the minister [of public instruction] and of his new colleagues. It would be truer to say that, because he is one of ours, his merit and the renown of his biological discoveries would be thrown into the shadows, while the places he frequented were made suspect, and that he would have to force open the doors of the microcosm of officialdom. And yet, we all feel it, I know, our dinner is not a stranger to the event that brings us together tonight. For was it not this dinner that brought together, for the past almost twenty years, the partisans of a common doctrine?

A little shout, among the indifferent and the hostile, in the disarray of opinions, has given, so to speak, a body to our ideas. It has contributed to creating the intellectual milieu that formed and affirmed the scientific convictions of Mathias Duval himself, and it has been a motivating center—the hidden, discreet origin of an unseen movement—the germ of an evolution, very slow for our taste and constantly being blocked but necessary and already fecund.

We should not overstate the significance of Lefèvre's comment that the regularly meeting dinner "has given, so to speak, a body to our ideas." We do not really know if that "so to speak" meant he was winking about scientifically measured human bodies *embodying* their ideas. Still, given the decor, the accomplishment being honored, and the company in general, it is a comment worth noticing because it suggests an interesting hint of self-consciousness. Lefèvre then gave a history of the dinner, which began with the founding of the freethinking journals and their suppression under the empire. He quickly passed to the role of Royer: "Madame Clémence Royer had just interpreted Darwin and, whether he liked it or not, she had pushed the master's doctrine to its final consequence. Moleschott and Büchner, also recently translated, had reanimated among us the memory of the French precursors of scientific materialism and transformism: d'Holbach, Diderot, Lamarck, and passing above the fictions of the concordat and the bastard eclecticism [of the empire], we came to seize again the heritage of our eighteenth century." This crediting of Moleschott and Büchner, even in the small role of "reanimators" of memory was extraordinarily rare. From the earliest days of *Libre pensée* and *Pensée nouvelle*, and their myriad articles on materialism, through to the freethinker's anthropological writing, when the Germans were mentioned, or included, it was generally as esteemed colleagues, not leaders. Even here, that tendency is well evident. To continue with Lefèvre's toast:

> Such was the object of our efforts. It was in order to sustain us in our goal and in our hopes that several of our colleagues, dead and alive, Coudereau, Asseline, Assézat,[150] Letourneau, Thulié, Yves Guyot, de Mortillet, Issaurat, Hovelacque—but we would have to name them all. It was to search for modes of propaganda and to prepare our next campaigns that we took on the habit of meeting together every month in a relatively secret room. . . . The terrible year did not separate us, we refound almost all our number at the Ligue des droits de Paris, between the victors and the vanquished.[151]
>
> Let us move on.
>
> The truly admirable energy of Asseline reformed the group and extended it. Without an organ of our own title, we spread our ideas in journals and reviews, and the dinner returned to its regular meetings.

It was at this point, Lefèvre explained, that the freethinking group joined the Société d'anthropologie and began publishing its scientific journals and several series of books. He also reminded his listeners of the great role they played in the organization of the centennial celebrations of Voltaire and Diderot.

> These results and the personal successes of our friends, aren't these the sure gauge of life for a dinner, where one meets with professors, councilmen, deputies, doctors, writers, linguists, and scholars who honor our country? A little joy is well permitted to us in these days of doubt and worry.
>
> Permit me to bring us together in a fraternal toast, our faithful quaestors Gillet-Vital and Issaurat; our profound moralist whose success gives all his friends a reason to rejoice, Letourneau, president of the Société d'anthropologie; and, finally, the hero of this party, the doctor Mathias Duval, professor at the Faculty of Medicine of Paris.
>
> Friends, I drink to you all.

There were cathectic bonds being formed here. The power to forward the various life goals of the group's members was imputed to the doctrine itself, yet a sense of secrecy and danger was made to adhere to the group as well. That speech along with the menu and its deceased presenters and understood within the context of the work in which these anthropologists were engaged, and their "profaning" behavior in particular, we are left with an elaborate description of a religious enterprise. It translated the individual's needs for community and for identity—needs that had once been mediated by religion—

into a nominally antireligious project, but it met those needs through religious behaviors. Years later, Clémence Royer would look back on the passion and ritual of the society with great affection and with strikingly explicit reference to the religiosity of the endeavor: "We were a little church. I love little churches; they are life and liberty. . . . In our little church we worked with fervor, we struggled valiantly, shoulder to shoulder, hoping to discover great truths, founded on ever more numerous facts and more general laws."[152]

CHAPTER THREE

∞

Scientific Materialism and the Public Response

The freethinking anthropologists grew into the roles they had taken on: they became scientists and were well respected among the scientific community at home and abroad. They wrote and published an extraordinary amount of material on their own and found several publishers for their scores of books. A great many of these books went into second and third printings. Their writing was blatantly, even evangelistically, materialist. As I will show, some of their audience celebrated this, some ignored it, and some spent a terrific amount of effort deriding it. Yet before entering into an analysis of their work and the professional response to it, I must touch on the question of the general public. Clearly, people were consuming all this literature, purchasing books and journals, and reading the professional reviews. But did the general public agree with the anthropologists? Those who joined the Society of Mutual Autopsy did, but what about everyone else? Were they fellow travelers? Were they being converted?

It seems that, despite their book sales, the materialism of the anthropologists never represented anything like a majority of French men and women. Instead, they were supported by the newly secular French government and also by a broad swath of the population that was eager to have a few materialist arguments in their personal arsenal and a bunch of materialist titles on the bookshelf. When a people seeks to describe its own moment, especially if it feels striking in its particularity, they cast about to name the quintessence of the new ideas that, in some attenuated form, have affected their lives. The freethinking anthropologists were kept in the public eye by both friends and enemies of their ideals, because they brought the hottest issues in France at that time to a thrilling extreme. They served as a sort of North Star of irreligious materialism: you did not have to go to Polaris, or even get close, or even be going north, to find the star an immense service in navigating the cultural terrain. The men and women who joined the Society of Mutual Autopsy were bundling up for the frozen North Pole of materialism. Even most other devout republicans were not going to follow the doctrine that far, but republican or not everyone seemed to check its position now and again and to refer to it as a reliable beacon in a complicated field of thought.

SOLDIERS OF ANTHROPOLOGY

The freethinking anthropologists often wrote that they intended to change the world through anthropology. They eventually developed a plan for doing this, but at first, and in a sense always, the idea was that knowledge about human beings had heretofore been mediated through religion and social hierarchy and was therefore corrupt. Philosophy had made some inroads in reinvestigating human nature from a secular and egalitarian standpoint, but it was mere conjecture, weak against enemies because it had no material proofs. What's more, its mere existence seemed to condone religious thinking because both availed themselves of feelings and unempirical concepts. Anthropology was going to do better. Without metaphysics or prejudice, anthropologists would collect facts about humanity. These facts might be taken from a very wide range of subjects: contemporary Parisian bone lengths, flint shards from an archaeological dig in North America, breastfeeding practices in Tunisia—any objective thing or behavior that could be described, sketched, or measured. The freethinking anthropologists did some traveling and digging themselves, they borrowed a lot from published sources, and they also deputized many amateur travelers. Before setting off on a voyage—on holiday, in the army, or in some other colonial enterprise—many people would come by the Anthropological Institute in Paris to be outfitted with extensive questionnaires to be put to the natives, as well as measuring devices and a list of what to measure. This may seem a strange way to collect scientific data, but back in 1865 Broca had said that "man . . . is not any more difficult to observe than a plant or an insect; any doctor, any naturalist, any attentive and persevering traveler can measure him, describe him methodically, without having to prepare by special studies, because the information to be gathered has to do with exterior characteristics that anyone can ascertain."[1] The freethinkers wrote a bunch of these questionnaires with Broca in the 1870s and without him in the 1880s. Most were published in journals as well as singly, in pamphlet form.[2] The travelers mailed in specimens and data, and the anthropologists published whatever they considered to be of interest. The measuring must have been fun—it certainly allowed for a good deal of otherwise unlikely human interaction and an odd intimacy. It seems to have felt like important work to everyone involved, but there was often no real attempt at synthesis or any interpretation that went beyond the most localized question.

Other sciences, astronomy, for instance, had collected innumerable facts that served later syntheses, so it seemed reasonable to keep amassing details even when there was no apparent point. Now and again the anthropologists articulated this, but they tended to claim more caution than was actually employed. Wrote Letourneau, "Social science is still in its infancy, to formulate laws is beyond its power, but scientific laws do not burst forth by spontaneous generation, one must prepare for them by pulling out of the chaos of details a few general facts—we hope to have succeeded in this."[3] What they really did was to turn "the chaos of details" into orderly piles of detail, without much rigorous sense to the piles. In fact, just before the positivist Eugène Dally rejected the materialists in the society as "trying to found a kind of church council," he railed against a committee-written questionnaire that followed Letourneau's classificatory system. As Dally put it, "I can't accept these divisions, not because they come from him but because they don't come from everyone, because they are not current, because if they were true they are not in our intellectual usage, but above all else because they are not true."[4] In the questionnaire to which Dally was referring, for instance, religion was grouped with dance and music, and we can understand Dally's frustration with this. He was explicit in his desire to standardize a single, international model upon which scientists everywhere could agree; the freethinkers, by contrast, were trying to change the way people saw the world.[5] Still, this work did process the material for general consumption, and it drew attention for a host of reasons. Some of it was rather titillating: sketches of a woman with a second left breast, tucked beneath the first; a lengthy description of a man who fathered children though he had lost his penis in an accident; accounts of cannibalism, hermaphrodites, the sexual practices of "savages." Articles about such things might be only a few paragraphs or a few pages, and though some were very long, the aggregate effect was a great crowd of singular, essentially disparate materials.

The freethinking anthropologists did, however, develop a theory into which some of these facts and descriptions could be pressed. The idea seems to have come independently from Mortillet working in archaeology and Letourneau working in anthropological sociology. As Michael Hammond demonstrated in his article comparing Cuvier's conservative politics and belief in the fixity of the species with the politics and ideas of these late-century anthropologists, the freethinking anthropologists believed progress was unilinear and followed the same stages in all places and cultures.[6] As atheists, they thought of themselves as the pinnacle of the history of religion and figured that the history of the rest of culture was also a story of sloughing off the primitive en route to an enlightened, utopian future. Mortillet had found a lot of evidence that pointed to a material culture at a time in prehistory when, according to his colleagues, human beings did not yet exist. He spent much of his life arguing that we must have already been around, and in the end he prevailed. But meanwhile he had to classify all his finds, and he settled on a notion of absolute progress in all places at all times and grouped the items accordingly, showing continuity between tools found unaccompanied by human remains and tools found near such biological evidence. Letourneau's situation was a bit less bleak and more overtly political. He wanted to write an anthropology of society, and his only theoretical claim was absolute progress toward his group's vision of human perfection; he rather straightforwardly lined his facts up to form an arrow pointing wherever it was he wanted to go. Mortillet and Letourneau both favored an animalistic vision of human beings in which various "hungers" determine human behavior, yet their understanding of the story of the human race extended into a triumphant and romantic future.

The other freethinking anthropologists followed this model. Whatever aspect of the field they studied, they classified what they termed early behaviors (specific to either prehistoric human cultures or to contemporary nonindustrial societies) as being natural and thus to be defended in modern French culture or as savage and therefore to be repudiated. The model was very similar to what Royer had suggested in her preface to Darwin's *Origin of Species*. The freethinking anthropologists determined whether a behavior was natural (good) or savage (bad) as follows: natural human behavioral characteristics were increasingly developed and perfected as a society progressed. Savage qualities, on the other hand, decreased with the passage of time.[7] It would be hard to think of a more manipulable system, and the freethinking anthropologists really just made their decisions by preference. In 1881 Abel Hovelacque wrote that through anthropology,

> we learn our origins and we see that our moral amelioration is tied intimately to the continuation of our organic evolution. We know all that is still present of the savage and barbaric in our modern civilizations: the priesthood, belief in gods, militarism, the subjugation of the weak and poor, the inferior condition of women, the cult of authority, respect for functionarism, suspicion of individual liberty, and social inequality. Such are the surviving traits from

which the development of anthropological science is called upon to liberate us. (314)

Not all the freethinkers agreed with his list of barbarisms—some were more concerned with class struggle than with women's rights, for example—but this is still an accurate depiction of the incredible list of charges they laid to the young social science.[7]

One item noticeably missing from the freethinkers' list of barbarisms was racism, and anti-Semitism in particular. The texts they wrote were often full of assumptions and claims about what they believed to be the temporary, but nevertheless real, biological limits of various non-European human groups. Still, they consistently defended the notion that all human groups, if encouraged or if simply no longer held back, could uphold the responsibilities of legal, economic, and political equality. As early as 1865 Eugène Véron wrote that he did not want to "imprison the entire possible development of a people in racial considerations," and this appeal to progress and possibility generally guided the freethinkers' approach to race.[9] Sometimes, though rarely, they went further. Mortillet, for example, argued strenuously against anti-Semitism, using anthropological arguments when he could think of any and otherwise simply trying to work out the socioeconomic origins of the prejudice.[10] The lacuna is still a bit odd. It is best explained in two ways: first, the timing—many members did not even live to see the Dreyfus Affair, let alone incorporate its issues into their work. Second, as I have mentioned, the written work of the freethinking anthropologists rarely mentioned concrete political questions of the day, despite many references to feminism, socialism, and egalitarian democracy.

Over all this reigned the freethinking anthropologists' notion that thought and emotion were physiological products, like sweat and urine, and that intellectual, emotional, and artistic desires were best understood as biological processes, exactly like hunger for food. This may sound like the naturalist literary movement championed by Emile Zola toward the end of the century, and, as I noted early, in the 1850s Thulié was part of a "cénacle réalist" and founded a journal called Le réalisme. (Naturalism and realism were distinguished primarily by the naturalists' devotion to objective description and their rejection of the moral commentary common to realism.) In fact, the connection is more profound even than that: Zola specifically credited Letourneau as having given him the information he needed to write his "natural history of a family," the Rougon-Macquart series. Among Zola's extensive early notes for the series are résumés of only two science books; Letourneau's Physiologie des passions of 1868 is one of them. Included in these pages are quotations from Letourneau that stand as singular elements of the materialist credo: "I sense, therefore I am"; "Of this vague formula is born the illusion of free will. Man, constantly solicited by numerous and simultaneous desires, obeys the strongest while being conscious of the others, and *that is why he feels as if he were free*" (The emphasis is Zola's).[11] Indeed, Zola reviewed Letourneau's book in *Le globe* and celebrated it for these qualities, writing, "Here we are fully in materialism, fully in experimental science. [Letourneau] has medically studied the passions, showing them to be born in the organs of the body and finding their cause to be in these same organs." Further explicating Letourneau, Zola related that what poets, philosophers, and theologians called "the unfathomable mystery of life" is nothing but a "phenomenon of organic assimilation and disassimilation." Humanity, "envisaged sanely and not through the tinted glass of metaphysics, is, like all other organized beings, nothing but an aggregation of histological elements, fibers, or cells, forming a living 'federal republic' directed by the nervous system and constantly renewing itself."[12]

The Zola-Letourneau connection is well known and often cited by literary scholars, but since they have heretofore had little or no information on Letourneau, the connection has not meant much. Still, one Zola scholar states that "[Zola] was particularly impressed by Letourneau's doctrine that man's moral and intellectual needs are as organic as his need for food, and that the need for food is the most imperious and indispensable of them all. He decided to make his novel [the Rougon-Macquart series] to a great extent a novel about different sorts of hunger—hunger for food, wealth, power, all the benefits of modern civilization."[13] Another Zola scholar notes his reliance on Letourneau and adds that the novelist voraciously read "the works of Darwin, recently translated into French," which suggests that Zola was under the influence of Clémence Royer as well.[14] It is surprising, from our present vantage point, to see that not only did Zola rely on Letourneau on a conceptual level, but he also saw him as capable of lending prestige and delivering a faithful and enthusiastic audience. In 1868 Zola exchanged a few letters with the naturalist author Joris-Karl Huysmans with whom he was planning to start a new journal. A note he received from Huysmans read, "Dr. Letourneau, a man well respected in philosophy, will cover scientific developments. This is good, I believe, as he can bring us the clientele that buys books at Germer-Baillière and Reinwald."[15] But the Zola-Letourneau conceptual link was strong in its own right, sales and clientele aside. In Zola's notes to himself in the Bibliothèque nationale, he mused that great novelists are supposed to have a philosophy; he resisted the idea but concluded that "the best one would perhaps be materialism."[16]

Not everyone was as positive as Zola, but in general the work of the freethinking anthropologists was highly esteemed. For the two decades between

1880 and 1900, anthropology in France was significantly defined by the journals attached to the Société d'anthropologie, the teaching at the Ecole d'anthropologie, the *Dictionnaire des sciences anthropologique*, and the books that were
published in the various "Bibliothèques." These included a "Bibliothèque matérialiste," a "Bibliothèque anthropologique," and a "Bibliothèque des sciences
contemporaines."[17] In order to map the science and politics of these works, I
begin with a brief discussion of the *Dictionnaire des sciences anthropologiques*, followed by a look at the Bibliothèque des sciences contemporaines (the collection discussed with such rancor by Topinard), and finish up with a discussion of
exemplary pieces from the Bibliothèque matérialiste and the Bibliothèque anthropologique. At the same time, I will consider the response these works generated from the scientific press. An analysis of the response from the philosophy and literary journals and the popular press follows thereafter.

THE *DICTIONNAIRE*

The *Dictionnaire des sciences anthropologique* was intended as a compendium of
anthropological information, with individual entries written by the most qualified expert—so long as he or she was a freethinker (or close). For the *Dictionnaire* was also intended as an ideological manifesto in the tradition of
Diderot's Enlightenment *Encyclopedia*. The work reflected Diderot's in significant ways. It was full of factual information, but many of its entries were
deeply political, and some of these communicated their politics through a
rather giddy sarcasm. Mortillet had been warned not to push this mood too
far. The advice came from a very interesting quarter: Carl Vogt, the German
materialist naturalist in exile in Switzerland with whom Mortillet had worked
during his own exile. Mortillet had asked if Vogt would contribute to the project, and, in a private letter to Mortillet written in the spring of 1880, Vogt declined, explaining that he had not written any anthropology in years and that
even were he to do so, the French press did not pay nearly as well as the German. Still, he took the time to warn his French colleague strongly against in-
dulging his usual ebullient atheism when putting together the *Dictionnaire*.
Vogt was no less a materialist, but he felt that the *Dictionnaire* would not make
any money if it were too insistent on the nonexistence of God and the soul. As
he warned Mortillet:

> If you want to be recompensed for your work on the Dictionnaire, do not make
> it too exclusively representative of our point of view. . . . If you want the
> dict. to be purchased, do not shake anything up, or leave anyone out, especially

anyone who has real merit but who does not go as far as you do, and do not be extreme in your manner of presenting things. *We are in an intimate minority*, it is incontestable, and except for a few rare exceptions, the majority is in control. Thus if you want to sell them your merchandise, do not print on it: Poison, prepared by several personal enemies of God. Otherwise you risk having your merchandise never leave the store. The nomination of M. Bertrand to the Conseil supérieur has proven to me that things have not changed as much as you like to say they have, that the Ac. des Sciences is still the same and the resistance of the Faculty of Medicine against pathological anatomy shows me that even in the groups that are said to be advanced, there is still an unshakable prejudice. So do not take the second step before you have taken the first.[18]

Perhaps Mortillet took heed of this advice; the *Dictionnaire* may have been originally conceived in an even more overtly atheist form than that of its final realization. As it was, the progressive ideology and anthropological methodology of the work were set out in a preface stating that "social evolution" was consistent in all societies as they progressed from tribes to castes to absolute monarchy. After that, the preface explained, "little by little the chains loosen, privileges attenuate or disappear, social inequalities increasingly raise public indignation." This pattern, exclaimed the anthropologists, "releases a great and powerful idea, the idea of progress, always necessary, always increasing in speed, despite the roadblocks it often meets."[19] Some entries stressed this notion more than others.

Letourneau's entry entitled "Femmes" is an excellent representation of the general approach (476–478). He discussed women in general and then moved on to a description of society's progressive stages regarding the relations between the sexes. Letourneau's understanding of these stages was as follows: women historically pass from being treated as domestic animals to being treated, progressively, as slaves, servants, and minors. "This gradation," he wrote, "is instructive; it obviously marks a direction, a slow work of emancipation and movement toward equality, which is not to be thought of as finished" (478). In his booklength study of the evolution of marriage, Letourneau had written that this movement had been hampered by the reversals "caused by Catholicism" and its submission of wife to husband. With the onset of atheism, "progress resumes its course."[20] In the *Dictionnaire*, some ambivalence about feminism showed through: he stated that he did not think women should be given political equality immediately, because they needed more education before they could handle the commensurate responsibilities. Despite his concerns, Letourneau wrote that he hoped and expected that "little by little, civil equality will be recognized, and a little later, political equality must follow."[21]

He concluded by tipping his hat to women such as Royer, who had moved faster than the rest of society by pursuing careers in fields that had previously been forbidden to women.

This juxtaposition of the capabilities of all women and the achievements of Royer was by no means casual. As freethinkers, this group of men was already inclined to champion the emancipation of women. As anthropologists, the inclination was significantly encouraged by their professional relationship with Clémence Royer. Letourneau's ambivalence about women's intelligence as well as his profound respect for Royer are evident in an address given at a banquet in Royer's honor in 1897. In it Letourneau elaborately praised Royer and admitted, with apologies to "the distinguished women" in the audience, that he had originally believed that "Cl. Royer" was a "pseudonym disguising a man and not an ordinary man but a philosopher, doubling as a vigorous writer." Letourneau went on to say that it was rare for either a male or female scientist to possess such a "virility of expression, rigorous logic, clarity of thought, and courageous penetration" as was demonstrated in Royer's famous preface to *Origin of Species*.[22] This language—from "virility" to "penetration"—demonstrates a strong desire to retain intelligence as a masculine feature, but it also ascribes that feature to Royer—and, by extension, to all women. As Joy Harvey tells us, Ernest Renan once praised Royer as "almost a man of genius," to which her friends and admirers replied, "Why almost? Why a man?"[23] The "distinguished women" in the audience probably included the several female scholars who had joined the society in the 1880s and 1890s.

Royer wrote the *Dictionnaire*'s extensive entry on evolution, which explained Lamarckian evolution and emphatically stated the French origins of the notion before discussing Darwin's contribution to the idea. Despite the Social Darwinism that Royer championed elsewhere, in the *Dictionnaire* she was rather leery of Herbert Spencer's assumptions regarding the application of Darwin's theory to human society. The entry amounts to a sober survey of the idea of evolution according to Clémence Royer, the most political aspect of it being the simple fact that a woman wrote it. Some entries took matters a good deal less seriously. Writing on fairies (the entry "Fées"), Lefèvre mentioned that in Christianity "one plays around with supposed apparitions of a Jewish girl who has been dead for eighteen hundred years."[24] Vogt's fear that the freethinking anthropologists would scrawl "Poison, prepared by several personal enemies of God" on the book was not far-fetched. Aside from Lefèvre's jokes, which were an obvious emulation of Diderot's (under "ANTHROPOPHAGY:" Diderot put "see EUCHARIST"), most entries were serious and scholarly. Many experts outside the inner circle of freethinking anthropologists had been called upon and had delivered dense, impartial definitions and descriptions. The freethinkers' entries, however, often followed their anthropological progress formula. Consider the entry on infanticide: "Today, the principal causes of infanticide are the shame attached to maternity outside legal marriage and the prohibition on searching out paternity. . . . Infanticide evolved like everything else. At first it was bestial, infanticide from need, then it became religious, as at Carthage, or social as at Sparta. As for infanticide in modern-day Europe, one could call it moral or legal, because it has as its principal causes the misunderstanding of public opinion and the injustice of the law" (610–611).[25] Hovelacque's entry "Nationalism" clearly placed it as a surviving form of a waning savage behavior rather than simply a bad modern development; in any case, it was a problem. He held that its suppression was one of "the practical goals of the anthropological sciences" (795). In one of the more obnoxious entries, Lefèvre described "apparitions" as "nothing more than hoaxes and simpleton traps, good for the little shepherds of La Salette or for the idiot of Lourdes" (108–109).

From his post at the Laboratory of Anthropology, Manouvrier was a fellow traveler of the freethinking anthropologists and was invited to write on many anthropological subjects for the *Dictionnaire*. Staunchly republican, materialist, and egalitarian, Manouvrier was also an excellent material anthropologist. He figured out a lot about the trick to deciphering fossil bones—that is, how behavior, conditions, and body size can affect a skeleton. Also, when people started finding bones that seemed to be from a missing link between apes and humans, Manouvrier was very early in his cautious but firm support for the theory of evolution. When Eugène Dubois showed up with a femur, discovered in Java, that seemed neither ape nor human, Manouvrier helped him measure four hundred human femora so that a careful comparison could be made. Manouvrier championed aspects of both Darwinism and Lamarckianism and brought much positive attention to the theory of evolution in general. He also pioneered the whole practice of estimating total height on the basis of individual bones. Unlike the more combative freethinkers, Manouvrier tended to be extremely cautious, considering anthropology to be politically useful but recognizing that it could be dangerously abused. As chapter 6 will show, he was very conscious of the political meanings of anthropology and never claimed to hold the science above social and political concerns. Still, of the many articles that he wrote for the *Dictionnaire*, only in one did he join the others in their combat stance. That was for the article entitled "Sexe," in which Manouvrier categorically dismissed female intellectual inferiority. "One can conclude," he wrote at the end of a lengthy analysis, "in sum, that there is no known anatomical fact indicating an inferiority of the female sex having to do with intelligence" (1000).

Mixed in with mainstream descriptions of phenomena and history were definitions that spoke of "universal atheism" as the goal of science ("Athée," 144–145), or explained that the family was progressively being phased out in favor of communal child rearing ("Famille," 469–472), or assured readers that within a half century, and without violence, workers would be in control of the means of production ("Industrie," 607–610). Positivism was accused of religiosity (918–920). The Société d'anthropologie was described and celebrated (1014). Thus the dictionary did serve as a manifesto of the freethinkers' "anthropology of combat," as promised. It also provided a generally scholarly compendium of contemporary ideas on the science of humanity. This is how its late-nineteenth-century readers responded to it, as is demonstrated by discussion of the dictionary in the *Revue scientifique*. This journal—also known as the *Revue rose*, for its pink cover—was the main science journal of the period, covering a range of applied and pure sciences and many other related topics (history of science and science teaching, as well as industry and "the military arts"). Each issue carried original articles as well as reviews of scientific books, conferences, journals, and exhibitions. Occasional contributors included Darwin, Pasteur, Marcellin Berthelot, Galton, Lombroso, and Vogt, as well as Broca, Quatrefages, Letourneau, Manouvrier, Royer, and Duval.

Book reviews in the journal were generally penned by its editor. In 1884 that was the doctor and medical researcher Charles Richet, who would later win the Nobel Prize for his work in serum therapy.[26] The dictionary's review in *Revue scientifique* was extremely positive, asserting that the "considerable importance of anthropology" was "increasing every day" and that its "numerous discoveries" had long demanded a source book such as this. Richet attested that the *Dictionnaire des sciences anthropologique* had been published under the direction of a "group of distinguished scholars and with the collaboration of a great number of anthropologists of the highest authority." The only problem with the work (aside from too many entrees that "really belonged" in a medical dictionary) was that the "tone of certain articles seem marred by a tendency to introduce politics—blatantly—into questions that absolutely must hover above politics and not leave the domain of pure science." Still, Richet cited no examples of this and quickly recovered his tone of praise, writing that, "this reservation aside, a great number of the terms have been magisterially treated . . . with all the developments that truly belong to them." The book was recommended for scholars and the general public, and readers of *Revue scientifique* were assured that anyone interested in the current research "will certainly rewardingly consult the dictionary on a great host of questions."[27] Within the scientific community, this was a very typical response to the freethinking anthropologists' writing in general: it was important work and good science, despite being problematically spotted with left-wing politics and evangelical atheism.

THE BOOKS

The Bibliothèque des sciences contemporaines was a great success. By 1883 eight volumes had appeared, and many had gone into a second printing. These included Letourneau's *La biologie* and *La sociologie d'après l'ethnographie*, Hovelacque's *La linguistique*, Topinard's *L'anthropologie*, Véron's *L'esthétique*, Lefèvre's *La philosophie*, Guyot's *La science économique*, Mortillet's *Le préhistorique*. In each case, physical anthropology—the biological nature of human beings—was the predominant focus and the source of all information. The soul was everywhere denied.

André Lefèvre's *Philosophie*, for example, was presented in two parts: one a history of thought from the earliest civilizations to the contemporary, the other Lefèvre's explanation of all things material and immaterial. The whole served the cause of science and progress with sedate, neat arguments and relatively controlled optimism. Only briefly did Lefèvre call attention to freethinking as such, and in doing so he did not mention that this was his project. Instead, in his history of philosophy, he wrote that:

> When, close to the end of the empire, an independent group, without anyone's backing, and without any compromises of any sort, raised up the flag of the *Libre pensée*, they were not walking in the steps of Virchow, Moleschott, Buchner, or Vogt (with whom there was, however, an alliance); they rescued a patrimony that had almost fallen into foreign hands. The disaster of 1870 and the lamentable discord that had followed it interrupted their work, at least apparently. The *Pensée nouvelle* did not live two years; the *Encyclopédie générale* has only three volumes. But, without a journal and without a visible corps, the doctrines that were represented in these two memorable publications continue to exist today as the only living doctrines standing out against vulgar metaphysics and refined idealism. Positivism, which is, by degrees, absorbed into this doctrine, will have the honor of having served to pave the way from the naturalism of the eighteenth century, to the materialism of the nineteenth.[26]

Letourneau's books were equally clear in their passionate materialism, often in the context of his insistence on the animality of human beings. This issue had become rather important in secular culture at the end of the century.

That Judeo-Christian ideology had so clearly and meaningfully separated humanity from everything else in the universe (including beasts and fish and fowl, over which humans had been given dominion), meant that Western atheism demanded a new formulation of the issue. The intimate relationship between anthropology and atheism heightened this demand because it suggested that the natural world of animals would be the primary source of new truths. Yet there was little consensus on whether humanity should distinguish itself from other animals (and if so, on what basis) or model itself on their "more natural" communities and behaviors. (This issue will resurface throughout the present study.) Letourneau's approach in *La sociologie* left little room for debate: "Man has long deceived himself with the idea that he was made in the image of the Divinity. It is now more than time to say and to repeat to this poor creature that he is animal in every fiber and in every particle of his being."[29] The pathos displayed in the epithet "poor creature" should not be overlooked.

La sociologie had a very clear projected telos. Throughout Letourneau's eighteen full-length books and innumerable articles he examined the evolution of morality, of marriage and family, of religion in various human races, and a plethora of other evolutionary tales including those of law, war, slavery, and the condition of women. In all these projects Letourneau aimed to show the development of human beings by systematically following "the progress" of human traits and behaviors from their "early" forms upward toward their attenuated or enhanced versions in "advanced" Western society. Letourneau mostly occupied himself with comparing past and present-day societies and merely suggested that certain traits and behaviors were on the wane. However, in *La sociologie*, his linear description of social evolution was projected into the future in a particularly detailed prediction. Letourneau wrote that "if, as in a fairy tale, some magician could display before us the tableau of the future, and maybe not too distant in the future, we would see the superior human races constituted into republican federations having profoundly modified their social organization." His vision of that social organization was sweetly tribal and democratic. "Confederated ethnic unions will then be little groups that administer themselves, by themselves, in everything that does not manifestly deal with the general interests of the republic. In each of these groups, social activity will be entirely absorbed by useful occupations." We will thus be delivered from fooling around and wasting time. The almost goofy optimism, the transformed nature of individuals, and the primordial feel of small, self-administered ethnic unions all combine to cast an eschatological tint on his predictions. In Letourneau's imagined future, "the greatest care will be given to the physical, moral, and intellectual education of the young generations; a great effort will be made, through appropriate training, to lessen organic inequalities, which will be the only ones that exist in this happy time."[30]

In order to facilitate the realization of this future, Letourneau wrote copiously, collecting and recording facts. He did not always claim to have an appropriately rigorous analytical methodology, and this humility led him to record endless data in the hope that future scientists could use it as the basis for their analyses. But no Kepler ever appeared. This did not much bother his scientist contemporaries. Charles Richet's review of La sociologie for the Revue scientifique was extremely positive, calling it original and profound and proclaiming that Letourneau "takes on, without prejudice and with an often happy audacity, the boldest and—if one may say it—the most frightening issues of the study of society." Richet was not kidding; he went on to explain that Letourneau's mastery was in demonstrating that "our society and this civilization, of which we are so proud, should not inspire such pure arrogance. Our costumes, our wars, our religions, our morals are compared without pity to those of savages." Like the Enlightenment writers who took readers back to the "state of nature," continued Richet, Letourneau was no less fearless in describing our faults. "It leaves one a bit disenchanted, even aggrieved, to see this tissue of folly and of cruelty that is the past and the present of man. Let us hope the future will be better." Letourneau may not have established a theory of sociology that would long outlast him (indeed, when Durkheim reviewed the history of sociology in 1900, he had little more to say than that Letourneau had "written voluminously"),[31] but he clearly did some important work in weakening the cultural arrogance of nineteenth-century Europe and helping its intellectuals to conceive of their culture as one among many historically specific and changeable discourses. Richet was also convinced by Letourneau's argument that "everything, or almost everything, is convention." Race eventually bowed to convention; more important for Richet, there were no transcendental absolutes, either. He also praised Letourneau's style as precise, elegant, and irreproachable. His only critique was that the anthropologist seemed rather casual in accepting travelers' descriptions as sufficiently perceptive and scientifically exact.[32]

The task that Abel Hovelacque set for himself in his contribution to the Bibliothèque, La linguistique, was to argue that the study of linguistics was the only means of differentiating between human beings and the rest of the animal kingdom. Like Letourneau, he saw humanity as fully animal. Hovelacque, however, asserted that humanity was distinct among the other animals but rejected the idea that religion or metaphysics was the distinguishing factor. He was also interested in elevating the importance of linguistics and confirming a materialist, progressive interpretation of the natural world. He claimed that anatomical differences could not separate humans from animals because some of these were more substantial between various apes than between apes and human beings. He also argued that the "inferior animals" were possessed of memory, imagination, shame, reason, pity, admiration, and so on.[33] As a result, he claimed, people had been forced to distinguish humanity by inventing "the argument of religiosity and the argument of morality. Their success has been poor" (24). Hovelacque explained that religiosity was just fear of the unknown and that morality was present among many animal communities (26). He then found a way simultaneously to support freethinking beliefs, defend scientific materialism, glorify Broca's memory, and promote the importance of his own specialization: he claimed that the only thing that truly separates human beings from the other animals is that the human brain has a speech section (21–37).

The *Revue scientifique* covered that book as well. At this point, the editor was Odysse Barot, and instead of a standard book review, the journal published a full-length article entitled "La linguistique moderne," subtitled "d'après M. Hovelacque" (according to M. Hovelacque).[34] Right at the outset, the article stated that Hovelacque claimed to have written his book to demonstrate the place of linguistics in the natural history of man but that "the author clearly had another goal, and that was to draft not a manual but a work disengaged from theological and metaphysical prejudice, composed following the principles of the current free scientific methods, as complete as possible and intelligible for all readers." Had he succeeded? "We think so," continued the review, "and we believe that this new science, linguistics, which bristles with so many difficulties and which had been reserved for only a few rare adepts, has never been more clearly put forward for the cultivated public. Hovelacque's book is a pleasing and interesting read; the views it contains are very clear and explained without the least bit of pedantry" (424). After a detailed discussion of Hovelacque's themes and arguments, the review resumed this paean to the book's elegance and the "inestimable service" it provided. What made the work so extraordinarily useful was that it was written in an "expansive, progressive spirit, disengaged from extrascientific prejudice" (428). Thus what some saw as the freethinking anthropologists' extrascientific prejudice against religion was here seen as a necessary and irreproachable defense: "We have here a book as learnedly drafted as it is freely thought; this is a double advantage for the reader, especially in an epoch when certain books on the science of language, coming from a noisy and overrated Anglo-Germanic personality and penetrated with a spirit of conciliation between the things of faith and the truths of science, serve as a kind of manual for everyone who begins to initiate themselves into linguistic studies" (428).

Eugène Véron's study for the Bibliothèque des sciences contemporaines was on aesthetics, a subject that may seem decidedly ill suited to the freethinkers' anthropological project. For this reason, Véron's work is particularly revealing of the extent to which these anthropologists subscribed to reductive materialism. Véron began his inquiry with the comment that "no science has suffered more from metaphysical dreaming than that of aesthetics" and asserted that throughout history art had been dealt with as "an amalgam of quintessential fantasies and transcendental mysteries."[35] His project was to tear down this "chimerical ontology" and replace it with the idea that "art is nothing other than a natural result of human organization, which is constituted so that man finds particular pleasure in certain combinations of forms, lines, colors, movements, sounds, rhythms, and images." Under this thesis, the principles of aesthetics could be explained, in large part, through optics and acoustics. Véron realized that the "explanation of the cerebral phenomena" was not sufficiently advanced for this, so until aesthetics could be a real science, it would have to content itself with "the statement and registration of facts, and to their classification in the most reasonable order" (vi). Véron also discussed the Society of Mutual Autopsy, the freethinker's greatest attempt to study the mind empirically. Here, too, he could not help admitting that "a large number of problems still remain unsolved" (vi).[36] This was the predicament of these champions of materialism. Given the example of the extraordinary advances in hard sciences in this period, they seem to have assumed that any empirical study would eventually pay off in some dramatically new, definitive construct. As for subject matter, Véron insisted that all art had to offer was the truth of its descriptions and the personality of the artist, so the only topic to choose was whatever small world one knew best.

This had an interesting consequence. In the United States, the naturalism literary movement took off after Theodore Dreiser published *Sister Carrie* in 1900, but there were a few major precursors, and one of them, Hamlin (Hannibal) Garland, was tremendously influenced by Eugène Véron. He touted Véron's call for authors to stick to realistic descriptions of their own corner of the world. Garland scholar Lars Ahnebrink tells us: "Garland's much thumbed copy of Véron's *Aesthetics* is full of annotations, comments and marginalia. Before the word, 'Introduction' (p. v), Garland had written in pencil: 'This book influenced me more than any other work on art. It entered into all I thought and spoke and read for many years after it fell into my hands about 1886. Hamlin Garland.'"[37] Quite a remarkable endorsement. Garland was, by the way, a "confirmed evolutionist" and an admirer of Darwin and Spencer. He called himself a verist, which he took straight from Véron's call for truth—and which also echoes the scientist's name.[38] As Garland explained, "Not being at the time a realist in the sense in which the followers of Zola used it, I hit upon the word verist which I may have derived from Véron."[39] Why did Garland reject

Zola? As he put it, "I thought to get away from the use of the word realism which implied predominant use of sexual vice and crime. . . . For the most part, the men and women I had known in my youth were normal. . . . Their lives were hard, unlovely, sometimes drab and bitter but they were not sexual perverts" (139). Elsewhere he wrote that the American followers of Zola seemed "sex-mad."[40] So verism was realism without the smut. Anyone who found that distinction unimportant did not have much use for the new term, and Garland became known as a realist. In 1928 the literary critic Régis Michaud wrote that realism was put to the American people as against sentimentalism, and that the choice had been theorized by Garland, who was citing Véron. Michaud tells us realism triumphed in American fiction until 1927, when Edith Wharton, tired of its provincialism, published a protest against what she called "the twelve-mile limit" in the *Yale Review*.[41]

What of Topinard's study for the Bibliothèque, *L'anthropologie*?[42] It is difficult to analyze considering Topinard's protestation that the freethinkers had edited it for political heresy before its publication. As it stands, *L'anthropologie* is a straightforward attempt at a physiological examination of the races of human beings. It presents some anthropological theory but is largely comprised of descriptive and quantitative racial portraits. Broca wrote the preface stating that Topinard's work was written as "an elementary treatise on anthropology—a systematic résumé" (viii). He explained that the need for a popular study of anthropology had long been felt but that an anthropologist "devoting himself to original research . . . is generally little disposed to employ his time in writing a work of a popular character" (ix). It does seem as if Broca, too, had a limited opinion of Topinard. The unilinear progressivism that was evident in all the other Bibliothèque volumes was absent from the book, but *L'anthropologie* did not contradict the ideals of the freethinkers. In its conclusion, in fact, it flattered these ideals enough to suggest that some lines had been added by the group. The text states that man's "spirit of inquiry is the most noble, the most irresistible of his attributes; and as M. Gabriel de Mortillet said at the meeting of the Association for the Advancement of Science, his special characteristic is here, and not in religiosity" (534).[43] Like *La linguistique*, *L'anthropologie* was reviewed in Barot's *Revue scientifique*, but this review was more sedate.[44] The work "does not disappoint our hopes: it is an exclusively scientific book." In describing the text, Barot took the time to list the amazing variety and number of skull measurements utilized by Topinard. He also reported the author's contention that there were at least three human species, more unlike each other than were dogs, jackals, wolves, and foxes. This was surprising enough to be repeated a few times in the review, but no commentary was offered on what this implied about race (334).[45] The ebullient response to scientific materialism was withheld in response to Topinard's neutrality on the question.

Of all of the studies for the Bibliothèque, Mortillet's *La préhistorique* had the most lasting effect on the discipline.[46] This is not too surprising, as it took part in the most enduring line of inquiry of nineteenth-century anthropology—does the fossil record prove evolution?—and argued, early, that it did. First published in 1883, *La préhistorique* summed up the theoretical picture for which Mortillet had been becoming famous since the late 1860s. Mortillet had argued the existence of "pre-Quaternary man," an earlier form of human being than had yet been conceded even by those who accepted the idea of human evolution. Mortillet called this precursor of humans "Anthropopithecus" and went on to divide his construct into further subdivisions. He also expanded on the notion of the Three Ages (Stone, Bronze, and Iron) with an original set of subdivisions for the Paleolithic period. Each was named for a French archaeological site (the Chellean, Mousterian, Solutrean, and Magdalenian), with the assumption that the types of flintwork found at each site were somehow limited to the era that the site represented. Mortillet saw these eras as necessarily successive: he expected all societies, all over the world, to follow the same course. There were those who had trouble with this notion, but many more simply rejected his work because they did not believe in this degree of human antiquity, and they would not accept material culture as proof without the further evidence of human remains. Mortillet's position gained acceptance as further archaeological finds were made, and, eventually, as further bones were unearthed. Mortillet was vindicated on the matter of human antiquity, and his fame grew. Yet, in the end, the greatest service this work provided was its formulation of a system for artifact classification. It was not, as I have hinted, a very good system, but it got things started. The archaeologist and anticlerical historian of religion Salomon Reinach was one of Mortillet's most critical eulogists, and he summed up the matter well:[47] "Mortillet efficaciously served the cause of truth, not only by his discoveries, which have remained perhaps less well known than his errors, but by his persevering—and, as a whole, fruitful—efforts to introduce order into an infant science that had been lost in confusion."[48] Reinach did not think much of Mortillet as an intellect but wrote that he had managed to create a discipline against great opposition. "After the cause of truth is won," offered Reinach, "let us not forget those who fought for it" (12). Mortillet's general construct was maintained by archaeologists for decades, and his nomenclature remains in use today though his terms have completely new meanings now.[49]

When possible, other freethinking anthropologists worked in support of Mortillet's ideas—sometimes without sufficient grounds. As Glyn E. Daniel writes in his classic *A Hundred Years of Archaeology* (1952), while Mortillet argued for pre-Quaternary man "very cogently" and with great conviction, "Abel Hovelacque, in an amazingly irrelevant argument, supported Anthropopithecus on linguistic grounds."[50] Though Daniel rightly critiques some of the conceptual flaws of Mortillet's system of classifying prehistoric human stages, he notes that "the de Mortillet system . . . became the orthodox system of prehistory until well into the twentieth century" (109). Daniel's work was in no way aimed at finding political meanings embedded in scientific theories. I quote it at some length below because it demonstrates how blatantly progressivist were Mortillet's conclusions:

> Its triumph was so complete because archaeology seemed to prove once and for all, and in an entirely unexpected way, the widely held doctrine of progress. The evidence of the geological and archaeological sequences seemed to show that man's story on the earth had been one of gradual progress from the primitive chipped flints of Chelles . . . to the ancient civilizations of Egypt, Assyria, Greece and Rome. "Impossible," declared de Mortillet, after conducting his tour around the Exposition, "de mettre en doute la grande loi du progrès de l'humanité." And this progress was clearly revealed in the archaeological stages: "Pierre taillée à éclats, pierre polie, bronze, fer, sont autant de grandes étapes qu'a traversé l'humanité entière pour arriver à notre civilization."[51] (116)

Mortillet's *Le préhistorique* was also eager to disprove the classic "argument of universal consent" that suggested that God existed because every culture believed, in one way or another. Without mentioning the specific argument, Mortillet contended that "paleolithic man . . . lived in peace, free and more or less wandering, completely without religious ideas."[52]

Along with the Bibliothèque des sciences contemporaines, the freethinking anthropologists produced an extraordinary number of books (some within other joint-venture series, some not) and articles (hundreds for their own various journals and an additional myriad for various other scientific, political, and literary magazines). Lefèvre's *La renaissance du matérialisme*, published with the Bibliothèque matérialiste, was reviewed by the *Revue scientifique* though the book was not really written for the scientific community (it was, as will be recalled, a compilation of Lefèvre's articles from *Libre pensée* and *Pensée nouvelle*). This was not lost on the reviewer. "Several times," he comments, "M. Lefèvre uses the expression: militant materialism. That is, in effect, the character of this work: it is a work of combat, made more for the public at large than for the scientific public." A few critiques were offered regarding the level of the science discussed, but the reviewer's overall attitude displayed great tolerance for this sort of endeavor: "Despite these critiques, one reads with pleasure this witty, alert book in which the talent of the author succeeds in couching hypotheses. We cannot, in any case, make it a crime that M. Lefèvre has not written a pedantic book. The greatest fault of an author is to be boring and that fault has been completely avoided."[53]

The scientist Charles Richet's review of Letourneau's *L'évolution de la morale* was less equivocal in its praise, claiming that the book, which was a résumé of Letourneau's courses at the Ecole d'anthropologie, was "certainly destined to hold one of the premier places in the work of reconstitution that will soon substitute itself for the old crumbling edifice of classical philosophy."[54] The review then offered a detailed and undisputed account of the author's system as it brought humanity from prehistoric savagery to present-day confusion, concluding with the wistful observation that "the ethic of the past no longer has any authority, and that of the future has not yet been formulated." Richet was unquestioning in his report that the "law of human evolution" indicates that "narrow egoism" will be bred out of humanity and concern for the whole society will take the place of concern for the self. The "goal" of this "law of human evolution" is "to create tendencies that are compatible with the greatest possible public and private happiness, which is to say, to make man more robust, better, and more intelligent." A big claim. Richet followed Letourneau in explaining away present-day criminal savagery by citing the effect of atavism, "immoral inclinations that are veritable specimens of the stone age." He further agreed with Letourneau in his claim that "every society has superior men within it, disrupters of the established order, contemporaries of the future, whose role is to prepare and hasten access to the next, superior, social level." As for the present "state of anarchy," Richet said there was no need to despair because evolution is most rapid in times that seem bad. "Letourneau's beautiful book ends with this consoling conclusion . . . everything confirms and affirms the existence of a great law: the law of progress" (536). It is a marvelous construct: criminals, Catholics, and metaphysicians are throwbacks from the savage, foolish past; materialist scientists are "contemporaries" of the wise and happy future.

Before moving on from the *Revue scientifique*, we must look at an extremely interesting and rather uncharacteristic review of Charles Letourneau's *La physiologie des passions*, a later edition of the book Zola had so admired.[55] The review was written in 1878, when the journal was under the care of two prominent liberal journalists, Eugene Yung and Emile Algave. As its title implies, *La physiologie des passions* was a naturalist explanation of all human passions, sometimes concentrating on physiology, sometimes on evolutionary development.

The review began by taking the work on its own terms, arguing with the specifics of its biological explanations—of music, for example, and of patriotism. But about midway through, the authors mentioned that "the philosophical system of M. Letourneau is materialism," and, after a few more biological discussions, the authors started to take issue with this general stance (1168). Letourneau, they wrote, "obviously cannot see in idealism anything but a fantasy of the imagination and an aberration of the mind. The theory of knowledge does not seduce him, he sees all discussion of the exterior world as a 'metaphysical refinement' and adopts the system that we call naive realism." Here Yung and Algave quoted Letourneau saying that if we do not trust our senses as "honest and sincere," we will be thrust into the "void of doubt," so trust them we must. Then they hit Letourneau where it would most hurt— and in a fashion rather strange for a journal of science:

> M. Letourneau is here much too much the metaphysician, he does not take into account the fact, and it is an obvious fact, not a sophist dream, that we know things only by our consciousness. . . . The workings of our consciousness always present themselves through the form of a rapport established between the subject and the object. The object could be either the not-me or an interior phenomenon, and we cannot conclude anything on the nature "in itself" of the subject or the object. Matter is nothing but an abstraction, like spirit; further, the word *matter* is an abstract term designating a specific part of our consciousness; the word *spirit* is another abstract term designating another part, equally determined by our consciousness. The materialists only attribute an objective reality to the abstraction called matter, the pure spiritualists objectify only the abstraction that is called spirit, and the great majority of people objectify both. Experience contradicts both these systems because it teaches us nothing of substances, not even of their existence or their nonexistence, and positive idealism, experimental and phenomenal, seems to be, in sum, the most supportable theory. (1169)

This is fantastic not only because someone has found a way to call Letourneau a metaphysician but also because it was an odd leak from the neo-Kantian philosophical school into the world of professional science. What the reviewers were talking about here is Kant's notion that our sensory apparatus and our inductive logic are both devices extremely peculiar to humanity and that it is unlikely in the extreme that a real world exists out there that somehow corresponds exactly to the kind of information that we are capable of gleaning. Anticipating the twentieth-century revolution in physics, Kant saw that our senses and our logic must also be responsible for the creation of time and space, and that suggests that time and space do not exist in "the real world," which he called the *noumenal* world, the world outside our perception. This implies that this noumenal, real world is totally different from our world, so no experiment could possibly give us any information about it. All this leaves for science to study is the entire world as we know it, the phenomenal world. Since that is where we live, it is a very worthwhile project. Kant was not saying to forget about science—quite the contrary—but he was insisting on the limits of its possible knowledge. As for religion, Kant figured that since consciousness and moral feelings are so different from most stuff of the phenomenal world, they may be emissaries from the noumenal.[56] Since we cannot know the noumenal, we are free to believe that it is the realm of God, if we so choose. Kant did believe. His division of the world was logical, rational, scientific; his religion subjective. Kant was being championed in France by Charles Renouvier, Durkheim's philosophical mentor, who, like other rationalist followers, made a big point of the fact that Kant's claim that the world we know is unreal was not a metaphysical claim, though it felt like one.

Normally, one could protest that experimental or descriptive scientists are so fully concerned with the phenomenal world that it is quite unfair to chastise them for ignoring our ultimate inability to know things-in-themselves. But Letourneau's brash materialism and utter disdain for Kant's transcendental idealism was awfully provocative when coupled with his tone of total mastery and unwavering conviction. Furious that Letourneau "willfully confounds *la critique* [Kantian idealism] with metaphysics," the review suggested "it would do him a great deal of good to be a little more skeptical." Letourneau's insistence on explaining the nature of thought and desire was particularly irksome. Yung and Algave's review insisted that "we do not gain anything by calling intelligence, thought, memory and will a property of the nerve cell instead of calling them faculties of the soul." The difference implied by the two names was conceded: "the soul" suggests that thought is an aspect of the workings of our consciousness while using the term "the cell" affirms "an intimate liaison between a psychic phenomenon and a physical action: a vibration of the molecules that compose the cell," but one can see how small a difference this was in contrast to Letourneau's claims. It is interesting to note that whereas physiologists might agree with this definition of their term (nerve cells produce thought), theologians would likely balk at such a rationalized description of the soul. That should remind us that despite this harsh review of materialism, the reviewers had more completely dismissed religion and the supernatural: *soul* was retained as an empty term to point out the weakness of the claims made in the name of *cell*. "The two expressions are legitimate to the same degree if one takes them to mean what I have just indicated, and they are illegitimate to the same degree if one attributes to the soul or to the cell an occult faculty or power, incomprehensible power, which explains thought."Thus, summing up, the review praised Letourneau for his abundant facts and his "occasionally original, often bold, always sincere" ideas but returned to the rather cruel accusation of metaphysics: "M. Letourneau doesn't use subterfuge. He's a declared partisan of the theory of evolution and of materialism, he defends his convictions ardently and endeavors to make them triumph. His greatest fault is to conserve the habits of his adversaries and to be still too imbued with the metaphysical spirit that he combats" (1170). What was being called metaphysical was empiricism itself: since sensory information is gleaned and processed by consciousness, it is less reliable and more of a phantom than conceptual reasoning—the earth does, after all, revolve around the sun, not the other way around, as sensory information would imply. As remarkable as this commentary was, coming from within the scientific community, it was par for the course among philosophers, as I shall discuss. Here, let me merely note that within my Polaris metaphor, Richet's *Revue scientifique* was located due north, though not actually lifting toward the stars. By looking to the freethinking anthropologists, Richet's readers could orient themselves on the cultural terrain, regardless of how far toward materialism they were personally interested in going.

RESPONSE FROM OUTSIDE
THE IMMEDIATE SCIENTIFIC COMMUNITY

In France in the 1870s and 1880s, there were three major periodicals concerned with philosophy: Ribot's *La revue philosophique*, Renouvier's neo-Kantian *La critique philosophique*, and the short-lived *L'année philosophique*.

Théodule Ribot founded *La revue philosophique* in 1876, when he was first making his name as an important positivist philosopher. His primary interest was psychology, which he wanted to base on scientific principles, "independent of all metaphysics," and he was frequently explicit in his insistence that "discussions of the 'real nature of the soul' have nothing to do with psychology." As late as 1926, general commentaries on intellectual life in France could claim that "Ribot's influence on present-day psychology in France is incalculable. Almost all the psychologists, in varying degrees, begin with his teaching."[57] Today he is often called the founder of French scientific psychology. The mission of the *Revue philosophique* was to create a forum wherein philosophers and scientists of the mind could exchange ideas and work toward synthesis. In the 1880s it published work by a range of philosophers, scientists, and other writers, including Darwin, Alfred Binet, Eduard Buchner, Henri Bergson,

Pierre Janet, Georges Sorel, Cesare Lombroso, Francis Galton, Gabriel Tarde, Paul Tannery, Jean-Marie Guyau, Charles Richet, Havelock Ellis, and Léonce Manouvrier.

On the face of it, this would have been a good place for the freethinking anthropologists to publish, for despite their horror of it, they wrote about philosophy a great deal, and they were purposefully trying to take over all questions ever conceived as philosophical. As for the materialist component of Ribot's project, the anthropologists fit right in. In psychologist Frédéric Paulhan's first review of a book by Letourneau for the *Revue philosophique* (there were three), Paulhan wrote that the work "was part of a collection [the Bibliothèque] already well known to the public interested in the progress of science and of philosophy and is destined . . . to popularize the results obtained by scholars in diverse branches of human knowledge."[57] Paulhan's reviews of Letourneau covered his *Evolution du mariage et de la famille, Evolution de la morale, L'évolution de la propriété*, and *Sociologie d'après l'ethnographie*. They generally recommended the works to the readers of the review, but with a good deal of reserve. Letourneau "wrote with erudition" and "offered an abundance of facts," but, warned Paulhan, "one may wish that there were fewer of them and that they were always well chosen." Paulhan also cautioned that the anthropologists seemed ready to understand all of psychology through biology, and, furthermore, "the theories presented by M. Letourneau were pretty much completely known before him." He expressed shock at Letourneau's claim that evolution would dispense with the family, religion, and property and with the anthropologist's evident desire for this to take place. "The state will eventually take the place of the family," Paulhan quoted, and this was good because "one would have to be blind to not see how many children are tortured there in their bodies or in their souls." Paulhan responded that while "it is possible that there is truth in all these claims, it is difficult to make a categorical pronouncement on such subjects." At the end of an extensive discussion of *L'évolution de la morale*—mostly critical, but respectful, and willing to discuss morality from a naturalist standpoint—Paulhan summed up by "recommending the book as a rich source of facts, well grouped and well organized, if too artificially, and as an interesting illustration of evolutionary materialism."[59] That is how it reads today. The fact that Ribot had Paulhan review the book, however, shows that Letourneau's book was more successful as an ideological work. What *La revue philosophique* did was to call attention to it while criticizing its particular manifestation.

Paulhan thought Letourneau's study on the family to be one of his best works ("the facts are abundant and well classified, the read is easy and interesting"), but it was still indicted for a lack of originality and a tendency toward unreasonable arguments.[60] Beyond the many small questions that Paulhan was willing to debate, he consistently took issue with the way Letourneau organized his facts by fiat. He was unconvinced that contemporary aboriginal peoples were "living fossils" of European prehistory. He also seriously doubted that evolution was necessarily progressing toward utopia: "Persuaded that evolution follows its course and emancipates man, Letourneau declares himself relatively satisfied, 'To dare deny progress, you would have to be blind or trapped in some chimerical system.' . . . It seems to me that Letourneau does not understand the question of pessimism."[61]

The journal's review of *L'évolution de la propriété* was furnished by Gustave Belot, a philosopher who would become well known for countering the socially based morality of Durkheim and Lévy-Bruhl with a defense of the traditional individual conscience. Belot shared Paulhan's frustration with Letourneau, for much the same reasons. In describing Letourneau's method, he assured his readers, "Don't fear, to attain his results [he says he relies on] the accumulation of facts and remains (at least in appearance) very sober in his generalizations and hypothesis," but in practice "the work rests on presupposed principles and hypotheses. The order in which we are presented the savage tribes, having nothing of the properly historical, is precisely determined by the same ideas that are advanced by his social evolution."[62] It should be noted that Belot still saw Letourneau's work as "important" and struggled with its individual issues and claims, but the piles of facts were really problematic. "We cannot but thank him for the abundance of his information, and if a few mistakes have managed to slip into this multiplicity of details, it would be puerile to make too much of it. But it does seem like one could wish that the facts could speak more precisely and more quickly to the point" (646).

La revue philosophique covered Thulié's *La femme: Essai de sociologie physiologique* by simply reporting its uncompromising—and for the freethinking anthropologists, totally uncharacteristic—claim that political duties would render women sterile and that their only function should be motherhood. The fact of "physiological sociology" was not itself a problem for the reviewer.[63] The journal was less sympathetic to Lefèvre's *La philosophie*, choosing not to go through, "one by one, all the arbitrary and hardly scientific affirmations that M. Lefèvre boldly produces in the name of science" and instead limiting itself to "remarking that [Lefèvre] had not understood the importance and the place of the problem of knowledge. Everyone who has worked on the thinking subject, from Socrates to Kant, incurs his critique."[64] To Lefèvre's surprisingly extreme empiricism, the review responded, "So be it!" happy to do battle with empiricism in this particularly simple form. It then proceeded to walk Lefèvre's contention through Hume's critique of sensory information and dis-

missal of the scientific "law." No amount of observation of particular phenomenon or causal relation can ever stand in for all observation. The reviewer then
demonstrated the progressive failure of empiricism's claims, as described by
Kantian critique, and concluded, "We thus arrive at a system more or less the
same as neocriticism, the idealist phenomenology of M. Renouvier, which is
to say, at a point of view of which we recognize the high value but which certainly is a far way from M. Lefèvre's materialism" (457).

It was again Paulhan who reviewed Véron's *La morale*, and he began by saying that Véron was so afraid of accidentally writing like a metaphysician that he
never got beyond generalizations. He went on to specify that this was an "application of the theory of transformist and utilitarian materialism to the question of morality."[65] Véron's central point in this book was that morality is but
an outgrowth of intelligence and that as humans evolve a higher intelligence,
our morality will also improve. Paulhan did not tolerate this well, writing that
he did not have to review the book systematically, since "readers of the *Revue*
can easily see what a precise mind can glean from 'evolutionism, utilitarian-
ism, and materialism'" (476), but he did caution against Véron's overweening
reliance on intelligence as the seat of morality. Paulhan also managed to critique Véron's logic as suffering from "a certain number of metaphysical tendencies, which do not accord well with the general spirit of his book" (477).
The problem was that "the notion of law, the subject of so much objection in
the spiritualist and critical theories, does not gain much in passing into a materialist theory" (477). Véron's conception was so "purely superficial" that it
never entered into the real philosophical problems of law.

Among all these bad reviews, a rather interesting development was taking
place. Ribot's *Revue philosophique* began publishing coverage of the Société de
psychologie physiologique, which was presided over by the famous psychologist Jean-Martin Charcot. These essays were all penned by a key member of
the Société de psychologie physiologique: Léonce Manouvrier.[66] Manouvrier
also wrote independently for the *Revue philosophique*, generally on descriptions
of brains that he had autopsied at the Laboratory of Anthropology. He was always careful to claim very little for these studies, repeatedly giving caution
against making too much of (even his own) contemporary brain studies. In his
detailed study comparing the brains of Louis-Adolphe Bertillon and Léon
Gambetta, Manouvrier explained that the brains suggested two different kinds
of intelligence; among other things, as we might guess, one was talkative and
gregarious, the other quiet and reserved.[67] But Manouvrier concluded by
writing that "all these findings have their interest, but we must insist that they
do not suffice to permit us to formulate a scientific judgment on the absolute
or relative general value of the two men, each illustrious in his various titles.

Imagine, in fact, a science sufficiently advanced that the preceding comparison could be done on fifty points rather than on four or five; it is impossible to say in how many points one brain might prove superior to the other or to the general average." He warned his readers that if they found "this or that psychological opinion" of either man confirmed by the study, they should realize that it was all "hypothetical." Manouvrier explained that, so far, scientists like himself were just trying to get the questions right: "The principal interest of isolated comparisons of this genre consists in the contributions they make to the precision of anatomo-physiological questions that are now able to be studied, to guide the investigations, and to raise new questions" (461).[68] He also took a moment to praise the Society of Mutual Autopsy. His work suited the *Revue philosophique*: it was cautious in its physiological claims but offered a good deal to think about and, especially in the above-cited case, it offered exciting peeks into the skulls of prominent Frenchmen. Like almost all the freethinking anthropologists, Manouvrier was an excellent writer, but, unlike them, he was not blinded by zeal. His work may have influenced Ribot (who was only about a decade his senior); in any case, it is clear Ribot admired the work. Articles that Manouvrier published in one of the Société d'anthropologie's journals were sometimes summarized and enthusiastically discussed in the *Revue philosophique*. Of his article "Analyses et comptes rendus: Manouvrier, Sur l'interprétation de la quantité dans l'encéphale et du poids de cerveau en particulier" (On the interpretation of the quantity of the encephalon and the weight of the brain in particular), the *Revue philosophique* reviewer concluded that "it is because he opens a large path of work of this type that M. Manouvrier's memoir will count among the most useful in the progress of positive psychological studies, while rendering an equal service to anthropological anatomy, in view of which it was specially created."[69]

With such praise, it is not surprising that, along with his special monographs and reports, Ribot welcomed Manouvrier to write reviews for the *Revue philosophique*. He reviewed his freethinking anthropologist friends' books almost exclusively. His efforts stood in sharp contrast to the reviews offered by psychologists and philosophers, but not because Manouvrier gave his friends that much credit; he simply wrote of these works as if they were primers of materialist science. As such, there was no need to chastise their lack of originality or their narrow, fervent materialism. Manouvrier even knew how to sell these works to people who might see themselves as beyond elementary books; he had an excellent understanding of the emotional component of the matter. Manouvrier treated his peers as serious believers of evolutionary science, not raised in the faith, who might very well regret the lost passion of their youthful conversions. They would welcome a revival. Further, he seemed to understand that many transformist coreligionists had never really read Darwin and never would; they had gleaned what they needed for belief from the cultural environment. They might now be too shy to buy a book called *Darwinism*, thinking that such a purchase would expose their ignorance and might prove too difficult to digest. These fears were easily alleviated by a deft pen. Manouvrier began his review of Mathias Duval's *Le Darwinism* as follows:

> Our generation was not brought up in the doctrine of transformism. Those of us who felt it was like a torch that we could not do without will not be unhappy to go back now and again to resubmerge their philosophical belief in the source from which they had drawn it. It is not that this new faith, which enlightens every day, is in risk of being shaken, like so many other opinions that have come later in life. But one only vaguely remembers the innumerable facts that, once given, had swept away and captured one's conviction, because more than anything one is impregnated with the general truth of what it expressed. One loves to reencounter these facts, from time to time, like a traveler loves to go back to see the stream where he once quenched his thirst. In any case, you will always find, in the enormous pile of proofs of transformism, a few facts that strike you in a new way, according to the present orientation of your mind, such that it feels like you are learning it for the first time again.[70]

Manouvrier went on to say that the old readers of Darwin, "still keeping his admirable books at hand," are always pleased to see published "the work of the master in a new form, generally less grim and forbidding than the first." Thus the theories of evolution and transformism "are henceforth going to penetrate young minds earlier, so that the ideas can be even more fruitful." Duval's book rendered an "especially grand service" in both reminding and teaching. What is more, Manouvrier asserted that in this new book "the proofs of transformism become much more striking thanks to their condensation and to the way they are organized. . . . They are corroborated by the addition of the latest scientific finds and made unassailable by the refutation of contrary doctrines. . . . That is why we do not fear to say that it will be able to convince certain stubborn spirits much more effectively than even reading the works of Darwin" (399). Duval's book was a compilation of his lessons at the Ecole, and Manouvrier did not fail to promote the school a bit in his review.

Manouvrier's concern with getting access to the next generation while they were young and with furnishing retorts and refutations further indicates that in 1887, in France, evolutionary doctrine was still serving as a point of honor and personal ideology for members of his own generation. The point is particularly striking in an issue of the *Revue philosophique* wherein Manouvrier reviewed the work of the venerable rearguard of anthropology, Quatrefages, as well as a book coauthored by two freethinking anthropologists. In discussing Quatrefages, Manouvrier did not actively disagree with the older man, but he gave his readers the information they would need to disagree on their own. Quatrefages argued that human beings formed a separate kingdom, outside that of the animal, marked off by our morality and religiosity, which he argued was nonexistent among beasts. Further, he discussed the human *soul* as opposed to the animal *soul*—both of which Manouvrier put in italics for effect. He then simply said that he would limit himself to registering these doctrines, "of which the readers of the *Revue philosophique* must have already read many discussions."[71] The review was not negative, but it was immediately followed by a review of Hovelacque and Hervé's book *Précis d'anthropologie*. This one, Manouvrier wrote, "seems destined . . . to present to the public doctrines absolutely opposed to those followed for ages by the wise professor of the Museum [Quatrefages]." Here were monogenism against polygenism (one primordial couple for all humanity verses one for each race) and species fixity opposed to transformism. Again, Manouvrier generously affirmed that such basic anthropological ideas would not be unknown to the reader of the *Revue philosophique*, but still, "many of them would not mind finding condensed, in a relatively short book . . . correct and clear, and well-written, matters treated in a great number of specialized works."[72] One of the problems Manouvrier rather charmingly pointed to in both works was that though Quatrefages would not admit transformism, he insisted that the single Adam-and-Eve couple was sufficiently malleable by environment to explain the various present-day races. Hovelacque and Hervé, on the other hand, were transformists who nevertheless insisted that no amount of transforming could have changed humans from one general race into the variety now available. Manouvrier did not say it, but the whole problem here was that one side of the argument was being faithful to the Bible (which has one Adam and Eve and no transformism—leaving the fact of human variation a bit of a mystery), while the other side was actively contradicting it. With his characteristic light touch, Manouvrier dispensed with the controversy by saying that "very happily, the solution to the problem is not very urgent, and we are allowed a little more time to make our choice between monogenism and polygenism. In any case, the first of these, it seems to us, has the upper hand for the moment" (326). In the end, he recommended Hovelacque and Hervé's book energetically and that of Quatrefages not at all.[73]

Such was the response to freethinking anthropology in Ribot's journal of science and philosophy: more critical than the purely scientific press but still very much interested in engaging materialists in discussion. The journal even welcomed into its ranks the most cautious and vigilant of the anthropological school, Léonce Manouvrier. The neo-Kantian *Critique philosophique* could not be expected to have so much interest or even tolerance. If the *Revue philosophique* was headed north by northeast, it can only be said that the *Critique philosophique* was headed out straight along the equator, steering clear of both materialism and religion. Even so, frequent, precise calculations regarding the position of our Polaris seem to have been necessary to guide their way.

Kant was a figure of the German Enlightenment, but in the first half of the nineteenth century his work had not been given much attention in French culture. It was Charles Renouvier who championed Kant's ideas, in a revised form, beginning in the 1870s. Renouvier had been an active member of the short-lived Second Republic, and when it was crushed by Louis Napoleon's coup d'état, he retired from public life and devoted himself to philosophy. During these years he grew famous for his *Critical Essays*, largely based on Kantian critique, and his *Science of Morals*, which attempted a logical, philosophical framework to establish morality on new grounds. It had to be outside both Catholicism, which he despised and understood as "an extremely organized absolutist international association, a threat to liberty, common rights, and public order," and outside materialism, which he liked no better because, for him, it could not support free will and innate moral feeling.[74] He rejected the term metaphysics, too, saying it stood for pre-Kantian philosophy in all its lack of rational rigor. Instead, he wanted to create a "republican philosophy in France [which would be] free from the speculations of an exhausted metaphysics."[75] Like Kant, Renouvier believed that the unknowability of the noumenal world allowed for the possibility of God's existence and that the human sense of morality and free will both emanated from the noumenal and were therefore real, natural facts. The phenomenal world was thus independent of God and ruled entirely by scientific laws: it was deterministic except in the case of humanity, which had free will. Unlike Kant, who held that the moral act was performed only for its morality, Renouvier conceived of a reward of immortality for those who led moral lives and a God who represented moral perfection. Unlike the positivists, the neo-Kantians were eager to discuss God, feelings, moral sensations, and free will and to rail against materialism with all the fervor of the spiritualist. In 1872, with the empire finally at an end, Renouvier founded the *Critique philosophique* with his friend and collaborator François Pillon, and in 1890 Pillon began publishing the *Année philosophique*. The journals were very similar, and the two men contributed heavily to both. One might not expect such reviews to have given any attention at all to a group of rowdy, materialist anthropologists, but apparently, they were too good a foil to forgo.

The two journals did not discuss the anthropologists as frequently as Ribot's *Revue philosophique*, and they certainly were not as patient in their response, but they did engage with them. A brief look at a piece published in the *Critique philosophique* in 1876 further clarifies its position. The piece is a half-page "prehistoric dialogue" (signed "X"—either Renouvier or Pillon had written it) between Darwin and God, in which an apparently noncorporeal Darwin requests that God create a cell:

GOD: Why? To put you in?

DARWIN: Me and all the others. I'll explain later.

GOD: Voilà. Is that all, can I go?

DARWIN: Well, it would be a gesture of your goodness if you would add the ability to produce, genealogically and by the struggle for existence, all the others who will be born down here.

GOD: I don't understand, and you ask a lot. In any case, I can't refuse you anything. It's done.

DARWIN: Now I don't need you anymore, you can go: as for the rest, indeed, I'll do it myself.[76]

And that was that. It seems an awfully silly piece for this sort of journal, but it was not silly then: it was a useful shorthand for the journal's complex position on philosophy, religion, science, and personal identity. Fervently against Catholicism, the editors were by no means against belief. Neither did their attack on materialism imply any disavowal of scientific power. They were also taking a middle road in genre, committed to explicate transcendental idealism and also to weigh in on matters of politics and policy. In 1880 the *Critique philosophique* published an article by Pillon entitled "La lutte contre le cléricalisme, ce qu'elle ne doit être et ce qu'elle droit être: Il ne faut pas que la politique anticlérical soit une politique d'irréligion" (The struggle against clericalism—what it must be and what it must not be: Anticlerical politics must not be a politics of irreligion). The reason it must not be was primarily because of the needs of "a numerous part of the population who do not know the ideal . . . and can only taste it in the form of Catholicism."[77]

André Lefèvre's *La renaissance du matérialisme* was taken somewhat seriously because the title alone was a threat to this audience. As the reviewer mused, the simple fact that anyone had created a "bibliothèque matérialiste" suggested that such a renaissance might actually be taking place.[78] Nothing about the book was praised; the review was essentially an alarm signaling Lefèvre's complete disposal of God and religion and his claim that materialism is "as well as the superior doctrine, the emancipatory doctrine." To the reviewer's distress, it was "a very easy read for someone who had never opened a book of philosophy. In writing it, the author—out of modesty no doubt—kept himself from showing the philosophical intelligence that he has; he rarely brings up the important problems, and, when he does, he doesn't discuss the arguments presented. He merely bats around the authors who are his adversaries, and then he formulates the truths that one must believe" (398). In the end, the review managed several times to liken Lefèvre's self-certainty to the dogmatic attitude of the Catholic Church (399). The insult is reminiscent of other journals' fun in calling the anthropologists metaphysical.

Hovelacque's *Les débuts de l'humanité* was also panned but given a bit more argument.[79] The "new faith" to which Hovelacque wished to convert his reader was—and here the review quoted Letourneau (a rare confirmation that the group was identified as such by others)—"morality is purely training" (399). This was important because the idea of a transcendental morality was basic to French criticism. The reviewer worried about the readers who might take all this at the word of the author "because it satisfies their rancor against the clergy and because it is a 'scholar' who explains the work of other 'scholars,' speaking of facial angles and cephalic indices" (399–400). A reasonable concern. Also quite reasonably, the reviewer insisted that *Les débuts de l'humanité* was ethnology, having nothing to do with prehistory, but he did then proceed to argue with the anthropologists in a rather naive way: both Lefèvre and Hovelacque insisted that some primitive people did not believe in God, but Hovelacque had written that these irreligious "savages" were just barely human, and the reviewer countered with the possibility that they were irreligious because they were not yet quite human after all. Anyway, "if universal consent cannot prove the existence of God, the lack of universal consent can't prove his nonexistence any better" (400).

Eugene Véron's *La morale* received a similar treatment: the reviewer was horrified by his claims that morality was hereditary, like any other instinct, and merely fortified by education.[80] He reported that, to Véron's mind, devotees of religion and of metaphysics were "both equally mystics and believers in fantasy." In this review and in the two aforementioned, the reviewer occasionally reminded the anthropologists that utilitarian doctrine simply did not explain all of morality: what about the difference between individual and public utility? What about freedom versus responsibility? What about group morality? But the reviewers knew that antireligion was the central issue here, not moral philosophy, and they pitched their criticism accordingly. It was also a matter of turf: Véron's potential reader was caustically instructed to "see, on the first page, the prospectus announcing the sale of volumes of the Bibliothèque des sciences contemporaines published with the cooperation of the most distinguished scholars and writers" (239).

The anthropologist who was given the most attention by the *Critique philosophique* was Clémence Royer.[81] In her book *Le bien et la loi morale* and elsewhere, she had introduced a new doctrine, "substantialism." It had a history: Spinoza's belief that the universe and God were identical had long been understood as atheism in light disguise (a God that *is* the universe itself, with nothing added, cannot intervene, or do or choose anything), but the formulation served some religious doubters as a way of visualizing life and the universe as unified and self-animated. Hegel's "world spirit" later arose as a potent source of secular explanations of life and the universe that still spoke in terms of overall unity and intelligence. By the late nineteenth century, "monism" was the most common term for such ideas (the term denies any duality of life, thought, and spirit on one hand and the material world on the other). Royer's substantialism was understood as a version of monism: everything takes part in the same energy and same moral law. As far as she was concerned—and they loved to quote her saying it—this ended the big feud between materialism and spiritualism: matter and thought are all there is in the universe and they are made of the same stuff, moving atoms; therefore, the universe thinks. *La critique philosophique* devoted considerable space to the rejection of these ideas, claiming both that Royer had not, in fact, invented this system (a technique that did not mark their dealings with the other anthropologists) and that she was wrong because she was stirring together the phenomenal and the noumenal. It was a full dismissal, and yet they devoted page upon page to the project.

L'année philosophique generally covered the more historical works of the freethinking anthropologists: Lefèvre's *Histoire: Entretiens sur l'évolution historique*, Letourneau's *Evolution de l'éducation dans les diverses races humaines*, *Evolution de l'esclavages dans les diverses races humaines*, *Evolution du commerce dans les diverses races humaines*, and *Evolution religieuse dans les diverses races humaines*.[82] These books were all taken to task for their blind materialism, their outrageously oversimplified versions of historical change, and their wild claims about the utopian future to which they would deliver humankind. And yet, once again, the aggregate was that the philosophers at *L'année philosophique* gave these anthropological works a great deal of attention and devoted to them considerable time and effort. They did it because the freethinking anthropologists represented an idea to which these philosophers were opposed, in a version that was rather easy to attack. In response to Letourneau's history of religion, Pillon wrote, "Materialism is wrong to speak in the name of science; it is not science; it is a philosophy, but a primitive and inferior philosophy that one cannot take seriously when one has understood that the phenomena called *material* reduces to sensations, that is, to modes or products of spirit."[83]

A few more stops are necessary to round out this portrait of the contemporary response to materialist anthropology. First, the Scientific Society of

Brussels published a journal called *Revue des questions scientifique* that tended to run five to ten original articles, a similar number of book reviews, and a rather extensive review of contemporary scientific journals. The section on anthropology was covered by Adrien Arcelin, a respected Belgian naturalist. But this was no ordinary scientific journal: it had been founded specifically to demonstrate that science and religion could happily coexist, indeed, many of its scientist contributors were also members of the clergy. Because it was their particular goal to show that religion did not get in the way of good science, their work was essentially positivist in temper, and the most common reminder of their special project was that the table of contents was studded with abbés, priests, and the occasional bishop. Another indication of their philosophy was their mild obsession with the enemy. Arcelin devoted tremendous attention to the materialist anthropologists of Paris. Granted, they were in charge of the oldest and most extensive anthropological institute in the world; they produced several anthropological journals and a storm of books. So they may have been hard to ignore. Still, Arcelin plainly both respected these anthropologists' work and enjoyed their materialist antics. A review of Mortillet's *Matériaux pour l'histoire primitive et naturelle de l'homme* in 1879 occasioned a rare articulation of the journal's attitude. Having called *Matériaux* an "indispensable source for anyone who wants to follow the movement of archaeological and anthropological researches," Arcelin wrote that "assuredly, the spirit that presides over the direction of *Matériaux* is not that of the *Revue des questions scientifique*. One there professes a great disdain for dogmatic discussions, and the cause that we defend here, the accord between science and faith, there provokes an undisguised hostility."[84] Further, Arcelin articulated what seemed to be genuine wonder that the famed scientists would mar their work in this way, since nothing could be easier than a religiously neutral discussion of archaeological finds (319). In general, Arcelin endlessly described and (usually) offered sober praise for the work of Mortillet, Véron, Royer, Letourneau, Lefèvre, and the Bertillons. When, however, the group began dissecting one another's brains, Arcelin reported it with a kind of mystified glee and rarely failed to mention these exploits to his readers. Even here, however, Manouvrier came off well and was celebrated for his rationality and nonpartisan approach.[85] There were times when Mortillet, in particular, was treated rather harshly, but he was always taken as an expert—which, of course, he was. Even the original articles in the journal treated him as such: An essay on the recent find of a prehistoric skeleton, written by the abbé E. Vacandard, cited Mortillet extensively and on almost every page.[86] The abbé did not mention Mortillet's take on the religious nature of this ancestor but himself claimed that for the specimen's "almost immediate descendants . . . religious sentiment was already rooted in their souls, and belief in a future life . . . already oriented their lives" (122).

A few more mainstream journals also occasionally covered the freethinking anthropologists. First, there was the *Revue politique et littéraire*, also known as the *Revue bleu* for its blue covers (it was put out by the same publisher as the *Revue scientifique*, or *Revue rose*). This was supposed to be a review of politics and literature, but such articles as "La moral de Darwin," by Lévy-Bruhl, were not at all uncommon; morality and evolution were the questions of the day.[87] Equally of the moment, Alphonse Bertillon took the front page in 1883 with an article on recidivists and the anthropology of the born criminal.[88] Here I will focus only on a particularly savvy article by the well-known historian of religion Jean Réville. The article, entitled "Une histoire des religions par un adversaire de la religion: M. Eugène Véron," explained that only a few years earlier, the "history of religions" had become a part of the upper level curriculum; "this egalitarian rubric did not fail to cause a certain scandal among those who find it natural to submit the religions of others to critical and historical investigations" but who would not thus subject their own religion, because of its "sacred character."[89] Réville did not mention it, but he was deeply involved in this "scandal," having written several important naturalist histories of religion. Indeed, before Durkheim wrote *The Elementary Forms of Religious Life* (1912) he attested to being "unqualified to speak" on the question of religion and "inclined to accept the naturalist hypothesis of Albert Réville."[90] Here, Réville explained that this problem of resistance to the history of religion had been resolved because the religious started publishing their own histories; "unable to keep it from existing," they adopted the subject themselves. "In any case, the church, in making a history of religions in its own image, should not be astonished if its adversaries create an alternative version, conforming to their principles and impregnated with their attitude."[91]

That, explained Réville, was what Véron was up to, but he found it a bit much "to baptize the history of religions with the name 'natural history' if one is not going to use the same impartiality with which one describes an animal or a vegetable." But Réville knew this was no oversight: "M. Véron has a holy horror of religion, and he doesn't hide it. . . . The alpha and the omega of these two volumes is 'religion is absurd.' Véron is convinced that he could not do a greater service for humanity than to demonstrate the inanity of every type of religion. He has antireligious faith" (15). For Réville, there was no crime in this, but he did wonder whether such an attitude would not necessarily get in the way of an appreciation of religious events. "What kind of history of art would we get from a man who had a horror for art and who was inaccessible to all aesthetic sentiment?" (15). It was an excellent question, and it pertained equally to the anthropologists' forays into philosophy. Yet it was not a perfect metaphor for the anthropologists. In human history, some people have enjoyed thinking about the mysterious weirdness of life; some people have enjoyed proposing an explanation of this weirdness and repeating it in catechism; some people prefer to ignore the whole thing whenever possible. The freethinking anthropologists cannot be accused of ignoring the paradox of human existence or of being numb to the delicious pain and wonder that it can engender: they indulged in it all the time. So the best metaphor for them may not have been art haters writing about art but rather zealots of a modern form of art cantankerously deriding all the "mistaken" art that had come before them. After all, as Réville noted, Véron had "antireligious faith."

Véron's utter disdain for all religion or spiritualism, and his particular hatred for Catholicism, began to wear on the otherwise anticlerical Réville: "Certainly the church, in the past and in the present, offers ample cause for critique, even the most severe, but it seems strange to us that in all the Gospel and in the whole history of the church one cannot find a single point, not even a single one, that we do not have to condemn. Done in this way, anticlerical history does not merit more confidence that clerical history, for which everything that happens in the church is admirable" (15). This was the central point, the primary use to which people put the freethinking anthropologists: the materialist extremism of their work allowed their audience to occupy a middle ground, that hallmark of the reasonable, sober humanist. Réville went on to point out Véron's gravest errors in assessing history, indicating that the war between the supernatural and the natural was a recent intellectual development and not at all stable and also suggesting that no religion can be understood as a "bloc" unchanging throughout history (16). Réville closed by worriedly mulling over how Véron would see him, were he to read this review. Réville wrote that according to Véron, "everyone who has not yet decided to intone the alleluia of atheism" refrains only because he is "obeying the intellectual habits that are imposed on him by atavism and reinforced since infancy by an education closed to the progress of science." Réville concluded that there would likely be no agreement between one who so completely believed religion to be dead and himself, "one who maintained its legitimacy so long as it progresses," and he conceded that they might as well each blame the other's "error" on his ancestors. "On that point, at least, no one will argue" (17). For his part, Réville was eventually rewarded for his careful secularism with a chair, created for him, in the history of religions at the illustrious Collège de France.[92]

AN ENGLISH QUARTER HEARD FROM

In 1879 Lefèvre's *La philosophie* appeared in English translation, under the title *Philosophy: Historical and Critical* and with a strange introduction by the translator, A. H. Keane.[93] Keane began by explaining that the author had not himself provided an introduction because the book's plan was so simple, its exposition so clear and so exhaustive, that the book could be allowed to speak for itself. It was so straightforward because the author was "a materialist of the most advanced modern school and as such, expresses his opinion in the most outspoken and uncompromising manner."What is more, "there are no abstract and but a few concrete matters that he has not had occasion to deal with, more or less fully, from the atheistic point of view." And yet Keane thought an introduction necessary, writing: "But this very circumstance would seem to call for a few words of warning to the unwary," particularly because its English incarnation was part of a popular series and "must necessarily fall into the hands of many readers who are apt to be carried away by a certain speciousness of reasoning, and who are not always possessed of a ready answer to a line of argument undoubtedly urged with great vigor and cogency" (xi) So the introduction set out to prepare the reader to reject the text. Lefèvre's head must have spun.

Given his proclaimed project, the translator asked the obvious question: "Why then publish such things at all?" The answer was the standard liberal notion of a free exchange of ideas, the notion of an index of forbidden books being "now everywhere happily abolished except, for obvious reasons, in the case of books injurious to the public morals." Keane's other argument was that since this sort of materialism could be found in many current educational works on the sciences, there was no reason to be more strict "merely because it calls itself Philosophy." He even pointed out that these scientific works were "freely placed in the hands of young students, [and] notoriously find favor with the promoters of 'the higher education of women.' " Of course, these were not the reasons at all: Keane, like many of those whom I have already quoted, was a man more often to be found arguing against churchmen and for the scientific side. After all, he translated and published Lefèvre! But neither could he go as far as the freethinkers. Once again, we find someone building an ideological nest for themselves by calling attention to the materialist extreme:

> Meantime, evolution, or as expressed by the distinguished French naturalist Prof. Charles Martins of Montpellier "the theory of evolution binding together all the problems of natural history, as the Newtonian laws bind together the motion of the heavenly bodies," is the great intellectual fact of the day and whether favorable or not to our personal views, cannot possibly be excluded from any intelligible treatment of philosophy. Indeed evolution, in some form or other, may now be taken as an established and almost universally accepted truth, being practically identical with that "progress from the homogeneous to

the heterogeneous," . . . from the simple to the complex, from unity to differentiation of functions and physiological division of labor, justly regarded by Herbert Spencer as the great law of nature.

At the same time it cannot be too often repeated that there are various theories of evolution. (vi)

So, Keane first told his readers that the theory of evolution is a gestalt explanation for all natural history, and then he said we have to consider it even if we do not like it. Here we may read: *you* have to consider it even if *you* do not like it—because, after all, he can not really share in this first-person plural when he goes on to assert that it "is an established . . . truth" and, further, that it explains not only bodies but social life as well. His back securely up against the churchmen, Keane could now attend to the present adversary. What he did was to argue that there were many ways to imagine evolution, and he listed several. In this list, there was no hint that the one he called "Wallace and Darwin's natural selection" was in any way the front runner. The final choice was "lastly, the crude materialist conception utterly eliminating the supernatural and preternatural elements, effacing soul and the Deity, . . . in short, the theory advocated in the present work" (vii). Keane then proceeded to argue a transformism that was guided by the Creator. Nothing in science, he argued, came close to explaining the transition from nonlife to life or from matter to thought. Nor did it seem possible that "matter alone, with the requisite amount of light, heat, moisture, electricity, &c., thrown in *par dessus le marché* [under the stairs]" could have possibly made a "single step" toward organizing and advancing life (x).

Lefèvre's determinist dismissal of free will and his use of the term "the thinking mechanism" also frustrated Keane, who wrote that "the writer here is at issue with the most profound thinkers of all ages" (though, of course, on similar issues, so was Keane) (xvi). Responding to these questions, Keane wrote, "How easily all these questions are answered from the idealist's point of view!" and, of course, "how impossible to solve them satisfactorily" when the mind is a machine and no guiding force maintains the universe (xv). Keane even took up the challenge of the third left frontal convolution of the brain and countered that even if a lesion there results in aphasia, "are we therefore in this case to say that the patient has ceased to be a human being? Assuredly not, because it is not the faculty of articulate speech that he has lost, but only the power of exercising that particular faculty" (an argument crucial to Bergsonian philosophy) (xvi). "The moral sense" also suggested God to Keane. Given all this, Keane concluded that "the dualists, the believers in mind and matter, have some grounds for holding that none of their strongholds have yet fallen . . . consequently that their conception of man and the universe is at least quite as philosophic as that of their opponents" (xviii). That "yet," coupled with his comment that this was the most "modern" materialism, suggests that he saw his position as rather more beleaguered than he was willing to admit. In the main, he was defending his right to see himself as a philosopher and as a brave and rational thinker. But he was also defending the things he wanted to believe:

> Hence the belief in a deity, in creation, in spirit as distinct from material substance, even in immortality, may continue to be entertained without rendering ourselves liable to the charge of superstition, prejudice, mental obliquity of vision, blind or inveterate anthropomorphism, and the other hard epithets flung about, often somewhat wildly by the eloquent and exuberant writer. As this belief is, further, quite as satisfactory, moral, and conservative of the social order, besides being a trifle more consoling, there can be no great harm in still upholding it against the atheistic theory of evolution. Evolution itself, as already seen, in no way necessarily excludes the theory of creation. . . .
>
> With these remarks it is hoped the present treatise—admirable in most other respects, and especially in its historical and critical survey of the philosophies—may be perused by the ordinary reader without much danger to the "faith that is in him." The religions, doubtless, receive some very rough handling, but they can probably bear it; and in any case, as the writer says of metaphysics, they must look to it. All of them, however, have made themselves at one time or another responsible for so many inanities in dogma and morals—belief in an impossible cosmogony; in a puerile astronomy; in the objectivity of certain Assyrian myths; in witchcraft; the efficacy and justice of the rack and the stake; intolerance, suppression of dogmatic error by fire, sword, and massacre of man, woman and child, predestination as understood by Augustine and Calvin; a personal devil presiding over an everlasting realm of material fire and brimstone; divine right of kings and the like—they could scarcely expect to escape without a few hard knocks in a work of this sort. (xviii–xix)

A remarkable sentence. The point of it, even the point of condensing all that critique in one sentence, was to contain, fiercely, the rejected authoritarian religion, on one side, and the rejected, mocking materialism, on the other. This did not leave Keane with much room in the middle, because both his opponents were likely to pounce on any phrase that might be used against the other. But note the tone of his separate arguments: it was as if he were begging victorious materialism to let him retain the "trifle more consoling" notion of God and yet still be modern, rational, and brave. In contrast, he spoke of Catholicism in a mood of vengeful triumph after a long siege.

MEN OF TODAY

My last source for contemporary response to the anthropologists is a rather odd one. In the 1880s the poet Paul Verlaine edited a biographical journal called *Les hommes d'aujourd'hui*. Each issue was four pages: the first being a caricature portrait of a famous personage, the rest, a biographical sketch. There were many such journals in France in the second half of the nineteenth century, but *Les hommes d'aujourd'hui* was easily the biggest and best known.[94] Subjects were drawn from the worlds of art and politics: the first featured was Victor Hugo; later issues were devoted to such figures as Stéphane Mallarmé, Jules Verne, General Boulanger, and Alexander Dumas *fils*. The paper had several editors, before and after Verlaine, but through them all, its politics leaned generously to the left: the republic was celebrated, the Second Empire resented, "decadent" artists were encouraged, and anticlericalism jubilantly supported. Over the years, *Les hommes d'aujourd'hui* featured Mathias Duval, Abel Hovelacque, Yves Guyot, Henri Thulié, and Clémence Royer.

Duval's portrait showed him at work hatching eggs—a reference to his research on the beginnings of life—but on the ground behind his figure stood an immense pile of skulls.[95] Guyot's illustrious political and journalistic career was detailed, including mention of a journal he had started with Louis Asseline, one of the freethinkers who had first drawn the future anthropologists together. His portrait showed him with a giant pen and bottle, the latter labeled "Democratic Ink."[96] Hovelacque's portrait showed him hanging a sign that read "Municipal Property" on a building that bore the word "convent" above its door. The text admiringly discussed his successful campaign to turn two Paris convents into secular schools. The Society of Mutual Autopsy was mentioned as well.[97] Much of the essay on Hovelacque was quoted (with attribution but no specifics) from something Lefèvre had written of his friend, and herein Hovelacque was praised because though "he had been raised in an ecclesiastic institution," he left it "not a skeptic but not a believer. . . . You say that this is not rare: happily! But neither is it so common that we can pass over it in silence." Again, these were Lefèvre's words, but the text quoted them with approval. The church, it continued, tortured and deformed many peoples' minds, and its powers of persuasion were not only negative; it could have made things very easy for Hovelacque. "Nothing could have been more simple for Abel Hovelacque, in the middle of the empire, to follow the regular channels to places and to magistratures. He could have been, today, one of the sweethearts of the moral order, fabricators and exploiters of social peril." But not Hovelacque. "Let us say that we are happy for him that he escaped . . . and he is no less happy for us that we may count among our ranks one more member. With no other incentive than the love of truth and of science, the young student of the church knew how to get out of the trap." All he kept from that education were "a few precious scraps of theology, which he would need to combat against it." Two "auxiliary powers" brought him his "emancipation": linguistics and anthropology (2). He learned Sanskrit and began attending the Société d'anthropologie.

> Thus by two parallel or, rather, convergent paths, he entered into the movement of contemporary thought. To analyze the elements of language is to grasp the mechanism of intelligence by the facts, to examine anatomical conformations, to measure the forms of skulls and cerebral capacities of diverse races, to follow from the prehistoric ages the progress of industries, of arts, of civilizations; it is to mark man in his place in the living series and to determine all the phases of his moral and social development. In both linguistics and anthropology, it is to apply to all objects of knowledge the experimental method. . . . But in habituating the spirit to rigorous processes and neat solutions, linguistics and anthropology did not dry his heart, they did not close it to love of beauty, to sympathy for suffering, to an enthusiasm for justice. Much to the contrary. There are hardly any sciences more large-hearted, because they embrace all of man with his physical and intellectual faculties in their rapport with the universe and with their fellows. They made of Hovelacque a philosopher and a citizen. (2)

That's no ordinary science. After the quotation from Lefèvre, the text went on to mention *Pensée nouvelle*, Asseline, Letourneau, Thulié, and Lefèvre, and more of their doings, but we have the idea. Lefèvre's biography of his friend reads like a spiritual conversion story in every point.

Thulié's portrait in *Les hommes d'aujourd'hui* showed him engaged in what looks like an autopsy.[98] He was generously praised for his "struggle against church and empire" in a manner that recalls the freethinking anthropologist's own lament over lost youth:

> He spent twenty years of his life in this deaf struggle without applause and without any other honor than that of having done the right thing. He believed, earlier than most, and rightly—and we have seen the proof—that the religious question was the Gordian knot of the political question and that clerical power was one of the most efficacious forces supporting the empire. He was one of the editors of a valiant newspaper, *La pensée nouvelle*, a courageous organ of freethinking that could not be very well supported by freethinkers, of which the number, at that time, was very small and which the doctor Thulié sees with joy to be so numerous today. (3)

There is a marvelous economy to these narratives: conversions and clawing to the truth or knowing the truth early and holding out, overcoming traps and prejudice, and teaching others.

On the cover of one last *Homme d'aujourd'hui*, Royer gazes out over a desk piled with books. She is wearing a high-necked, white, ruffle-collared shirt, cinched at the neck with a cameo brooch, and a black skirt and jacket. The books' spines before her say things like: "Theory of Evolution," "Maternal Heredity," and "Paternal Heredity," but the one she has in her hand, with her name on the cover, is called "The Philosophy of Hope." The text tells us that her childhood education was cut short at age eleven but that "reading our great poets and our modern novelists opened up new horizons for her," and a few scientific works provided revelations. Then, when she was eighteen, there was a political awakening: "The explosion of the Revolution of 1848 put before her a whole crowd of problems and doubts that she has felt the need to resolve ever since. Discovering she knew nothing, she recommenced her elementary education and quickly took her exams and got her diplomas." The few paragraphs left managed to sketch an extraordinary life of social and intellectual engagement, including a public essay contest on taxation in which she tied for first place, sharing the prize with Proudhon. The reader was also told that she had taught a course in 1859 that "defended the evolutionary theory of Lamarck against the Cuvier school just at the moment when Darwin revived it in England" and in 1864 had published a philosophical novel in Brussels that was banned from entering France. She had also "preceded [Ernst] Haeckel" and "had developed, even before Darwin himself, the consequences of the theory of selection relative to man and his mental faculties." A list of her publications and conferences was over a page long.[99]

So the freethinking anthropologists were famous enough to merit coverage in a popular one-person-per-issue biographical newspaper. They had become reasonably well known figures of republican politics, anticlericalism, and anthropology. They were a feature on the cultural spectrum, notable in part because they had lost so much of their lives waiting out the empire and in part because of their titillatingly gruesome preoccupations and their larger-than-life anticlericalism. What is clear from these portraits is that they were also beloved characters on the Parisian scene.

THE NORTH STAR

In the preceding chapter I reported how the freethinking anthropologists assigned new naturalist names and materialist meanings to buildings, bones, stationery, cemeteries, government personnel and their children, municipal conferences, anatomical miracles, funeral rites, and their own bodies. This chapter has shown how they worked to make mundane the previously metaphysical details of morals, marriage, philosophy, aesthetics, economics, linguistics, history, prehistory, and passion. The wider academic audience and general public at large may not have joined the anthropologists in all these deconsecrating efforts, but they did entertain the claims and support the work. Conceptual, epistemological work was getting done that was more interesting to contemporaries than any specific theory being proposed. Likewise, from our present vantage point, this secularizing work was more lasting than any individual argument the freethinkers presented. The anthropologists wrote well and in a casual style that seems to have charmed anyone not predisposed to hate it. They were seductively amusing, iconoclastic, and easy to read. What is more, they made some enduring—if impermanent—contributions to the development of their science.

In their moment, the scientific community (even its Catholic subset) respected the work of the freethinking anthropologists. Apart from the insistence that the whole project of anthropology added up to atheism, the scientific community accepted the work on the terms offered by the freethinkers, that is, everyone seemed to agree that this was important work that would become more meaningful as the facts were slowly and patiently gathered. In any case, much of it was interesting, peculiar, and unsettling and seemed worth the time to peruse. The greater part of the philosophical community was naturalist and scientific in temper, and, with a good deal more criticism, its members responded in a similar fashion. The members of the idealist philosophical community, by contrast, were rather scandalized by this stuff being called philosophy, aesthetics, and history, but even they did not ignore it: this work was neither below their dignity nor out of their cultural sights. In popular literature, reviewers also had problems with the anthropologists' more extreme claims, but they, too, found such claims to be a useful orientation point from which to announce their own, no longer extreme, position.

In the general culture, faith was advertised as stationary and science progressive, in a particularly tense formulation of modernity. The freethinking anthropologists helped people to question this dyad by taking it to an extreme. Also, it is clear that people felt an ethical need to register their disapproval of church malfeasance and cruelty and that a tacit agreement with the anthropologists served that need. In fact, the sheer presence of such white-hot hatred of the church may have given people the sense that their moral indignation was indeed being represented. Through the freethinking anthropologists served as a catalyst or inspiration for some, for others they likely served as a steam valve.

Most people who read and discussed the Paris freethinking anthropologists were surely not materialist atheists, but the changing role of religion in France was a major factor of their lives, and we may suppose they wanted that fact symbolized on their bookshelves and in their conversation. Of course, they also wanted to understand this change as well as they might, and reading this anthropology did shed light on the matter, whether one agreed or did not. To use the metaphor again, the freethinking anthropologists were a sort of North Star: some of those who referenced them were headed north—though perhaps not going all that far—and some were headed elsewhere; in any case, it was a wonderful beacon.

∞

Chapter Four

Careers in Anthropology and the Bertillon Family

By the 1880s there existed a second generation of atheist anthropologists who had trained with the freethinking anthropologists and then moved off in disparate new directions. The most important of these were the Bertillon brothers, Georges Vacher de Lapouge, and Léonce Manouvrier (this and the next two chapters will be dedicated to each of them in turn). This second generation did not form a cohesive social or political group; in fact, there were some major rows here. They each shared a great many common assumptions with the original group—they all joined or crucially enabled the Society of Mutual Autopsy—but they can be meaningfully distinguished from them in several ways.

The core group of freethinking anthropologists had written for *Libre pensée* or *Pensée nouvelle* in the 1860s, dedicated themselves to a wide range of freethinking anthropological projects, and cited the group as a primary allegiance. There were some exceptions to this: Louis-Adolphe Bertillon and Yves Guyot wrote for those early, atheist publications and took part in many freethinking-anthropology projects but were not as cultishly devoted to the group in their rhetoric and often had other things on their minds. (Bertillon had been part of the Société d'anthropologie before the freethinkers came on the scene; Guyot was an important political figure.) Also Duval and Hovelacque were younger and did not write for the early freethinkers journals, but became avowed and lauded members of the core group.

The second-generation figures were much younger and came to the freethinkers after the group had embraced anthropology. The original freethinkers were all about Broca's age: Broca was born in 1824; Mortillet and Louis-

Adolphe Bertillon were born three years earlier, in 1821; Eugène Véron was born in 1825; Letourneau in 1831; and Royer in 1830. Lefèvre, the poet, was the youngster of the original group—born in 1834. By contrast, Jacques Bertillon was born in 1851, his brother Alphonse in 1853, Léonce Manouvrier in 1850, and Vacher de Lapouge in 1854. Though these younger men differed immensely, all of them spent their entire adult lives dealing with the consequences of a godless world and bodies bereft of soul. And all of them found a way to apply the anthropometry they had learned from the freethinkers to political interactions with the French population and the French state. This took place on an intellectual plane that was supported by a social, political, and economic dimension.

CAREERS IN FREETHINKING ANTHROPOLOGY

The freethinking anthropologists managed to create elite careers for themselves—often with a considerable amount of imagination and hustle—in ways that brought a great deal of attention to their atheism. Most of them were able to make their living from a combination of politics and anthropology. They curated anthropological museums and exhibits, organized anthropological conferences, applied for government grants for fieldwork or to visit foreign anthropological museums, conferences, and so forth, and hosted short excursions to anthropologically interesting sites for paying amateurs. With Broca, they created the Ecole d'anthropologie, where many of them had professorships. They also earned income from their books and articles. All this served the dual purpose of allowing the group to live comfortable lives and assuring that their ideas received attention.

Consider, for example, the career of Gabriel de Mortillet. As I have already discussed, Mortillet was forced into exile after having published inflammatory socialist pamphlets during the Revolution of 1848. When he returned to Paris in 1864, he actively contributed to several scientific and political journals, including his own archaeological journal and *Libre pensée*. In 1865 he founded the International Congress of Archaeology and Prehistoric Anthropology, taking an active part in its increasingly prestigious meetings. In 1867 Mortillet was hired to organize the anthropological exhibit at the Paris World's Fair, and he published a pamphlet describing and explaining the presentation. (He arranged anthropological exhibits for the Paris World's Fair in 1889 as well, and both events brought tremendous attention to French anthropology and to Mortillet in particular.) In 1868 he was made conservator of the National Museum of Antiquities at Saint-Germain, which had been founded in 1862 by Napoleon III to complement the *Histoire de César* that the emperor was then writing. During his tenure at the museum—and after—he served as mayor of Saint-Germain (1882–1888) and retained his post at the museum until 1885, when he ran for a seat in the French parliament and was elected deputy of Seine-et-Oise, sitting on the extreme left. It was during his time as mayor and as deputy that he carried on his infamous local program of deconsecration.

Mortillet was by no means the only freethinking anthropologist to fill a political post. Abel Hovelacque was a Socialist deputy and was president of the Paris Municipal Council from 1886 to 1887. According to the *Grande encyclopédie*, in 1887 Hovelacque prevented Jules Ferry from being renominated as president by physically locking him out of the National Assembly.[1] Henri Thulié also served as president of the Paris Municipal Council, as well as the general councillor of Seine.[2] Louis Asseline was mayor of the fourteenth arrondissement of Paris from 1870 to 1871 and in 1874 served as representative to the Paris Municipal Council (from which position he was instrumental in procuring an annual allotment of twelve thousand francs for the Anthropological Institute).[3] Lefèvre served on the Paris Municipal Council as well. Louis-Adolphe Bertillon served as mayor of the fifth arrondissement of Paris.[4] Yves Guyot was the most politically successful as municipal councillor of Paris, deputy of the Seine (1885), and minister of public works in the governments of Tirard, Freycinet, and Carnot. He also served as director of *Le siècle* and editor-in-chief of the *Journal des economistes* (where he regularly featured Royer's work) and published extensively on economics and republican politics.[5]

Returning to the career of Mortillet, in 1876 he and Broca founded the Ecole d'anthropologie on private funds, having collected one or more subscriptions of 1,000 francs from each of its thirty-four founders.[6] Joined by Eugène Dally and Paul Topinard, Mortillet and Broca served as the school's first professors. By 1880 it was receiving 20,000 francs per year from the French state. The Institut anthropologique received a yearly 6,000 francs from the Department of the Seine, and the laboratory was yearly awarded another 6,000. These funds, along with the substantial gifts and donations, allowed the school to pay an annual salary of 3,000 francs to each of its seven professors. Professors at established Parisian institutions earned a starting salary of about 6,000 francs, which could rise to 7,000 (they began at 3,000 francs and went to 5,000 in the provinces), so the salary offered at the Ecole d'anthropologie was a healthy contribution to a scholar's yearly income in Paris at the time.[7]

After Broca's death, the possession and distribution of these stable positions were entirely controlled by the freethinking anthropologists. Not only was Mortillet able to support a family by these means, but he was also able to install his son in a similar position. By the 1889–90 academic year, there were eight regular chairs at the school: Mortillet's son, Adrien de Mortillet, held the chair in prehistoric anthropology; Mathias Duval took embryology and anthropogeny; Abel Hovelacque covered ethnography and linguistics; George Hervé was charged with zoological anthropology; Topinard had general anthropology (but would be dramatically ousted from his chair that year); Arthur-Alexandre Bordier taught medical geography; Charles Letourneau handled sociology and the history of civilizations; and Léonce Manouvrier received the central task of explicating physical anthropology.[8]

Whereas Durkheim was able to legitimate sociology by winning a chair at the Sorbonne, the freethinking anthropologists did not steward anthropology into any established institution. They could not grant degrees, and, largely because of this, anthropology remained slightly outside mainstream scholarship. But they avoided a great deal of rivalry and contention by forming their own semiprivate institute and filling it with excited fellow travelers. From all descriptions, it seems the place buzzed with a thrilled camaraderie as colleagues and specialized students worked together with a common methodology and spirit of action. As Royer said, it was a little church.

Over the years, Mortillet organized countless archaeological-anthropological tours. Each year the journals attached to the Paris anthropologists ran ads for these events. In 1885, for instance, L'homme announced anthropological day trips, inviting excursionists to guided tours of prehistoric sites a short train ride away from Paris.[9] Mortillet also frequently applied for government grants for anthropological conferences and archaeological investigations.[10] More often than not, he was awarded the sums he requested. He was given 1,200 francs to examine archaeological sites in the Midi in 1872 and was funded for a trip to a Stockholm conference in 1874 (600 francs). Broca, who was also attending, had heartily recommended Mortillet to the Ministry of Public Instruction.[11] In 1879 he was given 1,200 francs to attend a conference in Moscow after having written to Jules Ferry, then the minister of public instruction, explaining that France was the "terre classique" of anthropology, that for this reason more Frenchmen had been invited than any other group, and that they must attend in like proportion in order to maintain this reputation.[12] Following this trip, the Ministry of Foreign Affairs sent several published reports on the Moscow conference to Ferry, all of which celebrated Bertillon, Mortillet, Hovelacque, and Topinard as national treasures.[13] The freethinking anthropologists had done an excellent job of promoting themselves, and their discipline, within the contemporary climate of nationalist competition for leadership in the sciences.

Success bred success. In 1880 Mortillet received 1,000 francs to take part in a conference in Lisbon. An official observer of this Lisbon conference reported to Ferry, who by then had ascended to prime minister of France, that the king of Portugal had personally welcomed the visiting scientists and that the queen had invited them to an "intimate ball" that she held every year for her son's birthday. At the conference, the opinions of the French anthropologists were particularly honored, and in general "the scientific importance of the representatives of France . . . , the speeches and the toasts they had the opportunity to pronounce, have augmented the prestige of our nation."[4] France also played host to scientists from around the world. Foreign visitors to the society, school, museum, and laboratory were numerous and were given significant attention by those who ran these institutions. Some of the visitors were among the hundred or so foreign members of the Société d'anthropologie; they came to attend special conferences and regular meetings, to debate their positions, and to study at the museum and the school. Visitors of every background came to observe the many projects undertaken in the laboratory: the preparation of skull and plaster bone molds; the invention and use of new measuring instruments; the cataloging of human pathological specimens collected from the hospitals; the collection, characterization, and study of fossils; the dissection and anatomical description of primates; and the recording of thousands upon thousands of measurements taken from human subjects and specimens. Visitors also came simply to meet the authors of widely respected works of anthropology.

Mortillet published at least twelve full-length books on archaeology and anthropology, and his articles appeared in most issues of several anthropological journals. As I have reported, many of the freethinking anthropologists were extraordinarily prolific, a fact that is less surprising when one remembers they had originally come together not as anthropologists but as journalists and essay writers. Most of them supported themselves through careers that were quite similar to that of Mortillet: an amalgam of political, scientific, educational, and entertainment-oriented positions, with a good deal of their money coming directly from the state. Along with politics and anthropology, Letourneau continued to work as a doctor; Hovelacque as a lawyer. Lefèvre wrote novels and poetry and translated throughout his life. In these endeavors, members of the group continued to propagate their atheist anthropology and to transform the world from religious to scientific. Lefèvre's published translations, for instance, included a version of Lucretius's poem *On the Nature of Things*, one of the great essentially atheist works of the ancient Roman world, which itself owed its content to Epicurus, one of the great essentially atheist philosophers of the Greek world: a nice daisy chain of materialist atheism. There is little evidence as to Lefèvre's literary success, but Zola did review one of his novels favorably, and Lefèvre consistently published his poetry in the republican literary journals of the period. Some of these, like *La jeune France* and *La vie littéraire* were also happy to publish Lefèvre's (anti)philosophical musings and anthropological pronouncements. Some sample titles include "La vie philosophique," "L'histoire" (sonnet), and "Paléontologie intellectuelle."[15] I have translated the first three poems in a series of five that ran in an issue of *La jeune France*. The subject matter is typical of Lefèvre: the awe in the face of scientific revelation, the dismissal of the gods, and the wistful sadness and mannered courage with which he confronts the facts of life in an atheistic world.

The Retirement of the Gods

The gods in their youth lived on earth.
They were well nourished, free and vicious.
They stole the eyes from the forests
And the mystery from the mountain peaks.
Diminishing respect made their faces austere.
They showed themselves less: alas! they were old!
Finding man too close they took to the skies.
Copernicus chased them from the planetary world.
They climbed higher yet, toward the unknown. Their bodies
Emaciated, more hollow than the ghosts of the dead,
They made themselves a coat of the silver of the stars.
But to the eyes of humans, the skies were open.
Poor gods! Science has lifted all the veils.
They are gone. And nothing is missing from the universe.

Posthumous Rays

There are dead stars from which ancient light,
For millions of years, falls from the depths of the skies
And will bathe our night in a mysterious flow
Until its wave has run out entirely
And meantime even our dust will have perished.
Longtime, longtime after the extinction of our eyes,
In the void where once turned men and gods,
The river of light follows its career.
The poet and the star have the same destinies:
When the future receives their flame or their memory,
It comes from the depths of far away sepulchres.
Yes, the flash projected into the azure of history
Is the radiance of ancient extinct stars.
There is nothing but melancholy in our dreams of glory!

IMMORTALITY

Several millions of years! And the earth frozen,
In one block and naked, funerary flame
Will be seen engulfing it in its moving tomb.
Vegetation, life, and thought,
Nothing will rest of its eclipsed glory,
Nothing, not even a reflection of the True, the Good, the Beautiful.
Time will have taken back, to the last shred
Of the work of humanity, a page erased forever.
Nothing will exist anymore, nothing, not even the dead.
The world will be closed to immortal souls,
Because to be, they need to be in a body.
And those of the past? Alas! Where are they?
In the vague country of treasures spent,
Hopes accomplished, and futures past. [16]

Lefèvre was moved by these questions, writing historical verse on the origins of the secularist worldview and psalms to the mysteries of science. He was not just saying that the world was material and meaningless and so what? Rather, he was purposefully enjoying the invocation of disturbing scientific enigmas, staring boldly at the facts of death in a materialist schema and working to create a mood of deep, "religious" wonder.

In January 1881 Lefèvre published a short essay in *La jeune France* that had to do with a certain popular Christmas gift that year: the book was *Robinson suisse*, the French translation of *The Swiss Family Robinson*.[17] The book was great, asserted Lefèvre, but there was one big problem: it was brimming with prayers and other "religious banalities." "What was the author trying to prove?" asked Lefèvre, adding that if there is a Providence, it had thrown horrible trials in the path of this poor family, so why be impressed with it or thank it? Because of this, the prayers in the book seemed ironic to Lefèvre, who added: "What is worse is that they are as boring as they are superfluous. I will be amused, one day, to suppress from my copy all these superfluities. One would not believe how much of the text that would entail. Voilà the correction that we propose to an intelligent editor." Lefèvre cautioned that he did not mean that people ought to get rid of devotional books—we would need to study those in future—but to edit all "instructive and moral stories destined to the laic youth." Everyone in charge, "editors, authors, fathers of families," had to be strong and not let such things reach the children: "Never has such weakness been more foolish, more inopportune: it's a real absurdity. All its litanies, its invocations . . . are not only intolerable and nauseating; these literary pests are social pests. This is a solemn and terrible hour; it amounts to knowing if the world of science will definitively prevail over the world of tradition and of theocracy. And it is in this moment that they stuff nascent minds with twaddle and superstition!" Instead, he continued, children should be taught that nothing happens outside of "known, constant" natural laws that we "cannot escape" but can turn to our profit through our intelligence and courage and use to help establish justice, virtue, and happiness, "on earth and not elsewhere." In closing, he took up another scientific metaphor, warning that "we must not ourselves throw into their spirits, still vague and troubled, the germs of possible errors," since we have learned from Pasteur what "bad seeds" or "spores" can do. Such books should be edited out of respect for the future liberty of the children. Lefèvre also wrote a long essay entitled "Voltaire et les religions," in which he excused the philosophe for his deistic beliefs, reminding the reader that "if we go beyond Voltaire, it is thanks to him that we may do it. He cleared the path."[18] There was also an article on "man, according to the discoveries of anthropology," in which he discussed the accomplishments of all his friends at the Société d'anthropologie and told the story of humanity as an evolution of various races, some of which were destined to become extinct.[19] Of course, it was in modern humanity's religiosity that we most keenly showed the vestiges of these "earlier forms" of evolution (221).

As I have noted, the freethinking anthropologists managed to carry out a wide range of projects in pursuit of their goals. Some of this various work was more lucrative than it was influential, and some of it more influential than it was lucrative, but all in all it seems to have been a lot of fun and amounted to a career in anthropology at a moment when no clear pattern for such a career existed. Not surprisingly, the exception to this general success was Clémence Royer. As a woman, she had been barred from a scientific education, and throughout her life she was barred from most paying positions in science. As a result, she lived dangerously close to the edge of poverty. She was not ignored: she was well published, won prizes from such establishments as the Académie des Sciences Morales et Politiques, and was awarded the Legion of Honor, but the weight of economic penury took its toll. As she wrote in her will of 1895:

A victim of those prejudices that still are opposed to the intellectual development of women, I have worked all my life without pay to illuminate a blind humanity that has only created obstacles to the construction of my philosophical work by closing off to me schools, academic chairs, and laboratories. Everything that I know I have acquired after a great struggle, and I was obliged to forget everything that I was taught in order to learn everything by myself. I shall carry with me to the tomb useful truths that others will have to discover anew. Because I have had the bad luck to be born a woman, I have lacked all means to express, to correct, or to defend my thought, and I have done only the smallest part of what I could have done.[20]

A heartrending lament. Still, she wrote a great deal and managed to participate meaningfully in the creation of a new discipline. She was an active participant in society meetings and in a whole range of national and international conferences, and her theories on the nature of the universe and the meaning of life were publicly debated in the prestigious journals and institutions of her time. Moreover, while sometimes serving the discussion in important and imaginative ways, one reason she met with professional difficulty was that she was more racist and determinist than were her colleagues. In any case, all her other contributions aside, her preface and translation of Darwin's great work profoundly influenced the French reception of natural selection and determined the freethinkers' turn to anthropology. In their hands, it was becoming a viable profession, and since their work also amounted to an extremely multifaceted work of evangelist propaganda for atheism and feminism, merely by this she had done much to forward her ideals.

For another quick angle on who all these people really were, we may gain something by noticing the progressive and radical figures with whom they spent their lives—a defuse network of materialist, freethinking friends and family alliances. As a detail here I will offer only the most engaging and strange, a sort of train wreck of associations recorded in Edward Hallett Carr's *The Romantic Exiles* of 1933.[21] Alexander Herzen was a materialist freethinker, a professor of physiology at the University of Florence, and a follower of the materialist anthropologist Carl Vogt (who was friend to Broca and Mortillet). His father, also Alexander Herzen (who died in 1870), had been the famous Russian radical émigré and one of the founders of Russian socialism. Nicholas Ogarev, another key founder of Russian socialism, was a close friend of the young physiology professor. Both Ogarev and Herzen were married, both had mistresses, and Herzen's mistress was Ogarev's wife, Natalia Tuchkova-Ogareva. The linked families traveled together and in mixed clusters. Herzen and Natalia had a daughter, Liza, who in the early 1870s was in her teens and alone with her mother: her father had stuck with his wife, her mother's husband was living with his mistress. Mother and daughter were tense together, so Liza went to Florence to be near her father. She was living with her half-sister, Herzen's other daughter, Olga, who was acting as guardian, when young Liza fell madly in love with one of her father's friends: Letourneau. He had come to Florence to lecture at the university in 1874. Carr writes of him:

The other principal actor in it was a French *savant*, Charles Letourneau by name. He was already known as the author of a work entitled *La physiologie des passions*, now remembered chiefly as one of Emile Zola's sources of inspiration; he published during the year 1875 a text book of biology; and these two works were followed by numerous other treatises of scientific or philosophical character. He survived until 1902; and there was nothing either in his earlier or in his later life to suggest that his share in Liza's tragedy was anything but an incalculable and irrelevant episode in an otherwise ordinary career.

(350)

As for Liza, Carr tells us she was marked by imperiousness and shyness, that she was sixteen, and that she was the "most brilliant of Herzen's children" (348). Letourneau was forty-four years old when they met and was happily sharing his stint abroad with his wife and children. There were intellectual and scientific salons in Florence at the time that were open to women as well as men, and these families all seem to have taken part. Perhaps that is where Letourneau and Liza first spent time in each other's company. Letourneau tried to explain to Liza that he was thirty years her senior and married (we have their letters), but Liza was inconsolable and determined. Gabriel Monod, Olga's husband, stepped in to try to mend the situation. An important leader of the new French "scientific" history, Monod was another significant figure on the secularist scene, a founder of the *Revue historique*, and, later in the century, one of the earliest Dreyfusards.[23] Tragically, and even though Letourneau, too, made some serious efforts to discourage Liza's affections, the young girl remained heartsick and actually killed herself over him (chloroform, a towel, a note). The story takes on a strange weight when one considers that Letourneau had translated Herzen's *Physiology of the Will* from Italian to French, and Herzen had surely read Letourneau's work on the material basis of the passions, yet these students of the will and the passions had utterly lost their girl. The living arrangements of her family, the idleness of her hours, and the fact of Letourneau's status as her often-absent father's colleague—well, each explains enough, and the nature of the anthropologists' efforts to explain passion and will scientifically seems particularly maladroit in this context. In any case, this is a rare glimpse into the lives of these anthropologists and their social set; mostly we are able to know them only by their public writings and the slightly more personal material of their scientific societies.

The exception, again, is the case of Clémence Royer, whose private life with her lover, Pascal Duprat, a significant figure in the governments of both the Second and Third Republics, has been chronicled in much detail by Joy Harvey. Not least because she was a woman, and a feminist, and in a relationship with Duprat—who was married and with whom she had a son—Royer ran in different circles from most of the other anthropologists. Still here, too, Royer was part of a world of well-known progressivists, radicals, and active secularists. Harvey tells us that Royer and Duprat also spent time with Letourneau and Herzen in Florence, along with the materialist Italian anthropologist Paolo Mantegazza, and that the families of Broca, Mantegazza, and Letourneau sometimes met in summer by the seaside in Spezia.[23]

THE FAMILY BERTILLON

In turning to the second generation of materialist anthropologists, we must confront the profoundly overdetermined significance of the human body at the end of the nineteenth century. Michel Foucault's work along these lines is central to the present discussion but not the only source of this observation and attendant critique. In a broad sense, with hindsight, it would be hard to miss the sudden concern with the body—counting it, measuring it, and documenting it—that became an implicit part of the state's function by the early twentieth century.

The rise of the nation-state was attended by a new breed of experts, hired to guide and justify the state's bureaucratic and penal interactions with the body politic. The experts were self-invented at first and were trusted because of the immense prestige of science in this period. Through its chemical and technological service to industry, science had transformed the material world. In medical science, the identification of pathogens and the creation of vaccinations seemed likely to bring an end to all known disease. It was conceivable that even death itself might be vanquished. The movement toward democracy in many countries also seemed to be a sign that the Enlightenment vision of science would be borne out as all humanity progressed to a kind of rational paradise on earth. The social sciences were developed in this period and in this mood. Anthropology, sociology, psychology, and psychiatry all borrowed the confidence of recent scientific triumph, and as each competed for public attention, government funding, and university positions, the leaders of the new social sciences often made utopian claims. Governments were increasingly taking it upon themselves to cure the ills of their societies—often directly replacing the religious functions of the church—and the social sciences were called upon to act. Often, however, they profoundly overstated their knowledge and abilities. What they claimed was that they could make society happy, healthy, and normal. That is, they claimed they could identify and cure deviance and thereby eradicate poverty, crime, and social unrest. Psychiatry was going to do it through the mind of the individual; sociology was going to de-scribe and manipulate the patterns of society as a whole; and anthropology concentrated its attention on the human body, pledging to explain and regu-late our physical and social evolution.

There was always a temptation—as there is to this day—to define away so-cial problems by declaring some people to be less fit, innately criminal, bio-logically predisposed toward housework, or innately suited to poverty and squalor. It then becomes easy to argue that, for their own good and for the good of society, such people need to be kept in a certain role or excised from the population; the policies deporting criminals to far-off colonies were born in this logic as were a plethora of arguments about the role of women and the working classes.[25] Yet the vast majority of French social scientists examined in this study remained dedicated to egalitarian principles of race, class, and gen-der. With as much grandeur as exhibited by social pessimists, they declared that their social science would lead the way to an atheist, socialist, feminist utopia. Comparatively few of them, well represented by Georges Vacher de Lapouge, argued that the problems of the world were located in certain bod-ies and that utopia could only be achieved if these bodies were bred out of the population.

As we know, this approach would eventually be given a hearing. More palatable, environmental solutions were tried first, and governments busied themselves counseling individuals, changing curricula, and engineering public programs to enhance moral and physical hygiene. But deviance and poverty were not eradicated. As the rising eugenic movement of the early twentieth century earnestly advised men and women about whom they should marry and how they should raise their children, modern Western governments began to entertain the notion that the body politic could only be cured through sur-gery: metaphorically in the enactment of immigration quotas and quite liter-ally in the establishment of involuntary sterilization within the penal system.[25] The results were horrific. They were not, however, the only effects of the gov-ernmental shift from traditional political power to what Foucault called "pas-toral power," that is, the state's attempt to fulfill the historic role of religious care for the flock.[26] As Foucault put it, the modern state took it upon itself to "constantly ensure, sustain, and improve the lives of each and every one" (2:235). This recalls the kind of mission statement made by the freethinking anthropologists in the name of anthropology. The two claims taken together suggest a lot about how social services that had traditionally been negotiated through religious institutions and theories were reassigned in modernity. The monolithic state did not simply try to take more control of bodies in order to perpetuate its own power; sometimes individual people and communities of theorists began this process not only to collaborate with the state for the sake of prestige and reward but in response to philosophical crises and in an explicit attempt to fill the gap created by the rejection of religion.

The freethinking anthropologists did not make this move of helping the government in its attempt to control or manipulate the body politic. The practical use of freethinking anthropology to the French government was almost entirely contained in their deconsecration project. They served to support the general legitimacy of the new government by providing an alternate source of truth that could rival tradition and church. Also, in disavowing the soul and its philosophical equivalent, the freethinkers helped to establish the body further as the site for social change. In the anthropological work of the Bertillons, Lapouge, and Manouvrier, the markings, shapes, and affiliations of the body now became the key categories through which human beings tried to imagine their most authentic selves.

The whole project of locating the meaning of human beings in their physical form was partly initiated and profoundly augmented through Broca's work. After Broca died, anthropologists kept on measuring bodies, but measuring was never the central focus of the freethinking group. The freethinking anthropologists supported craniometry, and their work usually contained a good deal of it, but they did not generally base their conclusions on the measurements. They simply included such data as important descriptive information. They do not seem to have entertained the idea that the practice could translate into a powerful conservative social tool. Many of them had spent much of their lives as writers, and while they honored and supported craniometry as cleanly unmetaphysical, their own theories were primarily literary. In any case, whatever they measured, they were rarely making any other point than that there is no God, nothing is sacred, and religion is wrong.

The Bertillon family became interested in a very different set of problems, problems that served as an essential part of the new concern with the body politic. As noted, Louis-Adolphe Bertillon was a friend of Broca and one of the nineteen founding members of the Société d'anthropologie. As a young man studying medicine in Paris, Louis-Adolphe took courses at various institutes. He followed the lectures of Michelet at the Collège de France, and the two men began a long-term friendship.[27] Elsewhere, he studied population and statistics with Achille Guillard, an inventor and businessman who had christened the new science of demography (first publishing the word in 1855) and who had written a good deal on the subject. Bertillon and Guillard actually met in jail: between June 1848 and December 1851, Bertillon was arrested three times, Achille Guillard twice.[28] They were arrested amid the melee of demonstrations and police actions that accompanied the Second Republic and its fall, and into the Second Empire. It is harder to know whether they were involved as dedicated republican partisans, which the number of arrests suggests, or as physicians tending to the wounded on both sides of the barricade, which is what family members would later report. In any case, on one of these occasions the student recognized his professor. The older man was apparently able to arrange for somewhat more pleasant accommodations for both of them, which was lucky since they ended up spending some six months sequestered together in these hard circumstances.[29] After their incarceration had ended, Bertillon was welcomed into the home of his new friend and there met Guillard's daughter Zoé. They were married in 1850 and thus began a demography dynasty.

As I have intimated, one strain of the family's anthropological concern would become centered on natality, so it is interesting to note that Achille Guillard thought population changes were almost exclusively to be understood through death rates: natality did not have to be studied because it simply rose and fell with available sustenance. Also, he believed an automatic mechanism regulated the birthrate so that "when lives are short, there are lots of children."[30] Though he coined the term, demography was not Guillard's main concern; he was an inventor, ran a progressive school, and wrote on botany. His son-in-law, on the other hand, hammered out the basics of the science. Bertillon had earned his medical degree, but his friendship with Broca helped to lead him out of medical practice proper. He decided to join Broca in creating anthropology and took the impetus for his own specialty from Guillard.

According to a pamphlet written on Bertillon when he died in 1883, the whole thing was a kind of blind leap. Broca had bravely marched away from the Société de biologie and decided to create "a scientific society where one would have the right to draw all the philosophical consequences from one's observations. Few men at first had the daring to join him. M. Bertillon, naturally, wanted nothing better, but he feared being out of his element in such a society: 'I wouldn't be able to render it any service,' said he, 'as I don't know a word of anthropology.' 'Neither do I,' responded Broca with his usual swagger. 'All the more reason to learn it or, rather, to create it, because in truth, it doesn't exist!' "[31] When they got the society under way in 1859, Bertillon took a course with Quatrefages at the Museum of Natural History and studied craniology with Broca. Soon Bertillon was delivering papers on statistics and health to the illustrious Academy of Medicine, and in 1860 he assisted in the creation of the Société de statistique de Paris. In 1876 he became professor of demography and medical geography at the Ecole d'anthropologie. We can imagine that the first chair in demography would have taken a lot longer to materialize were it not for these relationships.

Bertillon was a freethinker. Under "Religion" on his sons' birth-certificates, he wrote "None."[32] He collaborated with the freethinking group before they joined the Société d'anthropologie, publishing in their *Libre pensée* and *Pensée nouvelle*. There he weighed in on issues of life and death in a godless world. For example, in his discussion of spontaneous generation, he reminded his readers that while Pasteur had proven that what we had thought was spontaneous was, in fact, a matter of contamination, there was no reason to assume that no spontaneous generation was possible, that is, that the creation of life required a God.[33] He was convinced that humanity would "very soon subjugate the living substance" as we had already begun to dominate the mineral world (171). The point of biology was "to vanquish illness and force death to recede! . . . That such a result might be despised by immortal gods, I can understand, but men?!" (172). The quotation is a lovely representation of Louis-Adolphe Bertillon's concerns: atheism, materialism, progress, medicine, and the scientific mastery of death. He saw the notion of gods as retrograde, and depicted them as wanting to hold down humanity, to keep immortality for themselves. Since he did not believe in God, he used the term "gods"—the plural suggested defunct deities but also implied that he was speaking not of God but of those who claimed to represent him. "They" wanted to keep humanity mortal, "they" actually "despised" the idea of progress toward immortality or even longer life, but "men" should want to take over the roles, rights, privileges, and responsibilities of the gods, especially the privilege of immortality. "Men" were those brave enough to defy the infantalizing coterie of the gods and their representatives and take the good stuff for themselves. He was optimistic about our ability to do this because so much had lately been conquered, why not death too? But especially in this community of avowed atheists, the question of immortality in a world without God was not only about stealing long life from the gods in heaven and locating it on earth; it was also about recognizing that one's own options for "living on" had been drastically reduced. His concern with death was as pronounced as that of the other freethinkers: a recent article on Bertillon by demographic historian Michel Dupaquier explains that he was "mostly interested in death: effects of vaccines, mortality by mushrooms, mortality in nursing infants, mortality of bachelors, statistics in causes of death, tables of mortality . . . , maps of mortality in France, etc."[34]

Bertillon's interest in "mortality in nursing infants" seems to have had a profound impact on the history of modern France. In 1858 Louis-Adolphe Bertillon addressed the Academy of Medicine with a paper entitled "Etude statistique sur les nouveau-nés," the general point of which was that urban French women were sending their newborns out to wet nurses in the countryside and the babies were dying in droves. Statistics made this visible by showing that infant deaths were highest in the thirteen departments surrounding Paris. Of course, many urban families who sent their newborns off for a few months or a few years knew full well that it was a dicey practice but were dependent on women's remunerative work to survive. When very poor families contracted with the very cheapest wet nurses, the practice amounted to semipurposeful infanticide. Better-off families might send the babies out, too, because they thought rest healthier for the mother and the countryside healthier for the child, but such families could afford to increase the infant's odds considerably by paying more for a reputable wet nurse—one who was lactating, for example.

Bertillon's announcement was not directed toward unsuspecting families but toward the nation. In his recent book *The Power of Numbers*, Joshua Cole has argued that Bertillon's paper led rather directly to the 1874 Roussel Law establishing strict regulations for wet nursing in France and thereby opened the way for a new kind of state intervention in the private lives of French families. Bertillon himself may have been primarily interested in showing that statistics could be useful; after all, the newborns were only one of his many concerns. But, as Cole points out, state control of the population was resisted at first because it offended ideals of individual autonomy. For this reason, early attempts to control the nation's individuals were discussed in terms of controlling families, that is, women and children, and thus preserving the ideal of the independent male citizen.[35] There is a considerable literature, into which this argument fits, that understands the welfare state as natalist in origin and, further, as being negotiated through women so as to preserve a semblance of autonomy for the male citizen.[36] It is logical that the first steps toward a new caretaking and controlling role for the state would be most easily tolerated if directed toward a group so variously disenfranchised and so plainly in need: barred from most assemblages that generate new, articulated ideology; easily silenced when it does manage to speak (remember Broca on Royer's study of natality); and unable to defend itself at the polls or in the legislature. The particular import of Bertillon's statistical observation, then, was that it started the French down the path to the welfare state. Of course, there were a lot of factors involved, but it is a persuasive argument.

The first-generation freethinking anthropologist Louis-Adolphe Bertillon did not make this issue his life's campaign. In fact, he never published the speech in question; his sons brought it to print in 1883, just after he died. At the very end of his life, he did become concerned with natality, but as late as 1874 he wrote, "Our fatherland is in need of workers and defenders . . . but I think that before studying the conditions of increase . . . it is urgent to discover the causes of our devastation, and in a word . . . it is better to conserve generations than to renew them."[37] Note that this was written four years after the Franco-Prussian War. I will have cause to return to this matter, but for now let us remain in Louis-Adolphe Bertillon's world in the 1870s. He was at the Ecole d'anthropologie hardly a year before he attracted an energetic disciple, the young doctor Arthur Chervin. Chervin joined the Société d'anthropologie in 1877 and launched the *Annales de démographie internationale* that same year. He published therein an eighty-page article by Bertillon. Chervin also organized a Congrès international de démographie in conjunction with the 1878 World's Fair, and Bertillon was co-president of this with the economist and statistician Emile Lavasseur. In 1880 Louis-Adolphe Bertillon became chief of the Bureau of Statistics of Paris, located at the Hôtel de Ville.

All this work involved a good deal of general population counting, but, again, its practitioners wanted statistics to be useful and engaged as well as merely informative. They devoted their studies to a range of social questions. Bertillon was particularly interested in the various mortality rates of the Parisian social classes, a popular topic at the time. He cooked up a formula for comparing life expectancy to hygiene but found there were many such formulas being used by different people and that made it hard to compare data for the various areas of France. Bertillon made it his particular business to champion the use of a single formula. The mess of competing doctrines had to be replaced with "a truly scientific method," a particularly difficult task when analyzing the mortality rate of, for instance, a hospital.[38] (He concluded it could only be done if calculated for the mean duration of stay.) As a modern observer has noted, while Louis-Adolphe is not always listed among the founders of the discipline, he "showed a remarkable perspicacity in figuring out a large part of the principal demographic problems."[39] He also received a good deal of approving attention in his own time.

When Louis-Adolphe died in 1883, his last words to his sons were remembered as follows: "I have always labored to serve the truth; You, my dear children, must do the same."[40] Jacques Bertillon served in the same manner his father and his grandfather had. He became a doctor, and in 1878 he joined the Société d'anthropologie. He had long been casually involved in demography, joining the Society of Statistics, which his father had helped found in 1879, becoming a member of the Permanent Commission on Municipal Statistics of the city of Paris that same year, and publishing *Statistique humaine de la France* in 1880. When his father died, Jacques took over Louis-Adolphe's professorship at the Ecole d'anthropologie, as well as several of his other professional posts, editing the *Annales* and heading up the municipal statistics office at the Hôtel de Ville. Throughout the 1880s and 1890s, the daily newspaper *Le temps* ran a regular column, "Statistique de la ville de Paris," which Jacques Bertillon provided and for which he was boldly credited. The popular column reported on the prior week's deaths, broken down according to death by violence, suicide, typhoid, tuberculosis, and so on. It also offered a much briefer catalog of the week's marriages and births.[41]

In 1885 Jacques helped found the International Institute of Statistics. For the Chicago meeting of this group in 1895, he was asked to provide an international nomenclature for causes of death, which the American Public Health Association adopted in 1897 and which remains the basis for the current international nomenclature. These included such caveats as, "In collective suicides there should only be counted those who have attained their majority. Minors ought to be regarded as the victims of assassination," and, for death by amputation, "Do not include Amputation of the breast; Amputation of the penis."[42] This was an attempt to create a uniform language through which varied localities could meaningfully communicate, even in an immense world full of weird occurrences, and it was a crucial part of the transformation of the modern state. The state was too big to form a useful, fulfilling community unless it could become visible, be tied together by likeness and difference, and speak of its members in unifying, overarching terms. As Bertillon noted, "The important thing is not that the classification be perfect but that the morbid unities counted by statistics be the same everywhere."[43] His classification was later modified, but he had started a movement with an important future. The World Health Organization describes the history of its mission thusly: "The history of the systematic statistical classification of diseases dates back to the nineteenth century. Groundwork was done by early medical statisticians William Farr (1807–1883) and Jacques Bertillon (1851–1922). The French government invoked the first International Conference for the revision of the Bertillon or International Classification of Causes of Death in August 1900. The next conference was held in 1909, and the French government called succeeding conferences in 1920, 1929 and 1938."[44] Bertillon made health visible in the population by the simple gesture of insisting that everyone use the same terms. In the history of modern France, another of Jacques's projects had an even more significant impact.

Despite his contribution of standardized nomenclature for causes of death, and unlike his father and grandfather, Jacques was not primarily interested in death. It was the other end of human experience that caught his attention. Cole has offered two reasons for the temporal gap between Louis-Adolphe Bertillon's 1858 paper on infant mortality and the government action that resulted from it in 1874: First, there was real contention over the meaning and use of statistics because they seemed to minimize individual clinical cases, which were all most doctors relied on and which they defended for ideological and professional reasons. A wider population expressed epistemological unease about seeing the nation in terms of these numbers; the many essays on statistics and health that followed Bertillon's study seem to have been necessary before the idea could influence policy. Second, the director of the Statistique générale de France between 1852 and 1870 was Alfred Legoyt, and he had been convinced, largely by the famine in Ireland, that French population "restraint" was to be applauded and encouraged. The devastating defeat of France in the Franco-Prussian War convinced a humiliated nation that Legoyt was dead wrong: he lost his job, and the nation turned to pronatalist doctrines so it would not be outnumbered in the future.

Jacques Bertillon was nineteen in 1870. The war must have had a terrific effect on anyone coming of age at that time, but we should not overstate the immediacy of the connection between the war and his obsession with the French birthrate: for most of the 1870s he lived as a general science writer. In the popular journal *La nature*, he described his father's statistical work, bone collections at the Museum of Anthropology, natural human "monstrosities," the importance of Broca's findings, the courses at the School of Anthropology, and many other efforts of his freethinking friends there. It was only when his father died and he became a full-time demographer that Jacques championed the notion that the French military debacle had been due to demographics. To explain this, he turned from the Guillard-Bertillon concern with death rates and announced that the imbalance in population had its origin in a dangerously low French birthrate. Even he claimed only that the natural rate of increase was slowing, not that it was actually declining. Still, the population in France was not increasing as quickly as it was in Germany or England. In most European countries, better nutrition and various other factors of industrialized modernity were allowing people to live longer lives about a half-century before social mores and bourgeois family ambitions slowed the birth rate. For reasons that still partially elude demographers, the situation was different in France: the death rate declined more slowly, and the "grève des ventres" or "belly strike" started much earlier.[45] Population decline was not the entirely French problem it was made out to be, but it hit a nerve for a sufficient reason, and depopulation anxiety became a central feature in fin-de-siècle and twentieth-century France. Of course, as intimated in the idea of a "grève des ventres," a lot of this had to do with controlling the "new woman" and blaming her for the perceived loss of national vigor.[46] Yet the population movement was also part of the ideological work of deconsecration taken up by the freethinking anthropologists and their students. Human sexuality and procreation had been monitored by religion and law. What the Bertillons inaugurated was a new, secular, even numerical way of talking about people's sexual and reproductive behavior.

France also supported an opposing doctrine that was also based in numbers, statistics, and science. Bertillon's pronatalist movement was in competition with the energetic and inspired neo-Malthusian campaign led by Paul Robin. A *lycée* teacher whose radical politics led to arrest and exile in 1870, Robin was a well-known revolutionary before he became an antinatalist. Karl Marx nominated him to the executive committee of the International Working Men's Association, but he fought with both Marx and Mikhail Bakunin and was expelled from the International by 1871. His exile in London brought him in contact with the founders of the British Malthusian League, Charles and Elizabeth Drysdale, and they convinced him that limiting the production of workers and soldiers was one of the real powers that could be wielded by the working class. He published *La question sexuelle* in 1878 and returned to France as the Third Republic came into the hands of republicans. The new government appointed him director of an orphanage (of which he spoke in his testament for the autopsy society). In a modern observers' words, he probably got the job "because his secular beliefs accorded well with the anticlerical bent of the government."[47]

Bertillon and Robin were competing openly in the late nineties: in 1896 Jacques Bertillon created the Alliance nationale pour l'accroissement de la population française, which would generate the great bulk of pronatalist propaganda for many decades. In the same year, he published *De la dépopulation de la France et des remèdes à apporter*, which would be followed by several other books analyzing the problem and suggesting solutions.[48] Robin responded by creating the Ligue de régéneration humaine in 1896, dedicated to shrinking the population, and founding the journal *Génération consciente* as the organ for his own propaganda campaign. The intention of Robin's group was mainly to impede war by slowing the production of working-class cannon fodder and to increase working-class wages by decreasing the population. But active feminists like the political organizer Nelly Roussel and the doctor Madeleine Pelletier joined his effort hoping to liberate women from the constraints of child rearing. Feminism and anarchism made this a prickly doctrine for some: Robin and Pelletier came under police surveillance, and it is no surprise that Bertillon eclipsed Robin in French public discourse.

In contrast to Robin's internationalist desire to reduce the working class, Bertillon championed a nationalist call for more French soldiers. It does not take a postmodern to spot the theme: In a 1938 study entitled *France Faces Depopulation*, Joseph Spengler listed a multitude of measures introduced in the French legislature in 1878–1895 but not adopted. They included such gems as "that every Frenchman (not a clergyman or infirm) aged 26–40 years be deprived of all electoral rights until he had contracted marriage," "that the state educate one child in each family of six or more," and "that medals be issued to parents of large families." In Spengler's words, "These programs were explicitly or implicitly based upon the assumption that each individual citizen was duty-bound to defend and contribute to the support of its government. Whence it followed that he who shirked the first [procreative] duty needed to bear more than his normal share of the defense and support of the state." Speaking of his own time, Spengler also noted that "all the principles and measures proposed and/or put into effect in the last thirty years were described and advocated in 1890–1913 by J. Bertillon and Paul Leroy-Beaulieu, each of whom believed that the 'normal' French family must include at least three living children."[49] Spengler went on to detail accurately the difference between those two theorists: Leroy-Beaulieu "urged that the attack on religion, especially Catholicism, cease, since its tenets and practices were favorable to natality." Bertillon called for an end to such efforts "to restore religious faith and sentiments," announcing that they were "illusory." Also, Leroy-Beaulieu was less convinced than Bertillon that people's life-and-death decisions could be manipulated by material gifts and punishments. Bertillon advocated, for instance, complete tax exemption for households of four or more living children and a tax increase of up to 50 percent for households of less than three living children; inheritance tax rates for families of three children high enough that the inheritance would have been the same for each child had there been fully twice as many heirs; limitation of the military requirement to one son per family; all government jobs, scholarships, and certain state loans reserved for members of families of three or more living children; leave with pay for women workers before and after pregnancy; suppression of all information on birth control; and celebrations for members of large families, designed to honor the children and encourage a sense of pride, responsibility, and national gratitude.

One of the more telling suggestions was "plural suffrage," which here meant "reform of electoral laws, providing to each voter an additional vote if he is married and further votes for each minor child" (234–235). The tense equation of rights and reproduction reminds us that the very word "proletarian" comes from the Latin for reproduction; they were proles because all they could offer the state was their reproductive power. The notion of celebrations for children from big families was very significant as well, and the Mother's Day celebrations that grew out of the eugenic, pronatalist concerns of fascist and democratic states between the world wars also served the celebratory needs and ideological dogma of the secular state. In 1919 Jacques Bertillon proposed the first of these for France: a "Journée des mères de familles nombreuses." The idea was accepted by the minister of the interior. Jacques Bertillon was charged with organizing the event, and the first was held on the May 9, 1920. At another event for mothers of large families later that year, a gold medal was awarded to Marcelle Comblet-Sue, mother of thirteen. The celebrations continued annually and were more officially instituted in 1941, under the Vichy regime. Historian Karen Offen has shown that in the face of this pronatalism, feminists argued that women were having fewer children because of the huge economic, political, and social burdens and humiliations under which they labored.[50] She explains, however, that Bertillon thought this was ridiculous; it was all about men and money. Interestingly, Offen cites Dr. Thulié's contribution to the debate because he was one of the few who championed women's own explanations for limiting family size, and he lamented their precarious and undignified position and the fear that went with it. But when it came to offering remedies, Thulié, too, suggested patterns of funding that would only increase women's dependence on men (648–649). As for Jacques Bertillon's personal feelings about women's participation in the public world, the first woman admitted to the Paris Faculty of Medicine, in 1875, was Madame Jacques Bertillon née Caroline Schultze. She was Polish and Jewish; her forward-looking thesis was on women doctors in the twentieth century.[51] Their daughter, Suzanne, wrote the booklength biography of her uncle Alphonse that is a major source of the family's history.[52]

Jacques Bertillon's works guided the Third Republic's active struggle against depopulation, and his generally economic approach dominated French policy for much of the twentieth century (234–235, 239). The Alliance Nationale organized conferences, published a journal, raised funds for various natalist programs, and lobbied the French government with an impressive array of pragmatic suggestions. It also served as a nexus for registering and responding to real fears about the future. As Jacques Bertillon put it in 1897, "In fourteen years Germany will have twice as many conscripts as France; then that people which detests us will devour us."[53] He was close with the date: seventeen years later he found himself serving as director of medical and surgical statistics for the army, a post he held for the duration of World War I. When the fighting was over, Bertillon founded the magazine La femme et l'enfant and generally resumed his propopulationist campaign. It was clearly convincing his peers: the Chamber of Deputies formed a natalist group in 1914, the Senate followed in 1917, and an official Conseil supérieur de la natalité et de la protection de l'enfance was set up in 1920.[54] Jacques Bertillon was an honored member of this significant branch of the Ministry of Public Health. Also in 1920, the National Assembly passed a new law on abortion and contraception, increasing the likelihood of conviction for both doctors and patients and mandating high penalties for advertisers of contraceptive methods. By the time Jacques Bertillon died in 1922, the propopulationist movement had eight national associations and sixty-two regional associations. A number of other institutions regularly paid out subsidies to larger families (128). In 1938 Spengler was still speaking of the Bertillon movement in the present tense, concluding that "since the war period, the collectivistic populationist philosophy of Bertillon has come to prevail in an ever increasing degree" (239). Several Bertillon economic schemes were put into action in the 1930s: the Family Allowance Act of 1932, for instance, and the Family Code of 1939.[55] Under Vichy, abortion was made a crime against the state. The Vichy government famously accused Marie-Louise Giraud of performing abortions; a thirty-nine-year-old washerwoman from Cherbourg, she was guillotined on a Friday morning in July 1943. The Alliance Nationale applauded the action.[57] Throughout all of this, by the way, the population rate did not respond.

Jacques's younger brother Alphonse took a different road but was equally concerned with reconceptualizing the body politic for the new French state. Apparently, in youth he was a poor student and a bit of a black sheep in the family. Years later, his niece, Suzanne, would attest that although Bertillon père, Louis-Adolphe, gave no credence to phrenology, he allowed one of his closest friends, "the biologist Letourneau, who was strangely fascinated" by the art, to analyze the bumps on his two young sons' heads. Letourneau announced that both had methodical and precise minds and that Alphonse in particular was capable of meticulous and orderly work.[57] This was not in any other way apparent for years. Alphonse was kicked out of several schools and seems to have gotten most of his education at home. It would have been an interesting one: His mother, Zoé, was remembered as an extremely intelligent and engaging person; for example, the family took a house at the seaside because "Zoé Bertillon wanted to rest on the beach and read the Ethics of Spinoza." Jules Michelet and his family were with them on the trip, and when Michelet heard of her reading material, he laughed at the notion of a woman understanding philosophy. "It was Zoé Bertillon who laughed last. In a few days the professor of history was to be found continually upon the beach arguing the merits of the systems of Comte and Spinoza with the young and elegant woman, the wife of his disciple."[58] In 1862 Madame Adolphe Bertillon and Elisa Lemonnier, wife of the Saint-Simonian turned Mason Charles Lemonnier, started a school together.[59] Called the Free Society for the Professional Instruction of Young Women, it aimed to help girls learn a range of employable skills. The school stood for "tolerance, respect for oneself and for others, devotion, sincerity, fraternity, and above all hatred of idleness."[60] Zoé must have been an interesting mother. According to her granddaughter, she was lean and graceful, and she dressed and decorated her home in simple republican good taste. To his distress, she was noticeably taller than her husband.

As for the influence of the Bertillon boys' father and grandfather, there are two early indications, one referring to extreme youth, one to the beach trip just mentioned. They are rather literary musings on the part of a biographer but were derived from unpublished correspondences among the Bertillons and biographical work written by the Bertillon family and are interesting enough to consider: "Dr. Bertillon and his father-in-law were deep in their statistical investigations. His room was full of calipers and gauges used for anatomical measurement. Even at the age of three, these mysterious instruments fascinated Alphonse and his elder brother. It was the beginning of an education in the guise of an intriguing game. In the course of a month of two they had measured with pieces of ribbon every article of furniture in the house."[61] Broca was also a common participant in skull measuring at the Bertillon house and was said to have been particularly impressive to the boys. Moving forward in time, while the rest of the family sunned itself with the Michelets, Jacques was at school abroad, and the Bertillon brothers wrote frequent letters to one another. Using the full Latin names, Jacques described the new methods of classification of rare plants that his grandfather was studying in Italy. "Not to be outdone, Alphonse retorted with grandiloquent descriptions of the marine plants he was collecting on the sea-shore. There was some mockery in this, but behind it was a real preoccupation with the Latin, and the botany, and the need to label the things they handled if they were to be recognized and understood" (36). There is a hint here about the nature of modern measuring, counting, naming, and labeling. After some category shift in theory, any collection of measurements or descriptive terms can suddenly become useless; but some of the use of these facts is in the experience of collecting them.

Alphonse eventually made up his mind to earn a medical degree and was doing well, but since he came to it late, military service interrupted his studies, and he never returned to them. While in the barracks he made "a metrical study of the 222 components which make up the human skeleton" (62). But according to his niece, when he returned to Paris, he infuriated his father with his ennui and indecision. His older brother, Jacques, was already well accomplished, and his younger brother, Georges, was passionately studying medicine, while he was without direction. At this point, his niece describes a moment when the prodigal son announced to his father that he had recopied a sentence from one of his father's publications; the sentence was about how science finds order within what seems to be chaos. The father "who listened with much interest, smiled affectionately and proposed to him that he become a member of the Société d'anthropologie, which Alphonse accepted with joy."[62]

It is significant that this society was the core community of Louis-Adolphe's life, to which he would long to bring his errant son, at last an active member of the fold. Alphonse began attending the Ecole d'anthropologie and tried his hand at writing, publishing *Les races sauvage*.[63] The book was not well received. At last, he asked his father for help, and in 1879 Louis-Adolphe was able to find him a low level job at the Paris Préfecture de police, recopying police-report descriptions of criminals so that repeat offenders could be identified even if they gave false names.

Recidivism was one of the great questions of the day, so the concept here was important, but the casual descriptions and muddled filing system made the project almost useless.[64] Like his father before him, Alphonse decided to make his endeavor "more scientific," and he pitched the idea of taking the anthropometric techniques practiced at the Laboratoire d'anthropologie (and discussed in every issue of the anthropology journals) and using them to regularize police descriptions. It was several years before they let him try out the system, but in 1882 he was given two assistants and three months to prove himself; as his niece would tell it, his future wife, Amélie Notar, joined them and did a disproportionate amount of the work.[65] When suspects were brought in, the group measured heads, forearms, fingers, feet, and a host of other minute bodily features. They also applied the anthropologists' precise calculations for eye color, patterns of the iris, nose and ear shapes, and forehead lengths. Making use of his father's statistical findings, Alphonse concluded that it would take eleven matching measurements to be sure that two sets of numbers had been taken from the same individual. Alphonse Bertillon took photographs along with his measurements, but, as Bertillon demonstrated, people could change their appearances easily from one photo to the next.

Yet though he downplayed the importance of photography in order to support the need for measurements, he would find new ways of using the camera for identification and for other police work. Because of Bertillon, in 1883 France was the only country in which a police department took identification photographs as a matter of routine (107). The measurements and photographs were cataloged using a clever new system.[66] Toward the end of the three-month trial period, a repeat offender did happen into the police station. The Bertillon team found him in their files, and when confronted the man confessed. Thus the "police identification" budget was extended. In the first year, they made 7,336 measurements and identified 49 repeat offenders. The next year, they identified 241, and in 1888 Bertillon was made head of a new branch of the police department, the "Service d'identité judiciare," at a salary of 3,600 francs a year.[67] Identity cards were standardized and printed with room for side and frontal photographs and spaces for a host of measurements. Such a card exists for Paul Robin. It was surely done for amusement, because none of the measurements were taken, and Robin smiles slightly in his double portrait. But it tells us something about the mood within the materialist movement that Robin was in so amiable a relationship with the brother of the man with whom he spent his life publicly disagreeing.[68] The fun here has something to do with the growing fame of Bertillon's endeavor. His techniques and his name would spread around the world, but even early on, in Paris, street lingo for an arrest was: "Un sourire pour le studio Bertillon"—a smile for the Bertillon studio.[69] The journalist and author Ida Tarbell described Bertillon's office at the police station in a piece on Bertillon for *McClure's Magazine*:

> There was a peculiar individuality about the place—the look which a room takes when the utensils of one's trade are scattered about it. They were odd enough—these utensils of M. Bertillon's trade—maps of France dotted with bewildering figures and marks; rows of photographs of criminals, some of them better looking than the most upright man; a chromatic chart of the hair of the head: huge cases of notes; queer measuring-appliances; pictures from the Russian prison service; volumes bearing the titles, "Anthropology," "Ethnology," "Criminology," and the names of Lubbock, Galton, Lombroso.[70]

She also tells us it was a bright little room, that in the corner stood a tall green palm, and that the view out the window was the chimneys of the Conciergerie, the Palais de Justice, and the Sainte Chapelle (355). Tarbell was quite serious of purpose—McClure's was an ambitious new general-interest magazine, and Tarbell had interviewed Pasteur for it the year before she covered Bertillon—but nonetheless, as an author, she capitalized on the exotic curiosities in Bertillon's studio. It was essentially the same exotica of secret knowledge one would find at the Laboratory of Anthropology, and it could still create a bit of a thrill in its new location and new role.

In the modern state there are many reasons for identification beyond criminality, and even in its beginnings "Bertillonage," as it was called, served a variety of needs. Consider a case that brought Bertillon early fame, in which a Madam Rollin complained of a missing husband. Friends visiting the Paris morgue for fun (apparently a somewhat popular idea) suddenly recognized a corpse as Rollin, and when they brought the "widow," she took a look at the bloated body, thought of her missing man, and confirmed the situation with tears and lamentation (103–104). Bertillon looked up Rollin's name in his files and found a card for him; he was no criminal but had been measured when arrested for drunken disorder. A photograph was attached, and everyone confirmed that the picture matched the corpse. Amid the sorrowful company, Bertillon was remembered as having shouted out: "The same man! Look at the ear!" and indeed the ears were very different (104). Bertillon was right. Rollin later turned up on his own. He had gotten drunk, fought with a policeman, and spent seven days in jail. Apparently the ex-widow was embarrassed but pleased. Photographs Bertillon took of the corpse "Rollin" and of the real Rollin have survived and confirm that without Bertillon's trick of comparing ears, anyone might think the two men were but one. A whole range of Bertillon cases remind us that the populace was much more slippery in his century. He won a medal from Queen Victoria for helping to identify ten of fifty-seven bodies found after the shipwreck of the *Drummond Castle* in 1896 (130–138). There were few comparative materials for Bertillon to work with, but his precise observations allowed some relatives to recognize and claim their dead.

Bertillonage was adopted all over the Western world, and though it was rather quickly superseded by fingerprinting, Bertillon had a hand in popularizing fingerprinting, too, and was mistakenly known as its inventor in the United States and elsewhere. Evidence of Bertillon's fame is legion; consider, for example, the testimony of Gallus Muller, a clerk of the Illinois State Penitentiary in 1889, who gave a detailed account of Bertillon's American converts and summed up by saying: "The Bertillon system is in a fair way becoming a fixture of permanent and universal usefulness in the United States and Canada." The existence of an "American Bertillon Prison Bureau" helps to confirm the assessment.[71] Photographers for American green cards (which confirm a foreigner's status as a resident alien) must still, to this day, insist that the subject's right ear be clearly visible, a convention set purely for the sake of Bertillonage.[72] In 1938 the retired royal commissioner of police in Dresden wrote that "Paris became the Mecca of the police, and Bertillon their prophet."[73]

A quick glance at the fortunes of Bertillonage in South America is presented by Alphonse Bertillon's biographer Henry Rhodes.[74] In the mid-1880s, Doctor Drago, an Argentine, visited France to study anthropometry in order to advise his government on the type of criminal identification system it might set up. The government was particularly keen on this because of the unprecedented immigration Buenos Aires was experiencing at the time: a full 60 percent of the city was foreign born and effectively unidentifiable, and petty crime was rampant. When Drago came back, he heartily recommended the Bertillon system, and 1889 saw the creation of the Anthropometric Bureau of Identification, attached to the Provincial Police of Buenos Aires. Juan Vucetich was appointed chief of the bureau, installing the Bertillon's system as he learned it from Drago and from the several textual explanations then available.

Only two years later Vucetich decided the system was too inconclusive (Bertillon blamed this on insufficient training) and began adding fingerprinting to his identification bureau. He dropped the rest of the identification measurements after 1895, claiming that these alone had not served to identify a single recidivist. Apparently, Vucetich was the first to secure a conviction with fingerprint evidence as the only clue: a Francisca Rojas left prints that identified her as guilty of infanticide—an interesting first case, considering contemporary concerns.[75]

Despite the demonstrated power of fingerprinting, Bertillonage continued to stand for measuring and identifying the population. In fact, when Vucetich publicly expressed his desire to fingerprint the entire population, there grew up Bertillon and anti-Bertillon parties in Argentina. In 1916 the government passed a law establishing a General Register of Identification to pursue Vucetich's plan.[76] Vucetich was made its director, and he announced that everyone was required to submit to fingerprinting. Large sections of the population refused to comply with the ordinance, there were arrests, the windows of the General Registry were broken, and serious riots ensued. As an immediate result, the law ordaining the General Register was repealed. Not only that, all the identification records that had been compiled—according to the Bertillon method and the simple fingerprint method that came after—were burned by order of Don José Luis Cautilo on May 28, 1917.[77] Reportedly, Vucetich never recovered from the shock.

Kristin Ruggiero's recent essay "Fingerprinting and the Argentine Plan for Universal Identification in the Late Nineteenth and Early Twentieth Centuries" confirms Rhodes's account of periodic assaults on identification files.[78] Ruggiero shows that Vucetich's motives were grand; he envisioned a world of peace and safety through identity knowledge. Everyone would feel "seen" and would therefore be their best selves, and he imagined this on an international scale. She also tells us that right from the beginning of this movement, in 1889, the Buenos Aires police department asked the Court of Appeals to force judges to let them measure people in their custody and the court had said no. "The court held that 'Bertillonage [involved] the mistreatment of people being prosecuted and that judges should not authorize it,' because of the view that measuring and photographing were intrusive and were like calumny and slander, and damaged reputation" (186).

It is interesting to note that the eclipse of Bertillonage measurement by fingerprinting may turn out to be a temporary lapse of a hundred years or so: computers will likely make facial measurements the basis for identification. Compared to biometrics, handwriting recognition has been more constant in its success, and it was also a major part of Adolphe Bertillon's program. This had some dramatic and dire consequences: in 1894 Major du Paty de Clam, "expert in identifying handwriting by the Bertillon method," along with Bertillon himself, concluded an investigation by these means.[79] The result that both announced was that Captain Alfred Dreyfus was, in fact, the author of a note that pointed toward espionage.[80] Dreyfus had been framed because there had been intelligence leaks to Germany, and because he was Jewish he made a convenient fall guy. The affair became famous for its ability to divide families and friends. Anti-Semitism was a defining issue of the day: the secular nation state could not be defined by monarch or deity, so the state *was* the people, and citizenship had no educational or property requirements: French birth and a Y chromosome would do. Taken together, these facts made the legitimacy of France and its elected government nervously dependent on the sense of a cohesive, *true* population. Anxieties in this regard were often negotiated through attitudes toward the Jewish minority. So Dreyfus was convicted, and, after an appeal, reconvicted, for reasons understood as protecting the French army from scandal and thereby protecting forces of authority and order. Half the population fought his conviction because they were defending the rights of the individual, even the outsider, in a just and decent state. But if the Dreyfus affair meant many subtle things, it is also true that Dreyfus was arrested and convicted because scientists said the handwriting matched—the bodies matched—and, at least at first, almost everybody listened.

The weirdest part is that Alphonse Bertillon was certain that the document *had been written in a false hand* to look like Dreyfus's handwriting: repeated words were so "metrically identical" as to suggest they were traced from a single source. But because he thought that the method used was derived from a common method of constructing military maps, he believed an army officer must be at fault. Thus he contended that Captain Dreyfus had actually written the note, purposefully making it look like someone had tried to copy his handwriting so that he could later use this as his defense. This is what Bertillon proposed, rather than the much more obvious idea that someone had in fact forged his writing. The whole thing was pretty incredible and was certainly guided by the little bits of information that Bertillon had been given before he began his analysis. He does not seem to have known that the suspect was Jewish, only that he was strongly suspected; yet he did not change his mind once he knew the details of the drama, and in defense of his findings he came to speak of "the Jews" as somehow responsible for whatever subterfuge was going on. Given his background, it was a strange response. The Bertillons were an egalitarian, anticlerical family, and Jacques became an avid Dreyfusard. It is true that Jacques's wife was Jewish, but that should suggest that he never was an anti-Semite more decisively than it suggests that her influence shaped his response to the affair. Meanwhile, Alphonse thought that the very forces of decline that his brother charted were actually at work here in a conspiracy to undermine French military and governmental authority. The affair so came between the brothers that they did not speak for many years. It will not be missed that Louis-Adolphe Bertillon's injunction that his sons should "serve truth" had consequences not only in the inclusive, caretaking role of the nation state but in its self-legitimating, exclusionary behaviors as well: the male citizen and the fatherland were fabricated by the exclusion of women from full citizenship, and the nation found its self-image through the rejection of its Jewish minority. Alfred Dreyfus on Devil's Island and Marie-Louis Giraud at the guillotine have more in common than the name of their accusers. Jews and women, signs and wonders.

Along with this infamous connection, Alphonse Bertillon is best remembered for his police work, and despite all the measuring, the most important aspect of this had to do with photography. He got very good at it, standardizing mug shots and improving flash technology through empirical experimentation. In the United States, too, he is widely credited with having invented the institution of the mug shot. Also, he created the *portrait parlé*, or "speaking portrait," a book of charts made by cutting up photographs (often of himself and his colleagues), with which he trained policemen to recognize and name a variety of distinctive facial characteristics. In his *Memory of the Modern*, Matt Matsuda demostrates Bertillon's important role in creating modern institutional memory.[81] An article in *Le matin* on Bertillon's training courses tells us that Bertillon's philosophy was written in big black letters on the white walls of the classroom: "The eye sees in each thing only what it is looking for, and it only looks for what is already an idea in the mind."[82] Bertillon specifically asserted that only what could be named could be seen, let alone remembered, or, harder still, communicated across space and time.

Alphonse Bertillon also began bringing his camera to crime sites and was among the very first to do so. Most important, perhaps, he helped to popularize the idea that an untouched, recorded crime site could aid in police detection. On the cover of a popular journal in 1909, a caricature depicted Bertillon studying a bloodied wall with a large magnifying glass; the heading was "Bertillonnades" and the caption read: "Assassins always leave traces somewhere."[83] Arthur Conan Doyle had Sherlock Holmes contribute "two short monographs" on the distinctiveness of ears to *The Anthropological Journal* in the short story "The Cardboard Box." In this tale, Holmes takes Bertillonage as a proven fact, commenting, "As a medical man, you are aware, Watson, that there is no part of the body which varies so much as the human ear."[84] In "The Naval Treaty," Watson records a casual talk with Holmes as follows: "His conversation, I remember, was about the Bertillon system of measurements, and he expressed his enthusiastic admiration of the French savant" (2:183). Finally, in *The Hound of the Baskervilles*, we find the following dialogue between client and detective:

". . . I am suddenly confronted with a most serious and extraordinary problem. Recognizing, as I do, that you are the second highest expert in Europe—"

"Indeed sir! May I inquire who has the honor to be the first?" asked Holmes, with some asperity.

"To the man of precisely scientific mind the work of Monsieur Bertillon must always appeal strongly."

"Then had you not better consult him?"

"I said, sir, to the precisely scientific mind. But as a practical man of affairs it is acknowledged that you stand alone. I trust, sir, that I have not inadvertently—"

"Just a little," said Holmes. (2:7)

So Bertillon was cool enough to turn the great Sherlock Holmes a little green. Taken together, the Conan Doyle quotations demonstrate how completely the mythic *Anthropological Journal* shared a cultural meaning with "the man of precisely scientific mind." Amusingly, between the world wars, Bertillon's disciple, Dr. Edmond Locard, the founder of the Laboratoire de criminalistique de Lyon, was referred to as "the French Sherlock Holmes." Art copied reality, then reality copied art. As these cultural responses suggest, Bertillon did a good deal of detective work, but he reportedly preferred measuring as a central occupation. Friends commented that Bertillon "worshiped precision to the point of idolatry" and that his "love of precision, even for its own sake . . . was almost obsessional."[85]

Detective work and measurements created a different kind of world of identity. Ida Tarbell put it rather neatly back in 1893, after describing Alphonse as "a tall man of slightly haughty bearing," with a grave face, of long regular lines, "a dark, almost melancholy eye," with a bit of a squint and "a nervous trick of knitting his brow":

This was M. Bertillon, the originator of the modern system of anthropometric identification; the man who has so mastered the peculiarities of the human anatomy and so classified and organized his observations, that the prisoner who passes through his hands is subjected to measurements and descriptions that leave him forever "spotted." He may efface his tattooing, compress his chest, dye his hair, extract his teeth, scar his body, dissimulate his height. It is useless. The record against him is unfailing. He cannot pass the Bertillon archives without recognition; and, if he is at large, the relentless record may be made to follow him into every corner of the globe where there is a printing press, and every man who reads may become a detective furnished with information which will establish his identity. He is never again safe.[86]

Adding "how this infallible Nemesis, this mathematically exact identifying machine, is constructed, was what I had come to learn," Tarbell echoed her earlier hints that measurements and identity science were a hostile force. With Bertillon, she went through a whole course of measurements on a man who had been caught stealing rabbits and who claimed it was a first offense. After pages of description of all sorts of measurements, they went to look for a matching card in their files and, to Tarbell's surprise, found one: "the rabbit-man" had been caught before. For her article, Tarbell also examined the photography studio and herself posed for a Bertillon photo. The description she left again reflects her sense of the hostility of the event: "In order that the distance may be invariable, the chair and camera are screwed to the floor, and there is a perfect system of adjustment. The light is thrown into the face. The result is hard on the subject. One does not care to display his judicial photograph, but for the purpose they are admirably, brutally exact." Especially for the most troubled citizens of the nation, there must have been a profound change from living in a world in which one might always slip into the faceless crowd, no more connected to past deeds than any single pigeon in a flock might be held responsible for a crumb of bread that eyewitnesses, only moments ago, had seen it steal. Of course, near one's home there would always be people to recognize one, but in the wider world changing one's name essentially meant one could plead "first offense" every time. With the mug shot, the measurements, and the *portrait parlé*, the individual bodies became visible, and in ways that could be communicated at a distance.

Jacques and Alphonse Bertillon used the gestures of the freethinking anthropologists but took a leap into aggressive utility. The Bertillon brothers learned practices of counting and measuring at home with the freethinking anthropologists and within the specifically religio-atheistic context of the Société d'anthropologie. In that context, these practices had remarkably little practical function. They were published, taught, and discussed by a large section of society because the freethinking anthropologists included such measuring behaviors in their journals, books, school, and various meetings. The freethinkers themselves were more interested in piling up archaeological artifacts and ethnological details, but they also took thousands of measurements, and their many fellow travelers were heartily encouraged to do so as well. When the freethinking anthropologists used numbers, it was above all else a complicated secular gesture, and one so pragmatically useless, while so prodigiously meaningful, that we are justified in calling it a rite.

When the Bertillon brothers found uses for their quantifying behaviors and convinced the French state of their necessity, they branched off into practices that may have had equal passion but look familiar to us—as pragmatic endeavors and as the subject of present-day critique of the power of science. Neither brother completely ignored the social aspects of natality and criminal identification, but their concern with bodies would be difficult to overstate. They were discussing health and class by noting life spans, counting up bodies as the salient factor in assessing the nation, measuring individual bodies for personal identification, and encouraging bodies to produce more bodies in order to save the nation from decline. Each of the brothers became the head of a large group of acolytes and devotees; they had flocks and converts, students, and central texts. They were also using numbers to talk about sex and taking measurements to talk about identity. Whatever it was that they were saying, they were making scientific, public claims about subjects that had been understood as private and religious.

Jacques dropped out of the autopsy society soon after his father died in 1883. He lived until 1922. His younger brother, Alphonse, did not have his constitution and died in 1914. A tribute pamphlet published in that year tells us that Alphonse's brain, noted to weigh in at 1.525 kilograms, was handed over to Manouvrier at the Laboratory of Anthropology.[87] He thus joined both his parents in having a publicly materialist death.

Though it was not in their official job titles, what was most visible to contemporaries of the freethinking anthropologists, what was constantly and explicitly articulated, was that they were atheists. Their job was to argue that everything, absolutely everything, could be explained and handled without recourse to God, priests, religious homily, saints, the devil, Catholic history, traditional morality, miracles, unctions, incense, or prayer. As for the Bertillons, their job titles accurately describe that for which they were best known: they were encouraging the production of soldiers for an as-yet-imaginary war and learning to identify the French populace. They were still measuring and counting all day long, but because the Bertillons convincingly applied the measuring and counting to specific "problems" of the French state (competition with Germany, criminals at large), they transformed the measuring and reclassifying behaviors into techniques of the pastoral modern state.

CHAPTER FIVE

∞

No Soul, No Morality: Vacher de Lapouge

In the next century people will be slaughtered by the millions for the sake of one or two degrees on the cephalic index. That will be the sign, replacing the biblical shibboleth and the linguistic affinities that are now the markers of nationality. Only it will not have anything to do, as it does today, with questions of moving frontiers a few kilometers; the superior races will substitute themselves by force for the human groups retarded in evolution, and the last sentimentalists will witness the copious exterminations of entire peoples.

—Georges Vacher de Lapouge, "L'anthropologie et la science politique"

Georges Vacher de Lapouge initially presented this idea in a series of lectures held at the distinguished University of Montpellier in the early 1880s. He first published it in 1887, in Topinard's *Revue d'anthropologie.* The article, entitled "L'anthropologie et la science politique," contained Lapouge's first descriptions of "anthroposociologie": the application of anthropology to social politics. Phrases such as "slaughtered by the millions" and "copious exterminations of entire peoples" remove this quote from run-of-the-mill nineteenth-century eugenics, though Lapouge himself was not calling for copious exterminations. The statement was intended as a warning about what would happen if governments did not take rational control of breeding practices. Lapouge was not, it should be noted, angry at the inferior races, but he was very angry at "the last sentimentalists." He had been taught anthropology by the freethinking anthropologists in Paris in the years between 1883 and 1886, and at some point he became enraged that they embraced anthropology and atheism and yet remained egalitarian republicans. Lapouge believed that these and other atheists had stopped just short of the awful truth: no God meant no meaning and no morality. For Lapouge, this translated into a complete indictment of the existing society, culture, and government, based as they were on principles derived from deistic morality. "Here is why I have been speaking to you of the abyss and of a cataclysm," wrote Lapouge:

> It is obvious, to my eyes anyway, that if one eliminates the supernatural element from the universe, it is necessary to eliminate, at the same time, a number of fundamental notions—all of which were, in the past, deduced from supernatural tenets. All of morality and all of the ideas that serve as a base for law and for the political sciences, in their present-day conceptions, constitute a series of deductions of which the first term assumes the existence of a personal divinity. . . . Remove all validity from this source, and there is nothing left.[1]

He was not happy about this meaninglessness, except insofar as it allowed him to eradicate moralist barriers to a "selectionist state." He did, however, relish the amorality of his imagined future and its brutal rule of science.

Lapouge is of interest here because he studied with the freethinking anthropologists and then remained a politically engaged atheist anthropologist throughout his long and strange career. He, too, completely rejected philosophy as a source of meaning and comfort. In an attempt to replace religion (without philosophy), Lapouge took what he had learned from the freethinkers and created his own impassioned science. Because he spent most of his life in the French provinces—that is, in republican France and not even at its urban center—he did not have many local converts. He did have some, but what is more important is the extensive community he was able to create by mail—a relatively new possibility as inexpensive, reliable postal systems had only just come into being, changing the nature of group formation as much as the Internet has done today. Through his articles, books, and tremendous correspondence, Lapouge led a vast, thinly spread, and distant flock. Central to his project were the deconsecration of sex and its translation into a scientific ritual and the reordering of humanity on the basis of skull measurements. To his far-flung audience he delivered scientific sermons, dramatically detailing the paradoxes of infinity, eternity, and other scientific enigmas.

Lapouge noted the freethinkers' attempt to replace religion with science and mocked them for it, insisting that humanity would never again know religious comfort. Yet his own religious behaviors and intonations show how much he wished to re-create religious experience and how far he was willing to go in trying. Lapouge occupied an odd conceptual space: his rhetoric was as adamant in its atheism as that of the freethinkers, and as suggestive of a cultic science, and yet it was racist and antirepublican, thus beginning the collusion of science and conservatism that is the hallmark of the new right. Further, Lapouge created an employable science: there was governmental use of La-pouge in Germany and the United States both during his life and after; in France, such use was made just after he died in 1936. (The next two chapters will explain how it was suppressed in France until Vichy.) Like the freethink-ing anthropologists, Lapouge's work was impassioned by a fiery and agonized commitment to the disenchanted world or, rather, a commitment to a world reenchanted by scientific proposal and paradox. But he also offered pragmatic suggestions that became part of the technology of the state in the twentieth century. There is a direct relationship here to the origins of the Shoah that lends further consequence to these issues. I will begin with a quick look at what Lapouge was claiming, and then, because Lapouge responded to it, I will examine a feature of the limited but important French Catholic revival in the mid-1890s. I then move on to demonstrate the anguished atheism, explicit an-timorality, and deconsecrated human sexuality proposed by Georges Vacher de Lapouge.

ANTHROPOSOCIOLOGY

Lapouge saw humanity as divided into two races, each of which could be iden-tified by its "cephalic index": one dolichocephalic, the other brachycephalic. The index was calculated by comparing the width and breadth of the human skull: a low index, that of the dolichocephalic, meant a long, narrow head; the brachycephalic had a higher index—a round head. The notion of a cephalic index had been around for some time, originating with the Swedish anthro-pologist Anders Retzius. While Retzius clearly thought dolichocephalics were superior to brachycephalics (he had in mind a good Swede/bad Slav binary), Lapouge's innovation was to attribute a host of specific qualitative character-istics to these labels. Dolichos, as he referred to them, were fair skinned with blue eyes, temperamentally energetic, creative, adventurous, and refined. They were more often Protestant than Catholic and were the majority popu-lation in northern Germany, Scandinavia, and England. Later, Lapouge came to use the terms "Aryan" and "dolicho" interchangeably.[2] Because of migration and intermixing, all those who spoke Aryan languages were not dolichos, but all large groups of dolichos spoke Aryan languages.[3] "Brachies," (also his term) were much darker in complexion, tended to live in mountainous regions, and were more often Catholic than Protestant. They preferred to stay near their homes and chose constancy over change. In a word, they were good, honest people whose lack of imagination and courage marked them as mediocre. They upheld civilization but did not create or advance it. Indeed, explained Lapouge, brachies were more comfortable in positions of servitude than in positions of leadership. It was this mild and uninspired group that dominated the populations of France, Spain, Italy, all of Asia, and most of the Slavic countries. Interestingly, where Broca had seen a causal relationship between skull size and aptitude, Lapouge saw skull shape (not size) as a general marker of aptitude rather than as its decisive causal agent. He could thus declare that some African peoples were dolichocephalic, but that in their case this did not indicate superiority.[4]

Lapouge saw his dolichos and brachies as belonging to different socioeconomic groups, and while his depiction of the relationship between social standing and skull shape sometimes shifted, he was unswerving in his insistence that the aristocracy of the ancien régime had been dolichocephalic. Because the French Revolution had removed the aristocracy from control in France, the dolichos were overwhelmed by the masses of brachies. Without legal or financial power, the group was losing its identity, intermixing, and dying out. Meanwhile, the brachies were ruining the country. Capitalist democracy was a disaster because it selected for mediocrity, allowing the spoils to go to whomever could outcompromise, outlie, and outhaggle everyone else. Lapouge wanted France, redesigned as a socialist-selectionist state, to regulate its citizens' professional and reproductive lives accordingly. As he would frequently assert, "Liberty, Egality, Fraternity" had to be replaced by "Determinism, Inequality, Selection."[5]

In his major works *Les sélections sociales* (1896) and *L'Aryen: Son rôle social* (1899), Lapouge attributed another notable evil to capitalism, namely, that it was the economic system most favorable to the Jews. Lapouge classified Jews as dolichocephalic, which was precisely why he considered them dangerous. This is significant when taken with the knowledge that, first, the cephalic index was a very slippery set of measurements, and there are many instances where one scientist declared that a given people were dolichos and another claimed that those same people were brachies; prior assumptions guided these measurements. Second, even to the degree that Lapouge may have assiduously measured Jews to be dolichos, he could have dismissed this as not, in this case, implying intelligence, as he did with African dolichos. Hence, it is reasonable to suppose that Lapouge chose to denigrate the Jews as an intelligent and therefore formidable people within the context of contemporary anti-Semitism. He saw them as the only human group that practiced eugenics, in that they stressed marriage within the group and steered clear of mixing with the brachies. They were thus the only group that seemed to have both intelligence and foresight; still, he considered them venal outsiders. Through their clever breeding and conspiratorial behavior, he argued, Jews were quite likely to become the new aristocracy of Europe, and the danger of Jewish domination was especially pronounced in France because of the thorough displacement of the proper, traditional aristocracy. Countries with large populations of dark, Catholic brachies were described as being more susceptible to Jewish domination than dolicho countries, partly because brachies preferred to serve ("they have found the master for whom they had been looking") and partly because they tended to set up the mediocrity-producing capitalist-republican societies that the Jews were able to dominate.[6] Anti-Semitism was not prominent in Lapouge's early articles, however, and in terms of scientific theory, Lapouge's work was similar to most of the other essays in Topinard's scientific journal. The most striking difference between Lapouge and the other anthropologists was in tone, for already in these early essays Lapouge envisioned the world as being on the brink of a total revolution based on his theories of heredity.

BRUNETIÈRE AND THE CATHOLIC REVIVAL: RELIGION RECLAIMS MORALITY

In France in the middle of the 1890s, a Catholic revival was getting under way. Of its several manifestations, I will deal here only with the one most debated in the academic press and most prominently rejected by Georges Vacher de Lapouge and Léonce Manouvrier. These two ex-students of the freethinking anthropologists of Paris came to situate themselves in opposition to this rather academic religious revival. Lapouge and Manouvrier were intellectual and political adversaries, but they both stood up for science.

In 1895 Ferdinand Brunetière, editor of the *Revue des deux mondes*, penned and published an article entitled "Après une visite au Vatican."[7] The article said very little about the interview he had had with the pope, instead presenting an aggressive argument for the revival of Catholicism. This article has been mentioned in several histories of the Third Republic and was discussed at length in Harry Paul's article of 1968 "The Debate Over the Bankruptcy of Science in 1895."[8] Paul's analysis of the changing relationship between religion and science identifies 1895 as a moment of heightened tension between the positivist French state and Catholic revivalism. The concerns of the present study lead to a different reading of the "bankruptcy of science" debate, concentrating on its preoccupation with the loss of the soul and its attempt to find a reasonable description of humanity balanced between materialism and the idea of spirit.

Ferdinand Brunetière was a prolific and respected literary critic, famous for his multivolumed tomes on lyric poetry and the history of French literature. Before 1895 he was widely known as an avid republican, rationalist, free-thinking scholar, *maître de conférence* at the Ecole normale supérieure, and editor of the *Revue des deux mondes*. He used the pages of that journal to discuss Darwinian theory, anticlericalism, and philosophical materialism—all in the name of secular democracy. In his last decade, however, he became increasingly conservative and nationalistic, and when the Dreyfus Affair polarized the country, he joined the anti-Dreyfusards in their defense of the state and the forces of order. His shift toward the political right can be dated from his article "Après une visite."Vigorous reaction to his apparent conversion (including a banquet held in express opposition to the essay) seems to have pushed him still further to the political right.Yet it is crucial to note that in "Après une visite," Brunetière was neither announcing a personal conversion nor even a changed intellectual position regarding the nature of reality. Rather, he was calling for a change in social strategy. He feared that without the morality of religion, the republican body politic would fall into chaos. In defense of bourgeois security, he dramatically asserted that he had been mistaken in the past: science could not convince the mass of human beings to be good.

Brunetière suggested, but did not insist on, the bankruptcy of science. Scientists had invented and discovered a great deal, but so far they had failed to provide meaningful answers to the great questions of human origins, human destinies, and human values. In late-nineteenth-century science, observed Brunetière, questions of free will and moral responsibility were outrageously dependent on results garnered from physiology. The "theory of evolution," he asserted, "will never tell us where we are going," and "neither anthropology, nor ethnology, nor linguistics will ever tell us who we are."[9] When Brunetière needed a quotation to demonstrate the position of total antimetaphysics, he turned to André Lefèvre's *La religion*. Lefèvre claimed: "Religions are the purified residues of superstition. *The value of a civilization is in inverse proportion to its religious fervor.* All intellectual progress corresponds to a diminution of the supernatural in the world. The future is science."[10] In citing him, Brunetière did not choose a scientific claim as "exhibit A" but rather a piece of evangelism of a level that even many secular republicans found a bit trying. Brunetière insisted that the progress that was intended by joining the moral sciences to the natural sciences was not progress at all but rather a step backward: "If we ask Darwinism for lessons in moral behavior, the lessons it gives us will be abominable."[11] Still, his chief concern was not that scientific morality would be vicious. Rather, his main point was that the secular morality of scientists taught the same lessons as religion without the requisite supervision. He had become convinced that science could not provide a convincing social morality because without sanctions in the afterlife, no morality could be sufficiently imposing.

Brunetière directed little attention to philosophical attempts at a science of morals, though he did not mind the idea that Christianity might eventually be replaced by another moral system, so long as it was metaphysical and not physiological. "I dare to advance the idea," wrote Brunetière, "that if we ever establish a laic morality, a morality independent not of all metaphysics but of all religion, it will not be in physiology that we find its base" (111). For the moment, democracy required Christianity. The only question, according to Brunetière, was which kind of Christianity. As I mentioned in chapter 2, many progressivists had hoped that France might deal with its anticlericalism by embracing Protestantism and thereby keep faith but lose the onerous authority figures. In an interesting reversal of ideas, Brunetière advocated Catholicism, "which is a government," while "Protestantism is nothing but the absence of government" (113). He wanted the authority figures and was looser about the faith.

This whiff of utilitarian motivation pervades the essay. What did Brunetière actually believe? He insisted that even were we to accept that all our human emotions and instincts were of a purely animal origin—"which, moreover, one can absolutely refuse to admit"—that would not discount the fact that "the object of the last six million years of civilization has been to separate humans from nature," not to tie human beings to a "moral determinism" based on the physical sciences and natural law (116). Further on, Brunetière explicitly referred to himself as "a partisan of the idea of evolution" and seemed to grin as he excused the Catholic orthodoxy as "no doubt hav[ing] its reasons" for not subscribing to the argument that "ferocious, prehistoric blood" flowed in our veins. But, Brunetière continued, evolutionists such as himself could enthusiastically agree with the church that "virtue is nothing but the victory of the will over nature. Which is to say, without metaphor, that the will only determines itself in breaking with nature" (117). As I have noted, the question of whether humanity ought to follow nature or distinguish itself from the natural world was a central issue of the period. Brunetière's choice to distinguish humanity served to remove the natural sciences from the position of authority on human affairs and thereby to support the validity of religious claims to knowledge.

Brunetière warned that to keep people of goodwill divided on the question of morality because of differences of opinion that were based on "exegeses and geology" would be "the most unpardonable, stupid mistake." He challenged those who had been long and publicly committed to materialism to swallow their pride and reverse their position. "Suppose," he asked, "that social progress could be made only at the price of a transitory sacrifice, which did not cost anything of our independence, nor of our dignity, but only of our vanity. Hesitation would not be permitted." So in late-nineteenth-century France, not believing in God supported or allowed a certain kind of vanity? Yes, and independence and dignity, too: the denial presupposes the notion. Brunetière was willing to trade in these feelings for comfort and stability. His call was for the intellectuals to fix society even if it meant reliance on a realm of unscientific (even untrue) ideas. As he put it, "Sick people don't care about rules, so long as you heal them" (117).

Published responses to this article were numerous and passionate. Catholics celebrated it, if not with all of Brunetière's caveats. Materialists disagreed with Brunetière's proposition outright, arguing that science was doing fine and had no need to run back into the arms of a paternalist, dogmatic religion. The famed chemist and Unremovable Senator Marcellin Berthelot was among the most prominent in this role.[12] In his several studies on science and morality, Berthelot argued that "the laws of natural determinism" were the only rational basis for a moral system worthy of free, republican citizens.[13] Religious dogma, he asserted, never helped to abolish slavery or torture or helped to further respect for life, universal liberty, tolerance, equality, or solidarity.[14] He held that Christian charity was inferior to positivist solidarity, and he met the unsavory connection between scientific morality and the "ferocious egoism" of the "pitiless struggle for life," with a similarly unpleasant connection between religion and the "fanatic who desires to conquer and dominate the world in the name of his God" (466). Ethics and politics. The natural sciences, on the other hand, were offering a new morality through the demonstration of morality's instinctive origins. "The hereditary perfecting of these instincts," wrote Berthelot, "is the true basis of morality and the point of departure for the organization of civilized societies" (463). This was very much the rhetoric of the freethinking anthropologists. They were not often personally involved in the debate—they were aging, and many had already been dissected—but Clémence Royer did weigh in on the question. Stymied, she asked, "How dare they accuse science of being bankrupt at the end of a century in which it has renewed the face of the world and created a new humanity?" and she listed the great accomplishments of technology as proof of her position. True to her life's convictions, she believed that the problem lay with those who held "their science in one cerebral hemisphere and their religion in another."[15]

The other major scientistic response to Brunetière's challenge was issued by Nobel Prize winner Charles Richet. As demonstrated in the pages of his journal, *Revue scientifique*, Richet was an enthusiastic supporter of the free-thinking anthropologists, though he sometimes cautioned against their mixing of politics and science. His "La science a-t-elle fait banqueroute?" appeared in *Revue scientifique* immediately following the appearance of Brunetière's piece, which had appeared in his own journal: they were dueling giants.[16] Of course, Richet did not think science was bankrupt. For him science could not possibly fail in its promise to explain everything, because it had made no such promise. As we know, the freethinking anthropologists had made this promise, and, of course, Richet knew it, but he let Brunetière be the only one to quote Lefèvre. Second, Richet claimed that anthropology had, in fact, done a great deal to explain human origins. He recognized that there was unknowable mystery in the world but argued that though science could not unveil the "intimate nature of things" (34), it was not an oppositional relationship; in fact, science kept discovering new mysteries. Richet argued that science was so profoundly responsible for all aspects of modern civilization, it was silly to judge it singularly, in any particular limited function. Science could not discover laws of morality, but it had indeed improved morals by creating the context in which liberty and responsibility came to represent the ideals of society. "Without getting lost in the clouds of the questionable future that awaits us after this terrestrial existence" (38), humanity could now simply be just and good and kind. In a rather open spirit, he claimed that the morality of religion and science was the same, but, in the event, we had come to uphold the ideals of peace and justice through scientific, civilized modernity. As Richet put it, it was not religion that stopped slavery and torture, it was enlightenment: "The terms science and civilization are identical" (34).

As I will discuss in chapters 7 and 8, Brunetière's call for a return to Catholicism also drew a vigorous response from philosophers and sociologists, challenged by having been left out of this "science versus religion" discussion of morals and unwilling to agree with either strict materialism or pragmatic religiosity. Some of this vigorous response, denouncing anthropology and religion, was a direct result of the vicious, scientistic antimorality proposed by Vacher de Lapouge at the very end of the century.

RIGHT-WING MATERIALIST ATHEISM

In his lectures and written texts, Lapouge assumed his audience to be secular republicans who believed in Darwinism and in the inequality of human beings but were not brave enough to take these notions to their logical conclusion. "I take a malicious but vivid pleasure," he wrote, "in catching the myriad errors in a number of recent articles that have appeared in the socialist, anarchist, or so-called democratic journals. In Darwinism, or in more general terms, in scientific doctrines on the origin of the species of the world, they have seen, above all, an argument with which to oppose religion and, here in France, an argument with which to oppose the church, which is creationist and dedicated to the text of Genesis." This was certainly true of Clémence Royer and the other anthropologists of Paris and could also be applied to the younger Brunetière, among many others. According to Lapouge, they were all using science to disprove religion, but when science suggested a course of action that contradicted their own moral and/or eschatological universe, they refused to acknowledge it. "They have not understood," he wrote, "that Darwinism applied to human beings in their social existence excludes for the future all elements of nonscientific social explanations, which is to say, it removes all supernatural causes from the general causality of the universe. When I say 'they' I mean the freethinkers, or those who qualify as such, because from the very beginning the churches have seen the consequences of these new theories and have taken steps to denigrate them."[7] Lapouge used the word "supernatural" to imply anything spiritual or philosophical: anything that could not be proven by his science. Consider, for example, his discussion of human equality: "Some people, believing in the mystical principle of human equality, cannot bear it when one speaks to them of superior races. I am not even going to take the trouble to contradict them. It is perfectly useless to reason with minds that are thus turned toward the supernatural; only fictions have value in their eyes. I address myself only to those for whom facts have meaning, as do numbers, which are also facts, grouped and counted" (397).

The freethinkers were so devoted to materialism that they wrote books on the biology of aesthetics, pronounced philosophy dead, and dissected each other's brains, yet Lapouge accused them of mysticism because they were egalitarians, and he believed that the idea of equality had originated in Christianity (rather than, say, the other way around). Themselves troubled by the idea of unscientific ideals, the Paris freethinkers had created an inherently progressive evolutionary model as a secular basis for their political values, but Lapouge did not accept it. He thought they were just too frightened to admit the truth. As he saw it, the republican postulates of fraternity and equality were based on the Christian idea that all God's children are brothers; liberty and individualism were based on the idea that each human being has a distinct and significant soul, and charity and morality were God's commandments. Many devoted republicans had similar misgivings: the republic did seem to have been based on Judeo-Christian ideals and philosophical tenets, from religious morality to the idea of individual free will. Like the freethinking anthropologists, other republican social theorists found new ways of justifying these notions, either scientifically or philosophically. Lapouge, however, was as angry at democracy as he was at Christianity, so he tried to salvage neither one nor the other from their damning mutual association. Christianity, it should be said, was despised by Lapouge because of the ideas the religion promoted—ideas Lapouge considered to be false, patronizing, and dangerous to the race. Christians, as a group, however, were not despised by Lapouge. This was because he did not define them as a homogeneous racial group but rather as representing two major racial categories, dolichocephalic Protestants and brachycephalic Catholics. By contrast, Lapouge's discussion of the Jews was sharply critical of the Jews themselves and not particularly mindful of the religion's doctrines.

In pronouncing the failure of Christianity, Lapouge wrote that religion had originally served three purposes: it explained the origins of things, it provided a moral system, and it comforted sorrows and assuaged the fear of death. Now, he maintained, the mass of people believed in natural evolution and thus no longer required religion to explain human origins. That no one had an explanation for the origin of living matter itself, he claimed, did not bother "a great part of thinking humanity": "This problem, they imagine, will soon be resolved like all the others; it all depends on a laboratory experiment that might take place tomorrow." Religion's second charge, morality, was a more formidable problem. People had always thought that society could not exist without morality and that morality could not exist without religion. Now, Lapouge contended, there was a modern understanding of the "arbitrary character of our morals." Yet that had not really changed anything, since the Christian moral system was the only one that most people knew. This was a problem because Lapouge believed the Christian moral system to be "among the worst" since it was based on the existence of a life after death, which he saw as "infinitely improbable." The belief in an individual's life after death had precipitated a moral system that "sacrifices society to the individual and real life to imaginary mystical interests." While Brunetière and others were trying to find the most meaningful and coercive basis for standard Judeo-Christian ideas, Lapouge was not interested in keeping those ideas. "Our contemporaries have seen all this, the moral crisis that has thrown everything into disarray. The morality of yesterday is on its way out, and the morality of tomorrow has not yet been born" (508). Taking the notion a step further, he mused, "We are on our way, by new formulas based on social hygiene, toward the elimination of the idea of morality. It is an evolution that has its advantages and its inconveniences but that the progress of human knowledge renders inevitable" (509). Lapouge saw the political and cultural battle over Brunetière's Catholicism as meaningless in comparison to the war on morality that was to come in the twentieth century. Operating without a heavenly censor and burdened with scientific proof of racial superiors and inferiors, human beings were about to face some extraordinary questions. Among his many warnings of crises and revolutions to come, Lapouge wrote that "our epoch of apparent indifference is the beginning of the greatest crisis of religion and morality that has struck humanity since it has begun to think."[18]

Christianity's third and final purpose, to be a comfort "during painful crises in life and at the hour of death," was, according to Lapouge, completely invalid now that it was clear that there was no life after death. "Oh, the millions of mourners who have been consoled by the golden promises of Christianity! Oh, the millions in agony that it has soothed—up until the supreme instant of the fall into nothingness!"[19] These phrases express both disdain for those who believed such ideas and envy for their existential comfort. The freethinkers running the Society of Mutual Autopsy had managed a slightly mournful culture of optimism. Lapouge seemed to celebrate his misery, but there was nothing optimistic about it. "The great consoler is gone. If religion has done harm to society, it is also true that individuals will never again have such promises of happiness" (509).

Lapouge certainly was not offering promises of happiness, but his passionate sermons on the meaning of life were meant to cast a kind of religious thrall. The freethinking anthropologists imagined an edenic future, to replace heaven, and replaced reliance on God with an appropriation of his power. Vacher de Lapouge instead cultivated a parallel to hellfire, even within his plans for a relative utopia, and replaced pained religious awe with pained wonder at scientific paradox. He wore his pessimism as a badge of honor, arguing that his bravery in accepting such a dismal situation proved that he was honest and, by extension, correct. Lapouge billed his relativism as the quintessence of objectivity, but it was all in the service of a passionate description of the emptiness of life. "There is no such thing as superiority in and of itself," wrote Lapouge, "any more than there is a top and bottom of the universe, or a good and bad, but we are used to orienting ourselves in space according to certain conventions. Accordingly, we regard the courageous as superior to the cowardly, the active to the indolent, the free to the servile, the intelligent to the weak of mind, the man of character to the waverer, the far-seeing to the short-sighted" (398). It is a classically religious gesture to announce that the hierarchies and truths of everyday life are of no consequence in the immense, real world. Lapouge echoed these religious concepts as had other freethinkers, but he refused the usual conclusion that science would now save us:

Progress is a purely human conception. Evolution is happening all around us, moving forward, backward, to the side, progressing, regressing, turning and returning. It does not tend indefinitely toward the best; it tends toward nothing. It is, at the moment, made to tend toward whomever has the greatest consciousness of it, but that consciousness will be extinguished along with the conscious being, who must eventually die. There is no heaven, not even on earth. One must not ask science to give more than it can give. It can give man consciousness and power. It doesn't have a direct control over happiness: for that you have to go to a priest, a sorcerer, a seller of alcohol, of morphine, or, best of all, go to the gun shop—the seller of suicide. (512)

His books were not teeming with this kind of religious-philosophical pronouncement; they were mostly descriptions of head shapes, language types, bone shards, and material culture, with all discussion dominated by neologisms and numbers. But every hundred pages or so Lapouge began to lecture on the ramifications of his materialist position and to stir up a sermon on the cosmic pointlessness of even his own project within this schema. Such conclusions made it easy to believe that only cowardice prevented his contemporaries from agreeing with his anthroposociological theories. They certainly stroked the readers who nodded as they read. As I have said, he was especially contemptuous of those who understood Darwinian evolution and claimed to be freethinkers and yet still supported the republic. Indeed, he said that it was purely "an act of faith that allowed these freethinkers to escape the conclusions of Darwinian political science" (514). Lapouge did not specify to whom he | was referring; he may have been directing this primarily at the freethinking anthropologists or he may have had in mind Brunetière and his "bankruptcy of science." The general nature of his accusation implies that he was referring to both, as well as to an entire segment of society that was freethinking and yet believed in equality, republicanism, and morality. Lapouge described all these people as lost between two extremes: that of the orderly, safe, familiar world of the Catholic Church, which they knew to be false, and that of the unknown, frightening, violent world of the amoral future. In his own startling words:

They only had a choice between a return, pure and simple, to the theological doctrines from which they had come or the acceptance, pure and simple, of the scientific explanation of social phenomena and the abandonment of all philosophical principles on which their political doctrine had rested. It is not toward science that they went. Their psychology is that of men who in times gone by would prostrate themselves in churches and light heretics on fire, and we are not the least surprised, as they are their descendants. Already liberals, socialists, and anarchists treat Darwinians as barbarians. So be it! The barbarians are coming, the besiegers have come to be besieged, and their last hope of resistance is to lock themselves up in the citadel they were attacking. The near future will show our sons a curious spectacle: the theoreticians of the false modern democracy constrained to shut themselves back up in the citadel of clericalism. (514)

Clearly, Lapouge heard hypocrisy in this about-face and in the willingness to jettison materialist science in favor of peace and comfort. Brunetière wrote that if we allow Darwinism to dictate morality, the lessons it gives us will be "abominable." Lapouge championed the abomination. Yet it was not the barbarism of the Darwinian vision that turned Lapouge away from morality, just as it was not really the barbarism of Darwinian evolution that had frightened Brunetière. Like Brunetière, Lapouge did not invoke the brutality of the natural world when arguing about morality; he explicitly referenced the loss of God rather than the law of the jungle: "If one no longer wants the supernatural, it is necessary to be logical and to say: there is no good or bad, in and of themselves; nothing is just, or unjust. All our ideas of morality and law are due to circumstances of ancestral evolution and the social conventions that are the most severely sanctioned by opinion and by law have, in themselves, the exact same value as the rules to a game of cards" (514). As such, morality was not to be reestablished or propped up; rather, it was to be bravely rejected along with religion.

It is fascinating to see the position Lapouge took on the question of humanity's place in nature. Brunetière had supported Catholic authority by declaring humanity distinct from the natural world owing to its will, its morality, and its spirit. Hovelacque and the other freethinking anthropologists of Paris had supported republican secularism by arguing that humanity was part of the animal world but distinguishable in its material makeup (the possession of an enlarged third left circumvolution) and the resultant ability of speech. Lapouge supported his amoral selectionist vision by arguing that humanity was not at all distinct from other animals. He came at this in several different ways, negating both free will and human rights. "Man," wrote Lapouge, "is not a being 'set apart'; his actions are controlled by the determinism of the universe" (511). But if humanity shares in the materialist determinism of the universe, we also share in the violent self-interest of the animal world. "Man, in losing his privilege as a being 'set apart' and created in the image of God, has no more rights than all the other mammals. Even the idea of right is a fiction. . . . All men are brothers, all animals are brothers, to one another and to man, and fraternity extends to all beings, but that they are brothers does not get in the way of their eating one another. Fraternity, so be it, but too bad for the victims! Life is only maintained by death" (512). His vision was the worst of all worlds because he would not give credence to any feelings of meaning, morality, love, or transcendence since as far as he was concerned they could not be proven scientifically. He dismissed free will and human rights as not real. Their existence was not enough to convince him of their existence.

Lapouge felt rather isolated in coming to these conclusions. He had spent several years studying with the freethinking anthropologists, so he could not have felt alone in his atheism. But this idea of amorality was certainly outside their canon. The freethinking anthropologists understood evolution as inexorable progress. Lapouge was furious with them and the host of secular sociologists and philosophers who were attempting to devise a new moral system: "We have attempted many systems in order to maintain morality and the fundamentals of law. To tell the truth, these attempts were nothing but illusions. . . . The conscience is nothing but a particular aspect of instinct, and instinct is nothing but a hereditary habit. . . . Without the existence of a distinct soul, without immortality, and without the threat of the afterlife, there are no longer any sanctions."[20]

The end of the soul meant that every aspect of human behavior had to be reconceived. As soon as scientists managed to generate life spontaneously in laboratories, science would vanquish religion and finally assume its full authority over society. Lapouge did not think that science could replace the role of religion: "When the biologists have finished their work of destruction [of religion] on this point, the anthropological sciences will not have the ability to fill this lacuna, which will be eternal." Lapouge well understood the project of the freethinking anthropologists of Paris, for if "the anthropological sciences" were not trying to fill the lacuna created by the destruction of religion, then why mention that they would not be able to manage it? After this direct attack on his former professors of anthropology, he mused that eventually, "they will limit themselves, as they should today, to studying the processes of the formation of moral ideas (like all those that are generally referred to as common sense), and their hereditary transmission." Freethinkers tried to ground republican principles in biology in order to wrest them from Christianity and thus preserve them. "The rescue of those so-called political principles," wrote Lapouge, "is, on the contrary, not possible, not by these means or by any other. But though the current doctrines find no more mercy before science than do religious beliefs, biology does, at least, have the means to replace them" (144). There would be no replacement comfort, but there would be replacement principles: selectionist racism, anti-Semitism, and the state-controlled breeding of a superior, Aryan race.

Lapouge held that as long as humans were nothing more than a physical conglomeration of matter, the only possible way of relating to the past and the future was to foster biological continuity. He saw this creation of meaning as retroactive as well as future oriented, so to fail to reproduce was not only a crime against the future of humanity but a crime against one's own ancestors. Without God or soul to give an individual life meaning and a place in eternity, only "le plasma germinatif" could serve this role. "What is immortal," wrote Lapouge, "is not the soul—that unlikely and probably imaginary personage—it is instead the body or, rather, the germ plasma. . . . The individual who dies without leaving descendants puts an end to the immortality of his ancestors. He manages to kill his own dead."[21] With this statement, Lapouge revived his own dead, along with the deceased ancestors of his auditors.

Very rarely, Lapouge made use of a spiritual metaphor in describing his scientific vision. The metaphor had a lot in common with both monism and Bergsonian vitalism. Vitalism and monism were a bit mystical for Lapouge, but he agreed with their rejection of religion and their sense of coming drama. This is clear in a preface Lapouge wrote for a work by Ernst Haeckel, the most prominent monist in Europe. The book was *Le monisme: Lien entre la religion et la science* (Monism: The link between religion and science), and Lapouge translated it from the original German.[22] As I have noted above, many scientists were interested in such ideas because they seemed to make room for a rationalist concept of a unified, "spirited" universe. Some went further, seeing monism or vitalism as supporting a kind of group afterlife. Clémence Royer dabbled in monism when she began writing explanations of the world that emphasized its godlessness and its atomic and lawful unity. She was also interested in it as an alternative to the Western dualism that seemed to support the sharp division between masculinity and femininity. Because of their atheist materialism, neither Lapouge nor Royer was particularly dependent on the spiritualist concept of monism, but Lapouge, too, made some use of it in his writing. Several years before his translation of Haeckel, Lapouge wrote that "living beings are really the depositories of the conscience of the world. This conscience rises from inferior animals up to man and from man up to scholars. One could say that these are the cells of the cosmic brain."[23] "From man up to scholars," indeed. Among atheists in late-nineteenth-century France, there was pride in unbelief, and this pride was intelligible to the wider society; it was part of the profile.

With the universe empty of consciousness save that of animals and empty of intelligence save that of human beings, Lapouge saw the mass of living things as closest to traditional ideas of divinity. "This is why the sexual act is not only a creative act because it brings about a new being, the accomplishment of the absolute condition of immortality. It shines with a divine character because it is the transmission of the world's consciousness, it becomes theogonic. That is why the only absolute sin is infecundity" (306–307). Historian André Béjin has read this passage as an attempt to replace the Christian idea of God as simultaneously human and divine with a parallel idea of the germ cell as human and divine. [24] The literary flourish does not mean that much (because Lapouge was elsewhere and often explicit about there being no replacement divinity), but Lapouge's anthropology was certainly designed to perform religious functions—without any gods. He was saying that procreation is the only godlike thing because it attaches us to the eternal, such as it is.

When Lapouge spoke of immortality, then, it was within the context of an entirely atheist conception of the universe. His approach to the question reminds one distinctly of his anthropology professors. The freethinkers had written that the "only means of not dying entirely was to disperse to the four winds all that one could of the fire of one's heart and the light of one's mind," and they had augmented these intangibles by donating their corpses to the future of science. [25] They had enshrined intellectual and humanistic labor such that it represented all there was of the sacred and the immortal. Vacher de Lapouge referred to biological procreation in similar terms, writing that "the only immortality to which man could lay claim is not at all in the domain of theological dreams, it is that which he is assured in transmitting his germ plasma to future generations. Each of his descendants is a part of himself, indefinitely reproduced, transmitted and constructing new organisms." [26] His language even contained the same sense of posthumous dispersion.

Though Lapouge did not make it explicit, he seems to have seen himself as back in the position of the biblical Abraham. Abraham lived before the Hebrew development of the idea of an afterlife, and God promised him, as the greatest possible gift, that he would become a multitude. Lapouge positioned himself as living after the idea of an afterlife had expired, and he, too, sought meaning and immortality in procreation. "The ancient world had a vivid instinctual understanding of this material immortality by reproduction. The double cult of ancestors and of reproduction rested on a scientific base infinitely superior to that of all the spiritualist religions" (190–191). In what he perceived as an unfair, Godless, and socially atomized society, Lapouge could imagine only one way to attain a meaningful relationship with the past and the future: to have children and trust that they would have children, too. Yet no matter what arrangements were made for selectionist breeding, all that could be achieved thereby was a "relative immortality," which would last "so long as the rites of fecundation are repeated, this is the only immortality: all others are chimera" (307). Why mention it? Why was he so interested in some kind of immortality? The not-believing mission was a passionate, active faith: the wider meaning of existence was aggressively confronted, and faith was boldly enacted in the negative.

It is at this point that morality really began to get in the way. If the most crucial role of every human being is reproduction and if failure to reproduce wipes out the reproductive effort and the very essence of all one's ancestors, then there can be no true morality that prevents the individual from reproducing. "Selectionist morality," wrote Lapouge, "places responsibility-to-the-race in the supreme place—there where the morality of Christianity put responsibility-to-God" (191). Individuals were temporally finite and thus meaningless.[27] Properly understood, they did not even exist: "The individual is crushed by his race and becomes nothing. The race and the nation are everything" (511). Here was a glorification of the state and defense of social hierarchies that did not rely on religious dogma or tradition for its authority. The rites of fecundity would take on powerful new meanings in the twentieth century.

LAPOUGE IN CONTEXT: EUGENICS AND SOCIALISM

Lapouge's weird pessimism critically distinguishes him from later French eugenists. His brave new world was never described in glowing terms. It was depicted, instead, as the only possible way that civilization might survive, and as often as not Lapouge expressed more doubt than hope that it would ever come to pass.[28] According to him, the ultimate "law of anthroposociology" was that since prehistoric times the cephalic index had everywhere and constantly tended to increase (meaning that the population had grown ever more brachycephalic). In arguing this, Lapouge was drawing on an element of Darwinian evolutionary theory that most Social Darwinists ignored: the most complex, intelligent life forms are not necessarily the "fittest," that is, the most successful reproducers. Darwin had noted that wonderfully complex species had died out, while simpler, less impressive species survived.[29] Spencer and Haeckel, two of the earliest and most influential Social Darwinists, generally ignored this evidence that survival of the fittest could not always be equated with progress. Darwin wrote that a quick or cataclysmic environmental change, allowing no time for the species to adapt, results in particularly dramatic changes in the relative populations of species. The cataclysmic event that Lapouge referred to in order to explain the waning of dolicho dominance was the French Revolution, though he argued that the process had been going on since prehistoric times. He saw a delicacy linked to the refined qualities of the dolicho. This meant that all else being equal, the superior people would not necessarily rise to the top—an excellent rationalization for professional failure. It was also a strong argument for a controlled society, since a free market and a social meritocracy both rest on the notion that the best minds and best products have an ability to assert themselves as such. Other eugenists believed that the best would eventually triumph and that the important work was essentially to speed up a process that was already in motion naturally. Lapouge thought the natural process had to be stopped and put into reverse.

Another reason for the general optimism of most French Social Darwinists was that they were only marginally Darwinists. As discussed in previous chapters, the French held on to notions of Lamarckian evolutionary theory long after it was discredited elsewhere. Lamarck had worked out and published his transformist theory during the revolutionary decade, and the theory continued to be associated with the political left. The very notion that species could be altered was revolutionary, echoing political discourse on the ability of the individual to raise his or her social station. Lamarck's mechanism took this even further, for in it species change is enacted by individuals. Within this theory, when a creature spends its life striving to adapt itself physically or mentally (the giraffe stretching its neck to reach sweet but lofty leaves is the classic example), the personal improvements it attains are passed on to its progeny. Working one's way up was thus naturalized within a scientific framework.[30] By the late-nineteenth-century, Lamarckianism was no longer a revolutionary stance, but it still retained a strong measure of its leftist significance. Unlike Darwinism, it provided a mechanism for evolution in which environmental amelioration could be seen as having direct and positive influence on future generations, which meant that social welfare could be seen as promoting evolution. Educate this generation, and the next will be ever so slightly smarter and perhaps less dependent, criminal, or offensive.

Vacher de Lapouge did not believe that any significant inheritance of acquired characteristics was possible, so even if social reform and education could improve people, these improvements could not be heritable. In evolutionary terms, it would be a totally pointless endeavor. He came to this by interesting means: The soul doesn't exist, so you don't exist after your body dies. If everything you are is physical, however, you can pass yourself into the next generation by physical means. It was not as good an afterlife as one in which you remember the first life, but it was all we had:

> The soul and the body are one; the psychic phenomena are functions of the brain. The soul is thus hereditary, like the body. An individual's psychology is dependent on his ancestors. The fundamental inequality of individuals results from the difference of birth, and the inequality of birth is the only one that

cannot be repaired. The effects of education are essentially conditioned by heredity; they are not transmittable. Education can disguise the individual, but none of this lie will be passed on to future generations. One cannot enter a family or a nation by decree. The blood that one carries in one's veins at birth is the same all one's life.[31]

For it to be emotionally compelling to imagine living on in this way, the "psychic phenomena" have to be relatively inviolate. That is why an individual's psychology has to be "dependent on his ancestors," and that is why education cannot fundamentally change the soul. "Nations" (of Lapouge's defining) and families are more than groups that join individuals together: they are biological groups that share physiological "souls."

So evolution was not naturally progressing, nor was it able to be directed by anything short of a breeding project. Further, Lapouge believed civilized society, and especially modernity, were forcing evolution backward. Mechanized war drafted only the fittest and then further endangered the bravest, and capitalism promoted marriage for money rather than racial type (502). Capitalism was also responsible for a general abasement of refined values. Moreover, industrial machines caused injury and loss of life to some of the most hard working, and religion lured the most honest and generous men and women into lifelong celibacy. The only way to combat the decline of civilization was to practice scientific selection: the controlled breeding of dolichos. While other eugenists merely proposed reeducation regarding marriage choices, Lapouge envisioned eugenics as a fully controlled scientific process in which children would be conceived through artificial insemination of fit women with the diluted sperm of a few perfect dolichos.[32]

With all this talk of breeding, one might well expect Lapouge to have some very firm ideas about women's role in society. Certainly, his belief in the power of heredity over environment explains his lack of concern about some of the usual eugenic issues: the mother's behavior before, during, and after conception and the nature of the mother's relationship with the child. But Lapouge did not even say much about who the woman should be. His few references to women, as such, occur in his discussion of negative social selection. Catholic countries, he wrote, were wasting large numbers of particularly bright women by allowing them to be sequestered in convents. A cultural preference for the demure Catholic girl had made French women weak and ignorant. "Religious selection," Lapouge insisted, had created "the extraordinary French doll," whose concerns were confined to her toilette and to living a safe and quiet life with her husband—so unlike "the Aryan woman."[33] His public stance is not much augmented by his private writings, though he himself married a cousin and seems to have given his son something of a hard time about the eugenic quality of his future wife.[34] He clearly thought that both women and men contributed important hereditary traits to their children, but even when he spoke of diluting the sperm of perfect dolicho specimens, he said nothing about the women who would be making use of it. The mother is not the only one missing from this picture. After the moment of conception, Lapouge gives no practical description of the ideal selectionist family. This is a serious omission, because his plan of artificial insemination using "a perfect dolicho's" sperm inherently disrupts the existing family structure. Would fathers then exist as such? Lapouge didn't say. The image we are left with is that smart women were to be inseminated (apparently at their own request, for no constraint is ever mentioned) with the help of the male scientist, who would, essentially, stand in for all males. She would then raise the children alone or with the aid of her less-than-dolichocephalic husband. There was socialist precedent for reconfiguring the traditional family, but Lapouge never made reference to it.

The socialism of Lapouge's socialist selectionism is another odd aspect of his ideology. It was not, obviously, based on a love of egalitarianism but rather on a preference for the pre-Revolutionary social hierarchies and regulatory government. Capitalism and republican democracy were the most significant forces that Lapouge saw as accelerating the demise of civilization. He believed both to be based on the erroneous notion of human equality and accused both of favoring the social, political, and thus reproductive triumph of the brachycephalics and the Jews. Socialism would redress this problem in several ways: It replace respect for money with respect for intelligence and refined cultural qualities. It would actively support dolichos so that they could produce dolicho culture. Most crucially, it would control physical reproduction so that dolichos could someday outnumber brachies. His socialism was in the service of his anthroposociology, but he also argued that the relationship was the inverse, positing that socialism would only be possible after the population was eugenically manipulated. "Socialism, in any case," wrote Lapouge, "will be selectionist or not at all: it could only be possible with people made differently from us, and selectionism can make those people."[35]

Lapouge's anti-Semitic socialism may seem like an odd mixture of left- and right-wing ideologies, but in the mid-nineteenth century socialism was by no means an inherently humanitarian and egalitarian idea. Indeed, many early socialists, such as Charles Fourier, Alphonse Toussenel, and Pierre Leroux, identified capitalism with the Jewish people.[36] It is thus possible that Lapouge came to his notions of anti-Semitism and racial hierarchy through socialism rather than despite it. In any case, by the time L'Aryen came out in 1899,

Lapouge had largely rejected socialism, pronouncing it unfeasible and undesirable for the brachycephalic present but occasionally suggesting it for the
dolichocephalic future. As French socialists moved away from designating Jews
as the quintessential capitalists, Lapouge increasingly argued, in his correspondence and in his last major work, *Race et milieu social* (1909), that there
was no Jewish proletariat—in other words, the equation of Jews and high capital was as valid as ever.[37] But it still was not simple. For instance, the anti-
Semitic Edouard Drumont, editor of *La libre parole*, often quoted long passages
from *L'Aryen* in his journal. In one article, after recounting the Jewish influences on Dreyfusard scholars, Drumont argued the racial differences of Jews
and brought out Lapouge for authority, writing "on this point we have the witness of a learned man, a real learned man this time," and he also quoted
Lapouge so that the anthroposociologist seemed to praise Drumont.[38] Yet
L'Aryen actually robustly dismissed Drumont. Lapouge explained that for Drumont "and his friends," the term "Aryen" referred to the French, or even the
Europeans, as opposed to foreigners and Jews.[39] These brachies did not know
what they were. According to Lapouge, political anti-Semitism was just a
brachy bid for power against the Jews—the one dolicho group giving them
any trouble. It made sense for them, mused Lapouge, but he wanted none of
it. Brachy "economic anti-Semitism" was just a brachy "form of protection-
ism," and their "religious anti-Semitism," just a "form of clericalism" (464). His
chief political concern remained the biological creation of a hereditary ruling
elite.

Once Lapouge turned his attention to the social system most appropriate
for this potential hereditary ruling elite, he described a surprisingly democratic
utopia. Using the United States and England as his best models of dolichocephalic rule (though both were compromised by the influx of brachy immigrants), Lapouge proclaimed that dolichos loved liberty above all else.
Though each item on his list of rights and freedoms essential to the Aryan is, indeed, represented in the Declaration on the Rights of Man, Lapouge argued
that the brachycephalic French only speak of these rights, they do not respect
them. While the French citizen, "the supposed sovereign," only has permission
to express his thoughts by a voting ballot, "in silence," the United States and
England have huge and numerous political groups that are not only tolerated
but legal and encouraged.[40] Such groups, and other manifestations of political
liberty, "are always a cause of astonishment for French people landing in an
Aryan country." This astonishment was, according to Lapouge, not merely because of the novelty of the situation; rather, "they are missing something in their
brains that would allow them to understand it" (376). The French citizen does
not vote on questions but only on representatives, and these representatives are

not chosen by the citizens but presented to them. Once the elections poll is over, "the poor sovereign," that is, the French citizen, "does not matter anymore. . . . The elected ones are everything. . . . He does not retain any control or any right to express what he thinks if the elected ones betray him or exploit him" (377). Lapouge criticized the republic, but it was the brachy race that was to blame, and he considered no political change possible until biological change had been effected. "It is certain," wrote Lapouge, "that functionarism suits the brachycephalics. In France, in the last fifty years, the number of functionaries has gone from 188,000 to 416,000. It is the same in the other brachycephalic countries. . . . If we continue to require, as the premier quality of a subject, that they are perfectly inert and submissive to authority, the brachycephalics will end up having the last word" (482). All of which, in Lapouge's characterization, led to the French Third Republic functioning as a constantly changing and entirely arbitrary monarchy, while in Aryan England the monarchy functioned as a stable, egalitarian republic (377–378). So, in a sense, Lapouge had classic republican values; he just wanted to create new human beings who would be better suited to these values. His Aryans would make material what had been metaphysical because they would be bred as flesh and blood embodiments of past theoretical goals.

A CAREER IN ANTHROPOSOCIOLOGY

In the 1880s Louis Liard, the director of higher education, was in the process of reorganizing the entire French university system.[41] He is best known for this work, as well as for having promoted the careers of young, republican, "modern" scholars. He helped to secure a fellowship for Durkheim and in 1887 made him professor of social science and pedagogy at the University of Bordeaux, a post created especially for him.[42] It is thus rather intriguing to find that, the very next year, Liard intervened on behalf of Georges Vacher de Lapouge, arranging for him to give a *cours libre* at the University of Montpellier.[43] The odd match of Liard and Lapouge was mostly the result of youthful friendship—and enough physical distance that Liard may not have understood just whom he was helping. Lapouge was born on December 12, 1854, in Neuville, a large town in the department of Vienne. His father was a minor government functionary, apparently of mixed Catholic and Protestant heritage, who, in his son's words, "died poor."[44] In 1868 Lapouge entered the Poitiers lycée, where his professor of philosophy, Louis Liard, introduced him to the works of Spencer, Darwin, Galton, and the Genovese racist naturalist Alphonse de Candolle.[45] Happenstance, then, brought these two men together, but Liard was an intense young man who saw promise in the younger Lapouge and took him under his wing.

Liard's own youth is worth a brief look. According to the reports in his personnel file concerning his years as a professor at the Académie de Bordeaux, Liard himself took much criticism for his ardent anticlericalism, his efforts in developing a worker's library ("which got him noticed and made him popular in a 'certain' party"), "and above all a degree of confidence in himself and in his opinions that his young age alone (twenty-four) cannot quite excuse."[46] Soon after this report, Liard was transferred, at a large reduction in pay and prestige, to the Académie de Poitiers. He wrote to the minister of public instruction asking him to remedy this "disgrace" with another transfer but was instead left at Poitiers for four years, during which time he taught and befriended Lapouge.[47] Here, again, Liard was criticized for his anticlericalism and his "exalted" opinions.[48] Now, however, the censure was more severe, centering on Liard's alliance with a professor of natural history, M. Contejean, "a learned man but of paradoxical and often dangerous ideas."[49] This alliance and the opinions that were at its base do not seem to have outlasted Liard's youth. They do, however, help to substantiate Lapouge's claim that Liard greatly influenced his anthropological approach to history and philosophy. Liard's troubled personal experience may also help to explain his encouragement and toleration of Lapouge long after Lapouge had begun to express ideas that would seem anathema to the social project of the mature Liard.

Lapouge attained his *baccalauréat* in 1872, for which he received high honors in the concours of 1873.[50] He went on to become one of the Université de Poitiers's star law students; one of his studies, a 750-page volume, was recommended by the faculty to be published at the expense of the state.[51] When, in his final year, he applied for a position as the university's law librarian, the director of the school wrote to the minister of public instruction promoting his candidacy, expressing, however, his reservation that Lapouge would soon attain his *agrégation* and leave the position for an illustrious career elsewhere. The minister took this worry to heart, and the request for the position was denied.[52] Late in his life, Lapouge would write that at this early juncture he had thought to go to Paris to prepare for his *agrégation*, but, fearing a life of penury in the capital, applied for a position in the magistracy instead. As a young man, however, he wrote that he offered himself for this service out of love for the republic. In March 1879, only days after receiving his doctorate, he was installed as substitute *procureur*, a public prosecutor, at Valence. Within a week's time, however, he applied for a position nearer to his mother and in this request also demonstrated his feelings toward the republic. He worried that, were he to choose his career over his filial responsibility, "I would neither dignify service to the republic, nor to the magistrate, because a bad son could not hardly be anything but a bad citizen."[53] Despite such ardor and the extensive concern he expressed regarding the responsibilities of the position, Lapouge was not very successful.[54] At this point, private letters and published articles, by Lapouge and about him, confirm the future antirepublican racist as a passionately anticlerical republican.[55] In his actions, the young Lapouge behaved as the freethinking anthropologist Gabriel de Mortillet had behaved when newly empowered: Mortillet secularized street names and uprooted cemetery crosses. Lapouge took advantage of a revolutionary law forbidding ecclesiastic dress that had never been officially revoked and arrested the first priest he saw passing in the street outside his window.[56] The list of attempts to deconsecrate people, objects, and places now included the priest himself.[57]

Lapouge was soon made *procureur* at Blanc; at twenty-six, he was the youngest in all of France. His superiors at Blanc described him as friendly, intelligent, and extremely well educated but with too little knowledge "of men and of things."[58] As a result, he was not recommended for a promotion.[59] On the July 23, 1881, he was transferred to Chambon, a move the republican newspaper *Journal du centre* reported as an act of political persecution against an "earnest and militant republican" who had simply let people know his beliefs. "We are not in on the secret between the gods and the prefecture," ran the final paragraph of the article, "we simply pose this question: Why was the radical and anticlerical M. Vacher-Lapouge sent away in disgrace to Chambon?"[60] The minister of justice, however, wrote that Lapouge had created an impossible situation for himself at Blanc and had been transferred for this reason.[61]

Lapouge was quietly tolerated at Chambon for almost three years, but this ended when he accidentally shot someone in a gun shop. The victim—it was the shop's proprietor—was not badly harmed: the bullet had lost much force in ricochet before striking him in the neck. But the situation encouraged Lapouge's superiors, who had until then held their tongues "out of pity" for him, to speak up and report that the incident could only diminish Lapouge's "moral authority," which "was already as meager as possible—one can be assured that whatever little he had is now completely gone."[62] They fired him. It was at this point, between 1883 and 1886, that he went to Paris to prepare for his *agrégation* and took courses at the Ecole d'anthropologie and the Muséum d'histoire naturelle. He later listed Quatrefages, Milne Edwards, Manouvrier, Mortillet, Hamy, and Topinard as his anthropological professors and reported that he quickly developed a distinct dislike for most of them (especially Quatrefages, whom he described as "resolutely opposed . . . to new ideas").[63] His correspondence indicates he knew Hovelacque and Letourneau as well.

Only in Paul Topinard did he find an intellectual companion, and their friendship lasted throughout their professional lives.[64] (In his *Revue d'anthropologie*, Topinard published five long articles and about a dozen book reviews authored by Lapouge.) Despite Lapouge's relative silence on the matter, the experience at the anthropology school must have been extremely meaningful. Whether or not he already shared it, Lapouge must have been struck by the atheism of the freethinking anthropologists. They may have converted him. Also, Topinard's racialism may have colored Lapouge's thinking during his years as a student in Paris.[65] Lapouge himself credited Louis Liard as his chief influence on these matters.[66]

In 1886 Lapouge prepared a series of lectures on inferior and superior races, which Topinard encouraged him to offer at the Ecole d'anthropologie.[67] According to Lapouge, this did not happen because while in Paris he had married and had a child, and his precarious financial situation forced him to leave the city and find a steady job. It is true that he had married: accompanied by some terrific brouhahas with his mother (they had an extensive, often contentious correspondence), the twenty-nine-year-old Lapouge married his "little cousin," the seventeen-year-old Marie-Albertine Hindré, on September 4, 1883.[68] Yet it is unlikely that this is what stopped him from running a course at the école.[69] It is more likely that the freethinking anthropologists, by now firmly in control of the school, rejected his proposal. Also during these years, Lapouge entered into the *concours d'agrégation* twice. Both times he failed to win a professorship.[70] With a family to support, no steady means of employment, a miserable record as a magistrate, and few professional friends in Paris, Lapouge returned to the career manqué of his youth and applied for a job as a law librarian. This time no one worried that he would leave this career of expediency for bigger and better things. He was assigned to be the underlibrarian at Montpellier—the town that would soon become known as the site of Vacher de Lapouge's anthroposociological lectures—and given a salary of 2,000 francs.[71] Though he would move around a bit, Lapouge would spend the rest of his professional life as a librarian.

Lapouge did not really want to be a librarian. In 1887, his second year at Montpellier, he was judged to be a very honorable man with a superior education but "without a hint of good judgment." In his superior's words, "an excess of tact [was] not one of his great faults either."[72] Meanwhile, he arranged for Topinard to visit Liard in person and plead his case for a new position.[73] Liard did not relocate Lapouge, but in 1888 he did arrange for him to give his *cours libre* at the University of Montpellier.[74] In report after report, his superiors at the library said that he did what he was told, "but nothing, absolutely nothing else; it's just too bad for you if you don't give him instructions."[75]

However, as long as Liard was in a position to assist him and was well disposed to do so, Lapouge did not suffer excessively from his unpopularity at the university. In 1890 he wrote to Liard asking again for a raise and for an indemnity for his courses and his research, which he claimed were yielding, especially that year, "important discoveries."[76] He received both.[77] In later years Lapouge claimed that these courses were an enormous and popular success.[78] According to Paul Valéry, however, who frequented the lectures while studying law at Montpellier, the courses were never well attended. The young poet found them fascinating, however, and though he did not put much stock in the racial theories argued by Lapouge, he credited him with being a thrilling lecturer. Valéry even joined Lapouge in his laboratory, helping him to measure six hundred skulls taken from an old cemetery. Valéry later commented that he did not learn anything useful from this effort but that, "among all the things that I learned that were never useful to me, those pointless measurements were not more pointless than the others."[79] An interesting assessment.

In 1892 new rules regarding the *cours libres* at Montpellier placed almost full control in the hands of the university's professors, and within the year Lapouge's courses were banned. He wrote to Dr. Collignon, an army doctor, avid head measurer, and amateur anthroposociologist, informing him of his situation: "Here I am in the same situation that Topinard was in, the only difference being that I saw it coming a long time ago and so I was able to bring all the work and specimens of my last five years into my private laboratory."[80] As he later wrote in *L'Aryen*, "They can destroy scientific documents or allow them to be destroyed, they can close courses, interfere with the publication of a book, and suppress a scholar into poverty—but they can not suppress science."[81] This does not seem to have been empty paranoia; I do not know what the mechanism was, but it does seem that some kind of ban was placed on publishing Lapouge's books. Contemporaries complained that they were nearly impossible to find after about 1900, and copies are still quite rare.

It was around this time that Lapouge carried out his experiment in "telegenesis." He had decided that one perfect dolicho man could impregnate 200,000 women a year if the semen were correctly diluted and if sperm turned out to travel well. He claimed that to check the second variable he performed the first experiment of telegenesis, mailing a dose of human sperm from one town to another and there attaining a successful conception. There is reason to believe this did occur, that he used his own sperm, and that it was his mistress who conceived the child.[82] To this oddity must be added a rather strange event that took place in 1896, when Lapouge was reassigned to Rennes.[83] An anonymous letter was sent to the commissaire général accusing Lapouge of luring five young girls between the ages of ten and sixteen into his laboratory on the pretext of a scientific experiment. Lapouge took pictures of the girls completely naked and allegedly engaged in acts of oral and manual sex with them.[84] The photographs were definitely real, but when energetically questioned the girls rescinded their more damning accusations and conceded that Lapouge had measured their skulls and chests. This, remarkably, convinced the judicial representatives of the state and of the university that the nude photographs had been taken for scientific rather than lascivious purposes. All charges were dropped. Louis Liard, now director of higher education, was informed of all this, including the tidbit that Lapouge was actively hunting for his anonymous accuser and he "will not abstain from resorting to violence or even 'shooting him in the head.' "[85] After this incident and throughout the rest of his stay at Rennes, it is reported that Lapouge lived in almost complete seclusion, hiding himself in his anthropological laboratory.[86]

The strained nature of Lapouge's relationships with women seems worth noting: his personal correspondence with his mother was extensive and turbulent, full of pleas for understanding, bursts of anger, and petitions for forgiveness. Before his marriage, correspondence between Lapouge and Albertine show him counseling her on how to deal with her difficult aunt, his mother; afterward, they seem to have had as difficult a time with each other. Add this incident with the young girls as well as the conception of a child, with his mistress, by "telegenesis," and a picture emerges of a particularly overwrought relationship with family, with gender, and with sex. Lapouge's theoretical dismantling of the traditional family and his extreme social isolation contribute yet more to the picture. Historians of race and gender have described the way systems of value based in race and gender share the same metaphors and patterns of logic, social conduits, and familial enactment.[87] Here, I think it is enough to mention that these relationships were fraught with tension for Lapouge, whose alienation and bitterness increased steadily as the years advanced. No one in his daily life seems to have been much impressed by Lapouge's first book, *Sélections sociales*. It certainly never entered into any assessment of his professional worth, where deeply negative employee evaluations were offset only by observations that he was "very miserable in his intimate life."[88]

It was in April 1897 that Lapouge contracted typhoid fever and offered his brain to the Society of Mutual Autopsy.[89] After his recovery, Lapouge's requests for a transfer increased in frequency and tone. When a position opened up in 1900, he was relocated to the library of the Université de Poitiers, where he would stay until retirement in 1923. This was where he had so brilliantly distinguished himself as a student of law. Now he was back as a librarian, while as an anthropologist he was growing increasingly famous outside France, especially in Germany and the United States.[90] As I discuss below, he kept up correspondence with an extraordinary number of mostly foreign adepts, disciples, and fellow racialist anthropologists, who all treated him with deference. His *L'Aryen* had come out in 1899 and been widely panned in France— for reasons that will become clear—but was lauded elsewhere. When he and his son Claude (then fifteen) visited Germany in August 1901, he was celebrated as he never had been in France. According to the *Badische Press*, which referred to him (erroneously) as a professor and as a "famous French anthropologist," as soon as Lapouge arrived, admirers came from all over and bestowed on him a crown of laurels, "one meter in diameter," which bore a large ribbon in the colors of the German and French flags and read, "To the great thinker who has shown us the way."[91] When he returned to France he was once again a little-known and disliked librarian—apparently, he socialized with no one.[92] In a rare indication that his superiors knew the extent of his renown, a 1903 report on his library work notes that "M. Lapouge is the head (outside of France) of a School of Anthropology."[93]

When Durkheim was offered a post at the Sorbonne in 1902, Lapouge wrote to Liard presenting himself as a candidate for the post Durkheim would be vacating: the chair of social sciences at Bordeaux. Having offered his candidature, he explained to Liard:

> In doing so, I in no way renounce my opinion that no sociology is yet possible. I still believe that everything that is said or written in its name is pure metaphysics. . . . It is precisely to protest against the pretensions of those who want to teach what they do not know, against the metaphysicians who seek to model a fiction of social science on their spiritual and sentimental prejudices, that I pose my candidature. I do it as the most authorized representative in France of the scientific school, which relies on facts and on their evident and immediate consequences, excluding abstractions and arguments, which knows neither sentimentalism nor partiality.[94]

In this light "sentimental prejudices" start to look awfully attractive. When one takes into consideration that Liard's interest in Durkheim lay in the latter's commitment to establishing a republican morality, it is evident that Lapouge's proposal was hopelessly out of touch with Liard's project. Lapouge was not offered the position.

In 1909 Lapouge applied for the chair of anthropology at the Paris Muséum d'histoire naturelle.[95] He did not get it. One of Lapouge's disciples, the German language teacher Henri Muffang, wrote to Lapouge regarding the application, and he juxtaposed Lapouge's failure with the Durkheimian success. "I was thrilled," wrote Muffang, "to learn that you had put up your candidature for the chair at the museum. You should have had a chair at Paris long ago. When one sees such people as Durkheim and Bouglé at the Sorbonne, with their egalitarian sociological phraseology, and Lapouge still at Poitiers despite a global reputation, one sees a beautiful illustration of backward selection."[96]

INFLUENCE IN GERMANY

I have elsewhere demonstrated some of the direct lines of influence between Lapouge and the Nazis.[97] Very briefly, Lapouge's work first entered into Germany between 1885 and 1920, through the anthropological and historical work of Ludwig Schemann, Ludwig Woltmann, and Otto Ammon—the three men who are generally cited as the originators of race theory in Germany.[98] But Lapouge lived until 1936 and actively supported his science until the end of his life, so he had a personal influence on theorists of the 1920s and 1930s as well, most noticeably on the racial theory and career of Hans Günther, generally known as the chief Nazi race theorist. Here, I want only to demonstrate that Lapouge's atheist, scientific materialism had an effect on his racialist colleagues, though they rejected it in almost all cases, and to point to some of the religious tones of his wider ministry. Consider, for example, Schemann's response to Lapouge's second major work, *L'Aryen: Son rôle social*:

> Even though, as a Christian idealist, I was seriously saddened, not by the pessimism but by the materialism, not to say the nihilism of your final pages, I still read your book with the greatest interest for its first part and the most profound emotion for the second part. Your imposing erudition, your so universal penetration, the grandeur and the profundity of your views, and more than all of that the heroism of your truthfulness made the same indelible impression on me as your *Sélections sociales* did in its time. The more I know your works, the more convinced I am that they are destined to play the most remarkable role in the science of the future.[99]

Schemann was saddened by the materialist nihilism of the book, but that is what lent Lapouge his air of truthful "heroism." The power of Lapouge's stark pessimism, even for those who did not agree with it, must be appreciated. Schemann, Woltmann, and Ammon were also impressed by his rejection of the sanctity of fatherhood, but they explicitly refused to go along with it. Still, they were enthralled. Consider a few brief comments from Ammon to Lapouge: "One often says 'poet, prophet,' but in our case it is a man of science and not a poet who has predicted everything, and that man is called M. De Lapouge!" In the same letter, speaking of a work of his own that was soon to be published, Ammon wrote, "You will see material that will excite the jealousy of your colleagues, applied to the glorification of your theories."[100] The deference was deep. As Ammon later told Lapouge: "I always regard you as a student regards his master."[101]

In the twenties and thirties, Günther came under Lapouge's influence. Often, as in the case of his influential *Racial Elements of European History*, Günther cited Lapouge more frequently than any writer except Gobineau.[102] And Günther also specifically praised Lapouge for his scientism, widely publishing that Lapouge had written "the first scientific work from the racial historical standpoint" (257). Beyond Günther's praise and direct citations of Lapouge, his works were profoundly influenced by Lapouge's very particular paradigm, and his prose drew heavily on Lapouge's odd lexicon. Their correspondence was tutelary but warm.[103] Günther sent Lapouge his books and gratefully accepted the elder man's criticism. In his letters to Schemann, Lapouge referred to the young writer as "mon bon disciple Günther."[104] In 1930 Günther was interested in a post at the University of Jena, but the Deutschen Liga für Menschenrechte (a group of thirty-one professors from all over Germany) did not think that he possessed the base-level qualifications the university demanded of its faculty.[105] Lapouge intervened with a few letters praising Günther's work.[106] Thereafter, Wilhelm Frick, minister of the interior and education in Thuringia and the first Nazi minister of a German state, took matters into his own hands, installing Günther at Jena against the continued protests of the professorial senate and the league.[107] Günther's chair was in "anthroposociology," a term of distinctive Lapougian origin.[108] That Günther's appointment was a serious affair is evident in the fact that Hitler attended his inaugural address in the spring of 1933.[109]

A casual letter that Lapouge wrote to an A. Assire (Lapouge had arranged for him to translate Madison Grant's *The Passing of the Great Race*)[110] in 1932 lends insight into the period and bears an extended quotation.

It has been a long time since I have had news from Grant, who is quite old. I am expecting Lothrop Stoddard sometime soon. They created for Günther, at the University of Jena, a chair of anthroposociology under my auspices. It was imposed by Frick, with pressure from the Nazis. Notice that the Nazis are nothing but the German branch of selectionist monists and that their nationalism makes no sense in selectionist internationalism, but the contradiction does not worry them. Hitler's social program was patiently constructed from the facts and ideas of my selectionist publications over the past years—except the milk has turned, and there is nothing in the casserole but a sorcerer's brew. The obligatory work for all . . . the methodical multiplication of eugenic people, the exclusion of noneugenic people from the right to reproduce, all that was already in the aristocratic socialism of Woltmann and of Lapouge when they founded, twenty-five years ago, the *Politisch Anthropologische Revue*, and when we lost my lieutenant, his place was filled by Hitler and Günther. Pan-Germanism has been an idea of political philologists for the past century, which they have attempted to hold together with the equation: "Aryan" = "dolicho-blond" = "German" = "all that is of the kraut culture [*de culture boche*] and German language."These things have nothing to do with one another. Another ingredient, which is very poisonous, is militarism. This is how demolishing things and massacring people became marks of high superiority and pious work.

You certainly cannot have doubted, my dear Assire, when you were translating Madison Grant's book, that this American work, translated into German, would become one of the catechisms of the Hitlerian party—oh! With how many little changes!

I strongly believe that the Nazis will eventually come to understand my explanations and how much they risk in fighting with their neighbors, especially because in the pure doctrine it is not Germany that is the country of dolicho-blonds. But what still worries me is to see Germany—an essential part of the machine of the world—end up in the most horrible civil war without much advancing the progress of selectionist monism. And this is the philosophy of those to whom we speak, with Grant's book in hand.[111]

So Lapouge saw himself as creator of the new German agenda.[112] It was a bit of an exaggeration, but Günther was quite clear about his reliance on and debt to Lapouge, and historians certainly have credited Günther as Hitler's primary influence on racial questions—from *Mein Kampf* through the Final Solution. Lapouge's *L'Aryen*, translated into German and edited, was published in 1939.[113]

French collaborators were impressed with Lapouge as well. Consider, for example, an article in the *Cahiers franco-allemands* in 1942, by Edgar Tatarin-Tarnheyden, on "Georges Vacher de Lapouge: Visionnaire française de l'avenir européen."[114] The article's proclaimed goal was to demonstrate that the changes going on in France were not "merely a result of the war." Rather, the author asserted, there had been isolated French precursors who had invented race science. These precursors were Gobineau and Lapouge, but Tatarin-Tarnheyden was considerably more impressed with Lapouge as a direct source of Nazi doctrine. Lapouge, he asserted, "was the first to . . . have established exact anthropological types and to have proceeded to a systematic subdivision of the principal European races."Tatarin-Tarnheyden credited Lapouge's work as having a fundamental importance "for today's German researchers." According to him, it was due to Lapouge that the Aryan "became a precisely established scientific fact" (339).

Among other reasons, Tatarin-Tarnheyden cited Lapouge as more significant because where Gobineau "was still solidly attached to the church's theory of the independence of the soul," Lapouge recognized that "the essence of psychic substance was the hereditary plasma, the racial soul" (344). Gobineau, he reported, had even specifically rejected Darwinian theory because it was too materialist. Tatarin-Tarnheyden hit this point several times, insisting that Gobineau's work was weak because it was based only on intuition and citing scholarly attacks on Gobineau's lack of scientific facts. Lapouge, however, he found irreproachable because of the combined effect of his atheist materialism and his scientific exactitude. Wrote Tatarin-Tarnheyden, "It is on this point that rests the grand progress and is the true progress of Lapouge. He did not separate the body from the soul" (345). As long as the greatness of a human being was understood as somatic, one could conceive of this greatness as heritable, and one could design a state around encouraging that hereditary line. But consider how, even this late, and even from the extreme right, the Frenchman sees science and a new future as dependent on the rejection of the church: "The subjugation of even spiritual leaders to natural laws appeared in Lapouge's work in the decisive weight that he attributed to the need for great space for the political success of a people—an idea that was foreign to Gobineau. Lapouge also surpassed him when, in detaching himself from the constraints of the church, he pronounced these prophetic words: 'The morality of today is almost dead and that of tomorrow is not yet born'" (346).

With the Germans, one sees a celebration of science but not antireligious scientism. Günther was editor of *Rasse: Monatsschrift der nordischen Bewegung* (Race: A monthly for the Nordic idea), which published discussions of dolichocephalics and brachycephalics, blood groups, and other biological determinations—all presented in strictly scientific terms, replete with numbers and comparatively devoid of vitriolic eruptions. As one contemporary critic noted: "It makes racism respectable among the educated classes by having a dazzling array of Herr Doktors and professors among its editors and contributors."[115] The invocation of science and scholarship mentioned here is plainly meant as a contrast to charlatans and boors, not religion and tradition. In *Rasse*, Günther published several short pieces by Lapouge and penned several more celebrating the older man as the "founder of racial science." Günther did not mention God or the church much, but he did appreciate the pessimism.

When Lapouge died, Günther wrote a mournful obituary that he published in his journal.[116] In it, he cited Lapouge as the first to apply the studies of heredity and selection to the life of peoples and credited him with having "gone further, earlier, than Galton, Gobineau, or Ammon in predicting the downfall of civilization." He also celebrated Lapouge for having "based morality completely on biology." Günther attributed uncommon insight to his mentor, writing, "We will never forget Lapouge. His name belongs among the great names of northern racial theorists!" (98). An obituary in the journal *Volk und Rasse* praised Lapouge as the founder of all race science, stating that, "though Gobineau was trained in the natural sciences, it was Lapouge who was the first to apply the scientific studies to the theory of races."[117] Most of the many German articles about him, before and after his death, mentioned his scientism but skipped over his atheist materialism.

The *Volke und Rasse* obituary proclaimed that "the success of the development that his theories had in Germanic lands, especially in National Socialist Germany, must have given him the assurance, in his final years, that his work would carry on" (258). This was true, to a degree. Lapouge did on occasion celebrate the influence he had had on Germany, as in a letter to Schemann written in 1934: "Tomorrow I turn 80, and I no longer hardly hope to take up my work on social science based on biology again, but as far as that goes, the battle is won, and in Germany above all they work with ardor on political and social applications of selectionism!"[118] A year later, however, he sounded a very different note, writing that "the future will tell us if this great man's politics-of-the-bogeyman could end in any way other than in horrifying exterminations and the end of the best people. But who would dare, who could proclaim this truth?"[119] This note of worry and remorse, and the few other comments that resemble it, help to place Lapouge historically, but they do not undo the fact that he helped to create the situation. In writing the history of the Shoah, we should at least note that in the mind of this first-generation scientific racist, morality was actively rejected rather than merely ignored. Moreover, that rejection followed a logic outside of racial science, a logic that rested on the apparent consequences of the end of the soul when coupled with an insistence on the end of philosophy. We must also note that this proponent of antimorality and state racism lived and campaigned until his death in 1936.

THE EXTENDED FLOCK

Lapouge was extraordinarily industrious in the service of his science. Aside from the skull measuring and publishing, he kept up a tremendous correspondence with disciples, sympathizers, opponents, journal editors, and colleagues all over the world.[120] These included a host of obscure men and women, as well as very well known racial theorists, eugenists, and other interested parties. Among the prominent racists and eugenists who held extensive correspondence with Lapouge were such figures as Madison Grant, Margaret Sanger, Charles Davenport, Lothrop Stoddard, William Ripley, John Beddoe, Ernst Haeckel, Francis Galton, Gustave Le Bon, Charles and Bessie Drysdale, Luis Huerta, Angelo Crespi, Georges Chatterton-Hill, Carl Closson, Jean-Richard Bloch, and Charles de Ujfalvy-Huszar. These were among the best-known racialists of the time, and they had tremendous impact; they all wrote immensely popular books, and many also proselytized in lectures.[121] In the United States the two most popular racial scare books were Grant's *The Passing of the Great Race* and Ripley's *The Races of Europe*; in Britain, Stoddard's *The Rising Tide of Color* and John Beddoe's *The Races of Britain* preached xenophobic hatred and encouraged the Nordic movement there. Many of these figures held extremely prestigious positions in their respective countries. Ripley was a professor at Harvard and MIT and also lectured on anthropology at Columbia; Beddoe served as president of the Anthropological Society of London and of the Royal Anthropological Institute of Great Britain and Ireland and lectured widely; Davenport was founding director of the famous biological laboratory of Cold Spring Harbor, New York, and of the Eugenics Record Office; Closson lectured at the University of Chicago; Charles and Bessie Drysdale were the leaders of the British Neo-Malthusian Society and delivered frequent lectures on eugenics. In the Lapouge archives, there are hundreds upon hundreds of letters to and from these people, and hundreds more to editors of journals and amateur scientists, both male and female, near and far.

Lapouge kept up these relationships (both in letters and visits) over the course of six decades, energetically facilitating the exchange of information, translation of work, publication of new studies, and organization of conferences. He introduced his friends and followers to each other, arranged for them to translate each other's work, and encouraged their friendships. He sent these fellow travelers his collections of data and his photographs of "types," lent them skulls, annotated their work, wrote their prefaces, and, as with Günther, even helped them attain university positions. They all flattered each other a good deal, but Lapouge was unquestionably respected as the master— or even "my superior," in many cases. Carl Closson informed Lapouge that his course on "Social Selection" at the University of Chicago was based on Lapouge's *Sélections sociales*.[122] William Ripley's interaction with Lapouge began when Ripley wrote a very critical review of this rival race theorist's work. Lapouge, however, wanted cooperation, not rivalry; he wrote charm-

ing, graceful letters to Ripley, and the two became friends and consistent supporters of one another's work. Much of the correspondence between Lapouge
and his colleagues was filled with details of the quotidian, serving essentially
as cathectic ritual. Some disciples were in contact with several other members
of the far-flung network, and most regularly either asked Lapouge for news
about the others or supplied him with such information. In general, the letters
kept track of disciples, colleagues, and various racial theorists around the
world. Many of these colleagues knew each other's personal situations and issued congratulations for good selectionist marriages, for the arrival of children, and then for the betrothals of these children. Another large portion of
the correspondence was devoted to the dissemination of specific anthroposociological facts. In France, Germany, Britain, Spain, and the United States,
Lapouge's friends and disciples kept him apprised of their almost constant
head-measuring plans, expeditions, and results. They asked hundreds of pro-
cedural questions, and Lapouge answered them. They also worked to defend
and promote Lapouge.

With figures who were more colleagues than disciples, Lapouge carried on
more balanced relationships. His language changed ever so slightly for each of
these figures, though his general concerns remained consistent. For example,
Lapouge told Ernst Haeckel, author of the very optimistic, monist-racist *The
Riddle of the Universe*, that he himself was "a pessimist monist" because he understood that "cold, darkness, death, and unconsciousness constituted the final
reality."[123] Lapouge rarely if ever described himself as a pessimist monist (essentially, a materialist) elsewhere. In another letter to Germany (the name is
obscured, but Haeckel seems the likely candidate), Lapouge wrote, "You
are . . . like me, a medieval chevalier lost in the modern world," and he congratulated himself and his correspondent on their ability to maintain "a perfect
detachment." "I really fear that our great courage will not serve for much and
that we will finish like Don Quixote. The world, which was never worth much
anyway, has gone rotten. My opinion is that a reform will require a profound
moral disturbance, and where can one nowadays find a crowbar with which to
raise souls? There is no more God, no more duty, no more morality, no more
country." Lapouge further complained that "the biggest moral movement ever,
Christianity," had failed. The republican movement was even more of a disappointment. Lapouge told of those who had "sacrificed everything and even their
lives to found the republic, the political regime that was going to put power
into pure hands and create happiness for everyone. You cannot doubt the enthusiasm and the righteousness of the apostles of the republic at a certain moment: the dream of purity and of happiness came to an end with the Wilson
scandal, the Panama scandal, and the ruin of honest naive folk and the enrich-

ment of rogues."[124] Lapouge was still invigorating the anthropological cult by invoking the loss of belief—in God, in Christianity, and in the republic.

With Margaret Sanger and other advocates of birth control, Lapouge concentrated his language on the prophylactic aspect of anthroposociology. Not surprisingly, his ties with this movement grew as the discourse of birth control shifted from feminism to eugenics. In 1921 he was invited and welcomed to the White House along with other members of the International Neo-Malthusian and Birth Control Conference.[125] On this visit he got to know Sanger, and they subsequently exchanged a series of letters; when she requested it, he agreed to lend his name to several of her projects. In 1925 he wrote telling her to be assured that

> you will have in me, henceforth, a very resolute collaborator and that I consider the interests and programs of selectionism as inseparable from those of the Birth Control League. There is a formidable amount of work to accomplish in order to change opinion and modify ideas on morality in general and on sexual morality in particular, but the future and progress of the human species demands that this work be accomplished without delay or timidity. I do not know how to thank you and all the other women sufficiently for your kindness toward me—I was very touched. I return to you the note sent by Dr. Drysdale. It is very good as it is and can be published with my signature.[126]

Sanger's description of Lapouge in her autobiography conforms to the rather maladroit image of him that emerges from the archives: he was an example of things that went wrong at the conference, getting lost when everyone was picked up and then finally found on the pier, "whence all had fled save one inconspicuous, desolate man sitting on top of his luggage, reading, waiting patiently for someone to come for him—so unimportant-looking that no one would have suspected him of being a renowned scientist," and later burning himself badly in the shower because he didn't know how to regulate the "much advertised American plumbing." He nevertheless charmed everyone that he wanted to charm.[127]

Because of the eugenics movement and the new immigration laws in the United States, Lapouge grew increasingly optimistic regarding the country's fate.[128] In 1925 he expressed as much to Charles Davenport, a Mendelian biologist and America's leading eugenist. Writing that the situation in France was getting worse all the time, Lapouge told Davenport that he was "very discouraged" and that "the only hope for stopping the decline of civilization was in America. . . . The progress of democracy has made Europe a land decivilized." He praised Sanger and lamented that in France, neo-Malthusianism was "above all else a doctrine of people who wanted to retain the fun of sex and get rid of the fecundity." In America, by contrast, "it seems to be inspired by selectionist tendencies more than anything else. I have a very firm opinion on this: the sacerdotal act of fertilization must not be open to everyone and in all circumstances; reproduction is a social function that must be put under society's control. By the weakness of the state and of the church, which opens marriage to all comers, it has most often become an affair of personal interests, sometimes of love. . . . This situation has been very badly handled."[129] When Lapouge wrote of "the sacerdotal act of fertilization," he was furthering his religious atheism, proselytizing to an attenuated flock. For Lapouge, and for some of his audience, a passionate, communally binding, meaning-laden science was replacing religion in very specific ways.

The American immigration laws that had so pleased Lapouge were best represented by the Immigration Act of 1924, overwhelmingly passed by the House and Senate and quickly signed into law by President Calvin Coolidge. When still vice president, Coolidge had publicly remarked that "America must be kept American. Biological laws show . . . that Nordics deteriorate when mixed with other races."[130] The law of 1924 reflected these concerns quite closely, seeking to "keep" America Nordic by letting in fewer eastern and southern European immigrants. As for the forced sterilization of the unfit and criminal, eugenists such as Sanger and Davenport actively supported such measures. The first state sterilization law in America was passed in 1907. At the end of the 1920s, sterilization laws were on the books in twenty-four states, and there were people sterilized for little more than moral infractions. Frustrated by an opposition that called these laws unconstitutional, American eugenists brought a test case to the Supreme Court: seventeen-year-old Carrie Buck had been labeled a "moral imbecile" because she conceived a child out of wedlock. She, her mother, and her daughter "tested" below normal intelligence though there is no good reason to believe they were not all perfectly normal. According to Justice Oliver Wendell Holmes: "The principle that sustains compulsory vaccination is broad enough to cover cutting the Fallopian tubes. . . . Three generations of imbeciles is enough."[131] Carrie Buck was sterilized. By the mid-thirties some twenty thousand legally performed, enforced sterilizations had been carried out on men and women in the United States alone.[132] Lapouge's role in all this should not be overstated, but neither should it be ignored. There were several routes to this kind of racism and coerced eugenics, but Lapouge certainly invented one of them, and through epistolary and organizational efforts, as well as through his publications, he devoted his life to supporting the doctrine's popularity.

Amid Lapouge's long-lived correspondences there are a great many incidental interactions, many with interested novices and many with ideological opponents.[133] These tended to concentrate on head shapes but sometimes contained striking ideological and cosmological assertions. In 1915 Lapouge sent a long letter to the writer Paul Gaultier at the *Revue bleue*, where Gaultier had recently published an anti-Lapougian article.[134] The letter claimed that the ideas Gaultier "took pleasure in destroying" had the power "to convulse the world such as it has never been before" and would "inevitably supplant the civilization that you love and that I will miss." After several paragraphs of anthroposociology, Lapouge added an unusual coda, asking Gaultier:

> Since God has died, don't you feel at all that something has changed in the world? There always remains enough spiritualism for a doctoral thesis, but now that natural creation is a universally accepted fact, do you think that people are going to continue to believe in the existence of a creator? Now that the lawgiver of Sinai has disappeared, do you hope to keep people in the observance of prescriptions and the defenses that the wisdom of the church had deduced? Not one political principle, not one moral principle, can escape the revision.[135]

One long quotation will round out this portrait of the epistolary Lapouge. The following letter was written to Madison Grant in April 1929.[136] It begins with a page and a half of chitchat about the weather and his health and then mentions his latest work, "Der biologische Ursprung der Ungleichheit der Klassen," published in *Die Sonne*, and discusses the Norwegian translation of Grant's *The Passing of the Great Race*. Next, he praises "a zealous assistant" who reports from California that the Latin Americans are increasingly considered white, "especially the Mexicans."[137] Except in Chile and Argentina, "it is very clear that the European element is going to disappear." The remainder of the letter continues along these lines, touching on most of Lapouge's major themes:

> It's a very good operation to fix quotas of immigrants to be admitted, following the proportions of elements composing the nation around the end of the eighteenth century and not from the period up to the present time. But to avoid the Law of Lapouge and eliminate the dung of the recent immigrations, it is necessary to employ efficacious methods to encourage the descendants of these immigrants to return to Europe. It will require envisaging measures forbidding marriage with Americans of origin and also to preserve the quality of the citizen among the pure Americans. Thus, little by little, it will be possible to determine which foreigners to remove and to decant the American nation. Be particularly wary of the Jews. Germany is nothing but a Jewish state, as is France, under exclusively mammonist politics today. The Jew understands nothing but money; in social life he sees nothing but the economic costs. The

danger that Marxism has let loose on humanity comes because Karl Marx was a Jew. This is not appreciated these days. I knew Marx when I was in Creuse, where he had relatives, and I was at the time struck by the inability of this powerful mind to be interested in anything that couldn't be translated into riches and material pleasures. For him, brains didn't count; he only thought of bellies. That's why Marxism has had so much favor among the Badly-Descended-from-the-Monkeys [*Mal-Descendus-du-Singe*], for whom ambitions are located below the diaphragm, and why the Jewish banker plutocracy finds a way to make a useful alliance with the rebellious elements of civilization.

This particular critique of Marx is obviously a lot easier when one's own belly is full, but the important thing to note here is how adroit Lapouge was at combining two contradictory aspersions against the Jews: bolshevism and capitalism. To continue:

The most horrible corruption is developing in France and Germany, in all the classes and from all points of view. Mammonism devours everything. More and more, everything is for sale: women, judges, functionaries, the parliament, the clergy, the government. An honest man ends up scowled at, not only as an imbecile—an old idea—but as a latent danger for the success of the schemes of others. One after another, huge scandals kept breaking this winter, of which the most resounding was the Hanau affair. From the president of the republic and the directors of the major presses, all the way to the lesser functionaries, everyone is touched, but we have so easily extinguished the affair that apart from a few people of the second rank, no one will be condemned.

The "Hanau affair" refers to the Marthe Hanau financial scandal of 1927–1928 that did, in fact, bring down at least one major newspaper, *Le quotidien*.[138] There was a host of other scandals in these years, extreme enough to jeopardize the republic. Lapouge's comment that everything is "for sale" and that an "honest man" ends up being scowled at are older aspects of his thinking, folded in with these new disappointments in the republic.

You'll tell me that in all times moralists have said this. But the demoralization of Europe by the jolts of the war was so profound that its effects have been mounting instead of fading away, and all the resources of civilized life, after having served in the destruction of men and of things, put their formidable power in service of the destroyers of the civilization. I don't necessarily believe in the definitive triumph of the Badly-Descended-from-the-Monkeys. The entire world hasn't been affected, and even if it were and civilization re-

turned to a state neighboring animality, there would survive in a good number of places residuals of the best lines, to enact the recommencing of evolution described in my "Ursprung der Ungleichheit," and it will happen even
faster because they'll constantly uncover fabricated objects, and maybe in-
scriptions, and books to guide the animators of the new civilization. This is
what I say to those who, especially in Germany, doubt the final end. As long as
the sun shines and the descendants of man aren't entirely lost, evolution will
be able to start again. Whether it takes a hundred or a hundred thousand generations. Don't forget that chronology no longer limits the past and the future
of the earth to that of the incandescence of a cannonball cooling down and that
the study of the decay of radium brings us to counting the past by the hundreds
of millions of years; the future remains just about infinite. Evolutionists always
have to enter the factor of "time" into their calculations, because the duration
around the current time, the duration of the present civilization, counts for infinitely little. It's an idea that still escapes everyone and that must be spread
about because it assures the certainty of all hopes, even those that seem
chimerical today.

Here Lapouge enters a weird kind of thinking that "assures the certainty of all
hopes" but also assures the certainty of all failures. His reference to "a cannonball cooling down" is a particularly keen reflection of the anxieties about time
common to this historical moment. By the late nineteenth century, some religious people still maintained that the world was about six thousand years old—
a number derived by estimating the years described in the Bible. Scientists,
however, generally believed that the earth was millions of years old, following
Lord Kelvin, who had come to that number by estimating the rate of cooling of
the planet. Marie Curie's experiments with radiation led to a very new estimation: now geological indications that the earth was billions of years old could
be taken seriously, since the planet had its own radioactive heat source. Lapouge
was so stunned by the extensive amount of time available that he really defined
even his own project out of existence: the Aryan might triumph, but, with an
almost infinite amount of time, the Aryan might fall again, too. However, true
to form, he managed to leap back into his peculiar teleology:

> Notice that the profound demoralization of the present time is in any case a
> cause of selections that tend to compensate for it. All those deranged men
> and women who frighten us today will leave hardly any descendants, and
> their heredity will be snuffed out. The effect on the passive elements who fol-
> low them will subsist, but the number of factors of decadence will not be
> augmented.

The phenomenon of social disorder is due, above all, to the fact that the civic spirit is not developed among people to the same degree as the other faculties. Yet the base of societies is the spirit of abnegation and the devotion to social utility that neglects the interest of the individual, as one sees among the bees, the ants, and the termites. Individualism and the cult of me are at the base of the family life and social life of men. This is due to the imperfection of evolution and also very much due to Christianity and philosophy. Concern over a future life and the vindication of the rights of man put the interests of the collective into the shadows. One only finds the cult of the other developed in isolated individuals, by instinct, and among the adepts of the Great Consciousness, as with a religious base. The two of these together form only an infinite minority of humanity.

By "Great Consciousness" Lapouge means the composite intelligence that is humanity. Yet it is entirely unclear where the "mind" that Lapouge defended against "bellies" is supposed to find a place in an ant, bee, or termite collective. More important to my analysis are Lapouge's persistent, explicit rejection of Christianity, philosophy, and the republic and his continued delivery of scientific sermons on infinity, eternity, and the paradox of human meaning in such a vast expanse of time and space.

———

THE COMPOSITE EFFECT OF HUNDREDS and hundreds of letters to and from Lapouge is remarkable. Throughout the correspondence, there are many striking references to Hitler, Günther, and the racialist events of the twenties and thirties. The overall picture is that from the 1880s to the 1930s Lapouge trained and maintained an extensive network of disciples and colleagues, all of whom felt they were involved in altering the racial content of the Western world. They saw themselves as having directly affected immigration laws, changed the social and legal status of contraceptives, influenced the rise of enforced sterilization (especially in the United States and Germany), and actively helped to shape Hitler's ideas and put him in power. Except in France, where a concerted effort had been made to discredit him, Lapouge came to be accepted as a respectable representative of science.[139]

Like freethinking anthropology, this branch of the atheist, materialist, anthropological creed was marked by distress, wild proclamations of the future, and the division of the world between a new sacred and a new profane. Lapouge's science was pragmatic in a way, a powerful doctrine of exclusion, quite capable of being co-opted by governments for direct action on the populace. It adds interest and relevance to note that he shared the head measuring with Paul Valéry, shared the deconsecration of sex with Margaret Sanger, and, in his own assessment, counted Adolph Hitler among his disciples. Lapouge broadcast the tenets of his scientific cult with remarkable energy, in every direction, over the course of some sixty years and may have done considerable damage to the world in so doing. Throughout that long campaign, he continued to express awe and anguish regarding the challenges of late-nineteenth-century atheism, and he remained dedicated to a doctrine of antimorality and coleader of an extensive pastoral community that saw him as a scientist and a prophet.

3ᵉ volume. — Nᵒ 128.   10 c.   Un an : 6 fr.

LES HOMMES D'AUJOURD'HUI

DESSINS DE GILL

BUREAUX : 48, RUE MONSIEUR-LE-PRINCE, PARIS

HOVELACQUE

FIGURE I

Several of the freethinking anthropologists had a turn as subject of the biographical broadsheet *Les hommes d'aujourd'hui*, all during the 1880s, around the time that the periodical was edited by Paul Verlaine. Abel Hovelacque is the subject of this one, and he is shown taking over a convent for use as a secular school for boys and girls. Hovelacque was the linguistic specialist of the freethinking anthropologists, generating theories that would strike later historians of archaeology as "amazingly irrelevant," but which were well respected in his own time and helped to create a climate of acceptance for evolutionary theory. Hovelacque was also a successful politician: a socialist deputy and president of the Paris Municipal Council from 1886 to 1887. When he died in 1896, his wife became one of the first women after Royer to become a member of the Société d'anthropologie. She had probably always taken part in some events of the society's calendar but only had to join when Abel's death broke her formal connection to the group. *Source: Les hommes d'aujourd'hui 3, no. 128.*

FIGURE 2

Mathias Duval's issue of *Les hommes d'aujourd'hui*. In 1886 Duval replaced the illustrious Charles Robin in the chair of histology at the Faculty of Medicine, one of the most prestigious posts attained by the original freethinking anthropologists. Note the scientific dictionary, the book on anthropology, the pile of skulls, the jars of fetuses, and the eggs. *Source: Les hommes d'aujourd'hui 6*, no. 273.

FIGURE 3

Clémence Royer on the cover of her issue of *Les hommes d'aujourd'hui*. Darwin's first French translator, Royer was an evangelical freethinker and a controversial and prolific anthropologist. Her opinions were often considerably more deterministic—and racialist—than those of many of her French colleagues. She wrote many books on natural science, economics, and philosophy and an abundance of articles for scientific, leftist-republican, and feminist journals and newspapers. *Source: Les hommes d'aujourd'hui* 4, no. 170.

3e volume. — N° 144.   10 c.   Un an : 6 fr.

LES HOMMES D'AUJOURD'HUI

DESSINS DE DEMARE

BUREAUX : 48, RUE MONSIEUR-LE-PRINCE, PARIS

THULIÉ

FIGURE 4

Henri Thulié's issue of *Les hommes d'aujourd'hui*. He is pictured performing an autopsy. The letter accompanying his Society of Mutual Autopsy testament said that he had sent a will and a description of his mind, but it had been lost, and he couldn't bring himself to write it again. He instead confessed, "I was brazen then and will this time allow my friends to estimate my 'modest personality.'"

Source: *Les hommes d'aujourd'hui* 3, no. 144.

FIGURE 5

On the bottom, from the left: Paul Topinard, Gabriel de Mortillet, Armand de Quatrefages, and Paul Broca. The picture was taken at the Moscow Anthropological Exhibition in 1879. Broca would die in the next year. Later that same year Quatrefages would answer Jules Ferry's query about the Society of Mutual Autopsy, recommending approval, and a decade later Mortillet would have Topinard ousted from his chair at the School of Anthropology. Ernest Hamy, who would become the head of French Ethnology at the Trocadéro, stands at the far right; next to him leans the scientific-popularizer Gustave Le Bon, whose racism and sexism were rejected by the Société d'anthropologie. Le Bon dramatically resigned, several times, in 1888. The other members of the French delegation are Charles Ujfalvy-Huszar (left) and Ernest Chantre (right), standing, and Emile Magitot, seated at the far right. *Source*: Columbia University Library.

FIGURE 6

The brilliant egalitarian anthropologist Léonce Manouvrier (ca. 1915). By this time, he was famous for identifying what he called "prejudices of sex, race, caste, social class, or of profession that unconsciously make up a part of someone's mentality, even a man of science." A younger fellow-traveler of the freethinking anthropologists, Manouvrier was the best anthropologist among them and made important contributions to the way ancient skeletons are interpreted and deciphered. He ran the Laboratory of Anthropology from Broca's death in 1880 to just before his own in 1927. He also taught history of science at the prestigious Collège de France. *Source:* "Discours prononcés aux obsèques de M. L. Manouvrier" (pamphlet), 1927, Library, Muséum d'histoire naturelle.

FIGURE 7

The prolific freethinking anthropologist Charles Letourneau concentrated his efforts on a sociology based on progressive evolution. His books on the evolution of the family, education, passion, women, religion, commerce, and other grand themes all traced humanity's rise from brutes to a state beyond the present level of equality, socialism, feminism, and peace. Letourneau lived from 1829 to 1902 and was probably about seventy here. In 1868 he published *The Physiology of the Passions*, which gave Zola the materialist philosophy that inspired the Rougon-Macquart series of novels. Liza Herzen had killed herself over Letourneau in 1874. Durkheim would publicly dismiss his entire opus in 1900, but he lived his life as an esteemed atheist scientist and as the secretary general of the prestigious Société d'anthropologie de Paris. *Source: Bulletin de la Société d'anthropologie de Paris*, 5th ser., 3 (1902): 171. Columbia University Library.

FIGURE 8

The museum at the freethinking anthropologist's School of Anthropology. The drawing accompanied a youthful article by Jacques Bertillon in *La nature* in 1878; years later he would inaugurate the French birthrate scare and subsequent pronatalist movement. The wall of skulls behind the man in the stovepipe hat probably extended all the way around the room; the collection of skulls was in the thousands. Note the anatomically prepared (partially skinned to muscle and bone) gorilla and the skeleton of a giant. The skeleton of Jacques Bertillon's own father, Louis-Adolphe, would join these others in 1883. The museum of the Society of Mutual Autopsy would grow within a section of the museum pictured above, and the bones and brains of society members from all over France would end up here.

Source: Jacques Bertillon, "Le Musée de l'Ecole d'anthropologie." *La nature* 6 (1878): 39–42.

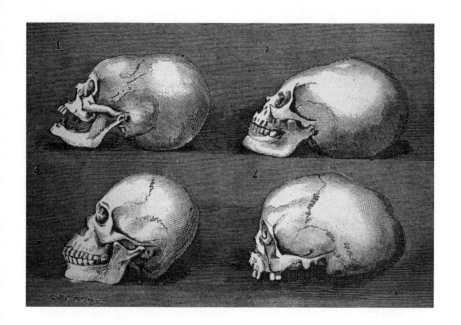

FIGURE 9

Artificially deformed skulls from the collection of the museum of the School of Anthropology. Bertillon explained in his article that Europeans who look down on such things should remember that "in the area of Toulouse it is not rare, still today, to see individuals given skulls elongated like a sausage by their wet nurses." The first of the skulls here is Toulousian, the second Peruvian, and the third and fourth are Mexican. The freethinkers enjoyed bringing such humbling observations to their compatriots. *Source*: Jacques Bertillon, "Le Musée de l'Ecole d'anthropologie." *La nature* 6 (1878): 39–42.

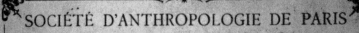

SOCIÉTÉ D'ANTHROPOLOGIE DE PARIS

15, rue de l'École-de-Médecine (École pratique)

CONFÉRENCE ANNUELLE BROCA

Le Lundi 29 Décembre 1890, à 4 h. précises

M. L. MANOUVRIER : LES APTITUDES ET LES ACTES

DANS LEURS RAPPORTS AVEC L'ANATOMIE

RAPPORT SUR LE PRIX BROCA DE 1890

Entrée pour deux personnes

Imp. J. Morien, 47, rue des Sts-Pères

FIGURE 10

The anthropologists printed tickets like this one for their lectures, scientific walking tours, and guided tours of exhibitions or museums. This one is for Manouvrier's address at the annual Broca Conference. Manouvrier published an article of the same title, "Les aptitudes et les actes dans leurs rapports avec la constitution anatomique et avec le milieu extérieur," which was an important essay in the campaign to debunk criminal anthropology. *Source:* Bibliothèque Musée de l'homme, Paris.

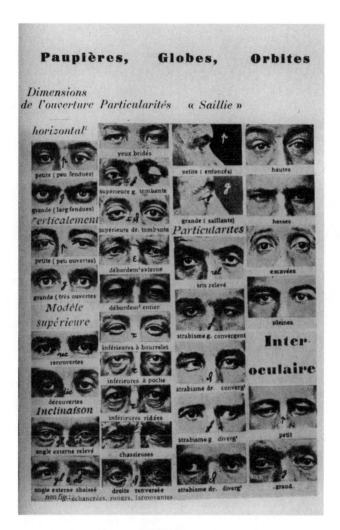

FIGURE 11

This is a page from Alphonse Bertillon's facial-recognition instruction books. Alphonse was the son of the demographer and freethinking anthropologist Louis-Adolphe Bertillon. Alphonse was one of the most important founders of criminal identification techniques, police photography, and modern detective work. He was internationally famous for this work and came to be internationally infamous when his techniques helped indict Dreyfus. Alphonse made the above *portrait parlé*, or speaking portrait, by cutting up photographs, often of himself and his colleagues, in order to train policemen to recognize and name a variety of distinctive facial characteristics. There were numerous pages on ears, naming outer and inner shapes and angles; noses with names for broadness and curve; chins; foreheads; as well as the texture and color of hair; and detailed drawings of the iris, with color charts; and a whole lexicon of descriptive terms. *Source:* C. Sannié et D. Guérin, *Eléments de police scientifique: Instructions signalétiques, l'anthropométrie et les antécédents judiciaires* (Paris: Hermann, 1938).

108

Les circonvolutions sont relativement grosses et assez simples. La deuxième circonvolution frontale gauche est plus large que la droite et subdivisée incomplètement en deux. La troisième de gauche, moins large que la droite mais plus ondulée, n'est pas subdivisée. La première circonvolution temporale gauche

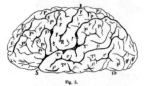

Fig. 5.

est plus large et plus flexueuse que la droite. » (Voy. séance de la *Société d'anthropologie* du 3 mai 1883.)

4° Léon GAMBETTA, homme politique, mort, le 31 décembre 1882, d'une pérityphlite, à l'âge de quarante-trois ans.

« Gambetta fut avant tout et par-dessus tout un orateur. Dans l'ensemble des hautes et brillantes facultés dont il était doué, celle qui dominait en lui et planait pour ainsi dire au-dessus de toutes les autres, instrument merveilleux et incomparable de son action et de sa puissance, c'était la faculté de la parole » (*Un cerveau illustre*, par le D' Laborde. Inédit).

Autopsie par le docteur Fieuzal et le docteur Laborde, membre de l'Académie de médecine, directeur des travaux physiologiques de la Faculté de médecine, vice-président de la *Société d'anthropologie*. — Description morphologique des hémisphères par MM. Mathias Duval et Chudzinski. — Moulage du cerveau par M. Chudzinski; de la cavité crânienne par M. Talrick.

109

« Le lobe occipital est extrèmement réduit, notamment du côté droit (fig. 6). Le lobule quadrilatère droit est très compliqué; il est divisé en deux parties par un sillon qui part de la scissure occipitale; de ces deux parties, l'inférieure est

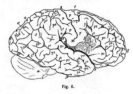

Fig. 6.

subdivisée en plusieurs méandres, par la présence d'une incisure à branches multiples disposées en étoile. On constate un développement extrème de la troisième circonvolution fron-

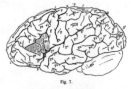

Fig. 7.

tale gauche, tel que jusqu'ici il n'en a été signalé de semblable (fig. 7), on peut même la distinguer sur le moulage de la cavité crânienne. Tout le monde sait que cette circonvolution est le siège de l'aptitude pour le langage articulé. D'une ma-

FIGURE 12

In 1889 the freethinking anthropologists published a book describing their projects for the World's Fair held in Paris that year. The first sentence of the book quoted Broca declaring anthropology to be "an essentially French science." In the section on the "Society of Autopsy," the anthropologists provided illustrations and discussions of six brains, those of Assézat, Asseline, Coudereau, Gambetta, Adolphe Bertillon, and Gillet-Vital. The text on Asseline mentioned that he was "an assertive combatant; all his life he fought for the destruction of error and superstition and for the propagation of the republican idea with all its consequences" (105). Here we see Coudereau's brain on the left and two drawings of Gambetta's brain on the right. *Source: La Société, l'Ecole, et le laboratoire d'Anthropologie de Paris à l'Exposition Universelle de 1889* (Paris: Imprimeries reunites, 1889): 108–109.

nombreux, c'est qu'il s'y trouve compris des crânes de femmes indigènes. Quand les Germains ne se trouvent pas superposés à des éléments de ce genre, l'indice oscille autour de 74 ou 75 et le type est franchement *Europæus*.

examen plus approfondi montre que ces sujets, fréquents dans certains cimetières de France, très rares en Belgique et dans la région du Rhin, n'ont aucun rapport généalogique avec

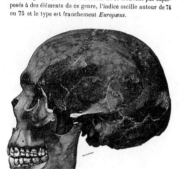

Fig. 26. — *H. Europæus.* Type germain.

Fig. 27. — *H. Europæus.* Type germain.

Les Francs présentent souvent une anomalie. La face est trop courte, le nez trop large et pas assez haut, indice 51 au lieu de 46. Le faciès a des analogies avec le *spelæus* pour quiconque ne connaît pas ce dernier d'une manière exacte. Un

spelæus. Les caractères particuliers de ce dernier, surtout ceux de l'apophyse orbitaire externe et de la base de l'occipital sont absents. La première est arrondie, au lieu d'être droite, la seconde est renflée au lieu d'être aplatie. Ces sujets anormaux

FIGURE 13

Georges Vacher de Lapouge studied with the freethinking anthropologists at the Paris School of Anthropology that they founded. He shared their atheism and their convictions about the downright religious importance of anthropology but broke with them when he began to see the world in racialistic and pessimistic terms. He believed there were two races in the world, and in France: one long-headed and superior, the other round-headed and mediocre—and then there were the Jews, against whom he grew increasingly vitriolic. Lapouge famously said the next century would bring the racialist extermination of millions. Here we see the long-headed "Germanic-type," the description of which Lapouge augmented with notes from anthropologists Broca, Hovelacque, Hamy, and others, as well as quotations from the ancients, including Horace, Juvenal, and Galen. *Source*: Lapouge, *L'Aryen: Son role social* (Paris: Thorin et Fils, 1899): 326–327.

FIGURE 14

In a sense the present study is on the posthumous life of Gambetta's brain. This is an image of a reading party at the home of Alphonse Daudet, April 11, 1882. That is Gambetta standing, smoking a cigar; he would die December 31 of that year. The round-faced man with beard and glasses, up front, is Zola, who had by then published many of his Rougon-Macquart novels. Next to him, in profile, is Charcot, who would start his Tuesday lessons that November. At far right, leaning head on hand, is Edmond de Goncourt. Edouard Drumont, not yet an infamous anti-Semite and champion of Lapouge, was also there and later published this image as a depiction of his enemies. The party was hearing a theatrical version of Alphonse Daudet's novel *Les rois en exile*. Drumont tells us Gambetta could not sit still and stood smoking the whole evening, yet he was attentive to the play's derisive treatment of "descendents of august families that had reigned over Europe" (564–565). He also mentioned that Gambetta's brain "only weighed 1,160 grams" (569). Drumont said Charcot "walked around the salon full of light and flowers," with a face of marble, "pensive and welcoming despite the sarcastic contraction of his lips"; he also referred to him as one of the "new laic confessors" (565). Léon Daudet was probably there as well. The woman playing with her fan is Charcot's wife, the artist Augustine Charcot, née Durvis. *Source*: Edouard Drumont, *La France Juive*, 2 vols. (Paris: Marpon et Flammarion, 1886).

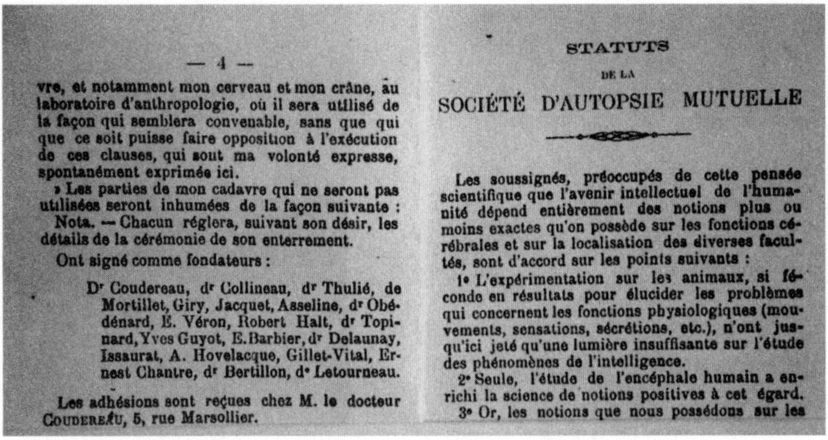

— 4 —

vre, et notamment mon cerveau et mon crâne, au laboratoire d'anthropologie, où il sera utilisé de la façon qui semblera convenable, sans que qui que ce soit puisse faire opposition à l'exécution de ces clauses, qui sont ma volonté expresse, spontanément exprimée ici.

» Les parties de mon cadavre qui ne seront pas utilisées seront inhumées de la façon suivante : Nota. — Chacun réglera, suivant son désir, les détails de la cérémonie de son enterrement.

Ont signé comme fondateurs :

D' Coudereau, d' Collineau, d' Thulié, de Mortillet, Giry, Jacquet, Asseline, d' Obédénard, E. Véron, Robert Halt, d' Topinard, Yves Guyot, E. Barbier, d' Delaunay, Issaurat, A. Hovelacque, Gillet-Vital, Ernest Chantre, d' Bertillon, d' Letourneau.

Les adhésions sont reçues chez M. le docteur COUDEREAU, 5, rue Marsollier.

STATUTS

DE LA

SOCIÉTÉ D'AUTOPSIE MUTUELLE

Les soussignés, préoccupés de cette pensée scientifique que l'avenir intellectuel de l'humanité dépend entièrement des notions plus ou moins exactes qu'on possède sur les fonctions cérébrales et sur la localisation des diverses facultés, sont d'accord sur les points suivants :

1° L'expérimentation sur les animaux, si féconde en résultats pour élucider les problèmes qui concernent les fonctions physiologiques (mouvements, sensations, sécrétions, etc.), n'ont jusqu'ici jeté qu'une lumière insuffisante sur l'étude des phénomènes de l'intelligence.

2° Seule, l'étude de l'encéphale humain a enrichi la science de notions positives à cet égard.

3° Or, les notions que nous possédons sur les

FIGURE 15

This was the standard Society of Mutual Autopsy testament, and it proclaimed that the signer was preoccupied with several "scientific thought[s]." Most people added liberally to it: the testament of Jeanne Véron, author of several educational books on natural history and the wife of anthropologist Eugène Véron, began, "Desiring to continue to work towards the diffusion of scientific ideas which I have professed for the last thirty years, I leave my body, and especially my brain, to the Society of Anthropology in Paris." It was accompanied by other letters and instructions, embellishing on her rejection of Catholicism and her devotion to science. Véron left behind much more evidence of herself than did the other freethinking anthropologists' wives, though many of them seem to have followed the Society's proceedings and otherwise shared its concerns and activities. Her paean to Clémence Royer, sent to the Society of Anthropology when Véron heard of her death, made it clear that for some of these women, at least, Royer was a real hero. *Source:* Extract of *Bien public*, October 24, 1876.

∞

CHAPTER SIX

Body and Soul:
Léonce Manouvrier and the Disappearing Numbers

"The theologians had asked whether woman had a soul. The scientist went further, several centuries later, and refused her a human intelligence."[1] Léonce Manouvrier wrote these words in 1903 in an article on "the anthropology of the sexes and its social applications." By then, Manouvrier had become a sort of a policeman for the entire discipline, reining in its excesses and publicly denouncing its more vicious doctrines. He did not campaign against the wild anthropological claims of his freethinking colleagues at the institute, presumably because their political beliefs were emancipatory and egalitarian. But a number of prejudicial, even violent anthropological theories had originated outside this group (they came from the past, from abroad, and from the provinces) and were growing in popularity. Broca's so-called proof of the inferiority of women was revived by a number of theorists, the Italian anthropologist Cesare Lombroso argued that criminal behavior was hereditary and that criminals could be identified by physical markers, and Georges Vacher de Lapouge argued the superiority of the Aryan race. All these theories were based on body measurements. They were repeated in popular literature and seemed to have been accepted as truth by many. These doctrines captured such attention because secularization had left an authority gap in the public world of French society and in the personal lives of some individual French men and women. As Manouvrier noted, it had been up to the theologians to determine whether women had souls, what kind of souls these were, and what that meant about the proper role of women in society. Religion had also been an authority on crime and sin, mercy and punishment, and here, too, its influence had been discredited. With the soul gone, the body became ever more important.

The hyperbolic claims of an intellectual discipline are often only limited through interaction with outside forces. Real reform often requires some sort of dramatic attack from outside, even if insiders later report that they knew change was necessary. Such an attack might come because the ambitions of one discipline encroach on the territory of another. Paradigms may also shift under the weight of political change. Theories may deflate under the attacks of individual victims, especially if a lawsuit is involved. Sometimes a journalist, primarily interested in a good story, can throw a stable doctrine into question or into chaos. It is more striking when a systematic discipline manages to police its own excesses from within. Yet that is what happened in nineteenth-century physical anthropology, through the careful and often brilliant work of Léonce Manouvrier. It is of particular interest because, in critiquing the excesses of physical anthropology, Manouvrier raised questions about the discipline as a whole and found himself struggling to defend his career and convictions. This important anthropologist has received little attention from historians—especially in comparison to scholarship about his mentor, Paul Broca.[2] Fame earned fighting against wild new ideas does not always last as long as fame earned proposing them.

Over a period of thirty years, Léonce Manouvrier developed an organized, and increasingly nonnumerical, attack on all anthropological systems of human-group inequality. This attack is surprising because it came at a time when certain human groups (women, criminals, and various races) were generally considered to be innately deficient in intelligence or character. The fact that the attack was progressively independent of numerical proofs is equally surprising, because it was initiated at a time when the world of French professional science was entranced by numbers. Historians have posited a number of ways to understand this.[3] First, establishing truth is difficult in a democracy, trickier still in a Godless world. Measurements gave experts a new authority, advancing their claim to objectivity and truth. Statistics made it possible to predict human-group behavior to a surprising degree, and many assumed that those who could predict could also manipulate. Second, social hierarchy was seriously threatened by the establishment of a government based on secular, democratic, egalitarian ideals. Without the authority of God and his representatives, and no longer able to use traditional abuse as a justification of continuing abuse, new proofs of women's natural inferiority served to preserve the social order. Third, the new, secular nation-state was conceived as a community but was too large to have many community features. Numbers offered a way to visualize the whole and to create noncorporeal subgroups: one could feel part of several national trends without ever seeing more than a few other trend members. Fourth, authorities trying to manage the new national community could speak to each other without worrying that local assumptions were skewing their descriptions. When ideas were translated into numbers, scientists, too, could speak to each other across great distances. Finally, the individuals doing the measuring may have had as much need to bolster confidence in their own abilities as the most suspicious audience. Numbers were comforting. Any system based on measurement was intrasubjectively verifiable; within the systems, some things were true, and some things were false. This created a very stable world of meaning, so long as one kept one's head in the system. But the thirst for valid claims was intermingled with semiconscious desires to keep the world the same and to keep personal privileges. Once the scientist had designed the conceptual framework, he or she could feel absolved of responsibility for the numerical results that were generated and the conclusions they supported. It is stunning, then, how Manouvrier was able to see past the compelling fallacy that equated quantification with objectivity and truth.

Manouvrier worked from within Broca's anthropological institute in Paris, holding a chair at its Ecole d'anthropologie, where he instructed and inspired a number of important future scholars, including the American anthropologist Aleš Hrdlička. Manouvrier also published extensively in the journals of the Société d'anthropologie, served in key administrative roles, represented France at a plethora of international conferences, and almost single-handedly ran the society's laboratory. From this base, he slowly demolished a fundamental notion on which the anthropology of his time was based: that a person's body could be "read" for information about that individual's abilities and characteristics.

In France, this biological determinism was tempered by a commitment to Lamarckian evolutionary theory, which allowed for the improvement of human beings through the improvement of environment. But so long as this improvement was conceived as biological (and generally very gradual), the idea that it was heritable offered hope only to future generations. This provided a strong argument for schooling the underclasses and cleaning the cities, but it had its limitations. Contemporary mature individuals could be measured, belittled, and classified as innately inferior in intellect or morality. Even deeply progressive branches of the eugenics movement tended to describe biologically defined human groups in a hierarchy of quality, while they discussed the malleability of each group's abilities over (generational) time.[4] There were many variations on such theories. Manouvrier stood apart from the whole debate, boldly declaring the at-birth potential of human groups (socially or biologically defined) to be essentially equal. As time went on, he further came to believe that levels of intelligence and character could not be read from the shape of a skull, weight of a brain, or any other biological datum.

Historians who have examined the theories of biological determinism of this period have consistently noted the remarkable nature of Manouvrier's critique. None, however, has taken Manouvrier as a central subject or given any attention to the scope of his work. As a result, he appears as a politically isolated figure, and the significance of his larger project is lost in seemingly discrete critiques regarding the nature of gender, criminality, or race. In truth, however, he was very isolated in his rejection of biological determinism but not in his politics. The freethinking anthropologists with whom he worked were more than willing to invent scientific theories to support their socialism, feminism, and atheism. Thus there was a lot of room for Manouvrier to develop and publish his far more empirically grounded studies. No doubt, his work also helped the freethinkers to believe that good science was indeed on their side. Further, Manouvrier's critiques were by no means discrete gestures. Rather, they formed a many-pointed attack on the wild claims of anthropology, all aspects of which cohered around his increasing suspicion that the quality of a human being could not be quantified.

MANOUVRIER AMONG THE FREETHINKERS

Léonce Manouvrier was born in 1850 in Guéret. Thus, when the future freethinking anthropologists started publishing *Libre pensée* in 1866, Manouvrier was still in school, and when the empire fell, he was twenty. Clearly, his life was not marked by the constraints of the empire, but, perhaps more important, he had entirely missed the democratic tragedy of the Second Republic: when the members of the freethinking group were about twenty, they were triumphant and very active revolutionaries; a few years later, their cause was lost and stayed lost for twenty years. By contrast, at twenty, Manouvrier joined up to fight in the Franco-Prussian War. France suffered a humbling defeat, but this loss led to the foundation of a republic—delicate, unsure of itself, and, after a few years of struggle, dedicated to secularism and democracy. The fiercely oppositional stance of the freethinking group did not make the same kind of sense for Manouvrier. Closer to home, there is some useful personal evidence. Manouvrier was the son of a doctor—the most left-leaning and anticlerical of professions. He studied classics in his home town until the war and afterward took his degree at the Paris Faculty of Medicine, receiving a *lauréat du prix de thèse*. At twenty-eight he was a leftist, egalitarian doctor in Paris, and

Paul Broca, by now extraordinarily famous, recruited him for the Laboratory of Anthropology. He never left. I mean this in several senses: he worked there until his death in 1926, but also, as all who knew him would attest, until his rather late marriage changed his routine he almost never left the lab. His manner seems to have been relaxed but neat and straightforward. He wore a Vandyke—gray in the photographs that have survived.

Manouvrier trained directly under Broca, but the tutelage was short-lived: Broca died only three years after Manouvrier came to work with him. No one anthropologist was considered capable of taking on the many roles that Broca had played in the anthropology institute, so after his death a general reassignment of roles took place. Jean-Baptiste Vincent Laborde was made director of the laboratory, but he took no interest in his new position: he was a doctor and medical researcher of some renown, an activist for a variety of causes (he was a major voice against absinthe, for example), and a member of the Academy of Medicine.[5] Thus Manouvrier, remaining a mere *preparateur* in name, became the functional head of the lab. He would eventually get the title as well.[6] So there he was, a young man in Paris, with a good job, conceptually supported by the immense prestige of his mentor but not held back by that mentor's actual presence.

In the 1880s the freethinking anthropologists took control of the society and the school, and Manouvrier seems to have gotten along extremely well with this older crowd. In this same decade, he came to rethink a number of Broca's conclusions, and to question the validity of his general methodology. Broca's craniometry was a system of quantifiable biological determinism that, as was well argued by Stephen J. Gould, appeared to Broca and many of his contemporaries as an empirical, numerical, and thus "pure" science.[7] But, as Gould has demonstrated, the ideologies of these scientists guided their measurements. Skull volumes were not easily measured, and when results clashed with accepted notions of gender or racial superiority, the numbers were either considered false and hence rejected, or theories were altered to change the meaning of the offending measurement. This subterfuge was generally unintentional, fueled by assumptions rather than by an articulated agenda. Manouvrier pointedly identified the preconceptions that guided many anthropometric studies of difference. It is difficult to know how he managed this, but the institute's archives have him requesting the purchase of books by such nondeterminist philosophers, sociologists, and psychologists as Alfred Fouillée, Jean-Marie Guyau, and René Worms; library records show that he borrowed these books as well.[8] He began his critique of biological determinism before the philosophers, sociologists, and psychologists had begun their dramatic shift to indeterminism, but they were clearly an influence. Manouvrier participated in the increasingly articulated suspicion that even for atheists the human world was not utterly material and was not ever going to find its Newtonian laws.

Manouvrier never protested against the freethinking anthropologists' use of anthropology. He certainly never sided with Topinard when the latter complained about their anticlerical, political anthropology. Though he kept some distance—he was not a very social person, and, in any case, the freethinking group was about thirty years his senior—Manouvrier cooperated with them extensively, contributing to their myriad publications, including the *Dictionnaire des sciences anthropologiques*, and performing many of the postmortem operations for the Society of Mutual Autopsy. As noted above, he also represented this and other freethinking projects to the world and made them look good. The freethinking anthropologists repeatedly elected him to the society's most important position, secretary general (held only by Broca and Letourneau before him), and enthusiastically supported his career at the school and the laboratory.[9] If the freethinkers were more adamant in their calls for political change than he was, Manouvrier was considerably more adamant than they in proclaiming that social hierarchies had no biological component. The freethinking anthropologists were egalitarian, yet they always assumed that intelligence and character had a direct correlation with somatic sites in the brain; they seemed unaware that this reification and quantification of intelligence and character had the potential of being used against them in antiegalitarian proposals. Their primary concern was always anticlerical and antisacerdotal, and they clung to materialist determinism because it was a description of human thought that negated the conceptual need for soul and spirit. It does not seem to have occurred to them that one could conceive of a nonspiritual mind-body indeterminacy. Manouvrier, by contrast, came to envision a nonspiritual but indeterminable human mind that had some innate aptitudes and tendencies but was effectively unpredictable. Thus, in critiquing a theory of human limitation, Manouvrier went further than the freethinkers, but it was their bombastic anthropological politics that had created the comfortable climate in which he worked.

This chapter will explore three of Manouvrier's great antideterminist campaigns and his attempt to redefine anthropology in the wake of his own critique. It will then briefly return to the question of Manouvrier's intellectual and political environment through a discussion of his collaboration with the great neurologist and psychiatrist Jean-Martin Charcot, famed as Sigmund Freud's mentor on hysteria and hypnotism. Examining the work of Charcot and his associates will also demonstrate that the enthusiastic deconsecrating work engaged in by the freethinking anthropologists was taking place in other sciences as well.

MANOUVRIER AND THE QUESTION
OF WOMAN'S INTELLIGENCE

One of Broca's craniometric conclusions was that women had lighter brains than men and were therefore less intelligent. He also showed that Germans were less intelligent than were the French. In order to arrive at this second conclusion, it was necessary to argue that Germans had proportionally lighter brains, for without factoring in a larger body mass for Germans, the brain-weighing test came out quite unfavorably for the French.[10] But, Broca argued, a direct comparison of German and French brains would not be equitable. Owing to the larger amounts of food and beer consumed by Germans ("even in regions where wine is made"), Broca believed that Germans were much bigger than Frenchmen—"so much so that their relations of brain size to total mass, far from being superior to ours, appears to me, on the contrary, to be inferior" (466). No such proportional considerations, however, were made in tabulating the brain-weight relationship between men and women. Broca recognized this discrepancy, musing that perhaps "the small size of the female brain depends exclusively on the small size of her body." "But," he cautioned, "we must not forget that women are, on the average, a little less intelligent than men, a difference we should not exaggerate that is, nonetheless, real. We are therefore permitted to suppose that the relatively small size of the female brain depends partly on her physical inferiority and partly on her intellectual inferiority" (153). Broca's data were widely cited, becoming a staple of books and articles dealing with the "woman question" and often producing even harsher interpretations than Broca intended.[11] The archconservative Gustave Le Bon took Broca's notion to great extremes, arguing that even "among Parisians, there are a large number of women whose brains are closer in size to those of gorillas than to the most developed of men's brains. This inferiority is so obvious that no one can contest it for a moment; only its degree is worth discussion."[12] In 1881 Manouvrier published the first of his many studies challenging these conclusions.[13] In that early article, published only one year after Broca's death, Manouvrier was quite faithful to Broca's style and general method. He set up graphs and comparative lists of data and did not question the overall anthropometric project. He did, however, announce very different conclusions than had Broca. Women's brains, Manouvrier demonstrated, were proportionately heavier than were men's.

A few years later, Manouvrier once again took aim at the issue of brain weight and group inferiority, and again he fought in the quantifying terms of the physical anthropologist. In an 1883 article, he reviewed and refuted the conclusions of the "illustrious" anatomist, Louis-Pierre Gratiolet, on the relative weight and proportions of women's and men's brains.[14] Gratiolet, reported Manouvrier, had announced that women's brains were inferior. Although this assertion "rested on no numbers whatsoever" (694), continued Manouvrier, Gratiolet's opinion had nevertheless become classic. Manouvrier then listed nineteen procedures that he had carried out comparing skulls according to sex, each of which involved measurements on anywhere from fifty to two hundred skulls. Manouvrier did not hide the fact that he performed these procedures in order to assure himself of something that past laboratory experience had led him to suspect: that "the proportional weight of the brain of woman is very much higher than the weight of the brain of man" (695). In light of this, his conclusions were rather tame. He made no overall statement about the respective intellectual capacities of men and women, and neither did he announce that their skulls were perfectly equal. Instead, he listed five specific conclusions regarding details of skull comparison; sometimes the men slightly exceeded the women, sometimes the women slightly exceeded the men.

Manouvrier presented his findings as if he were correcting a mere mistaken notion of anthropological science and made no political comment as to the significance of the mistake. In the following year, however, he published an article in the *Revue scientifique* entitled "Variétés: L'internat en médecine des femmes," which argued that women ought to be welcomed into the medical profession.[15] Women had been allowed to attend medical classes as externs and, obviously, to do various types of work in hospitals, but they had not been allowed to join the ranks of the interns studying to become doctors. In the 1880s they were petitioning for this right. An article in the *Revue scientifique* had recently argued against their cause, and Manouvrier took up his pen in their defense. Modestly, and somewhat teasingly, he claimed that it required much more cleverness to argue against the education of women than for it (592). In part 1 of his article, he asserted the feasibility of the women's demands. He emphasized that all the women wanted was the right to take the internship exam. His opponent had seen this and not made intelligence the center of his argument against the women. Wrote Manouvrier, "My wise adversary has too much wit not to recognize the wit of women, and he is not one of those men who seems to have had the misfortune of spending their lives among inferior women" (593). But what of the other necessary attributes: self-possession, sangfroid, good judgment? To this Manouvrier said that since it was true that the exam could not gauge such things, the only way to know was to let the women try. Still, he supposed that since these young women generally surmounted immense obstacles in order to take the exam, one might accept this strength of will as a measure of their character. In any case, these attributes are "pretty rare among the male interns." Put the women to work, "and you will see that these qualities, which are not measured by the size of one's biceps, and which female externs . . . have already begun to prove, really exist in women to the desired degree." As for women's horror of blood, it just did not exist, and in any case women were already working in hospitals. To those convinced that women would be disgusted or upset by some medical procedures, Manouvrier wrote, "Have you the right to block these distinguished women on the pretext of a sensitivity that has not heretofore upset them more than us? In the hospital, as in war, there are two types of courageous people: those who are not moved, and those who are very sensitive but who go to the front anyway." If the women were more sensitive, he concluded, they deserved to be honored for their courage. Professors of obstetrics often had to spend considerable effort instilling courage in their young male students. As for women: "Who has not had the occasion, in so many circumstances, of being astonished by their sangfroid and their presence of mind" in difficult situations? A common late-nineteenth-century argument against women working in responsible positions was that their menstrual periods would disrupt their jobs and exhaust their health. Manouvrier referred to this delicately, writing that some insist on the "physiological subjection that monumentally—or periodically—alters and enervates the resources of the female organism." But, except in "pathological cases," this just was not a real problem. In any case, wrote Manouvrier, "how come no one ever invoked this reason to keep women from the most punishing and constraining professions? They can be washerwomen, ironers, girls of the baths or the restaurants, workers without a day of rest, and yet people are afraid they can't support the quite moderate fatigues of service as an intern! . . . The female externs know the chores of the internship, and, more than we, they must know if they are capable of acquitting themselves" (593–594). Manouvrier charmingly addressed the "bad mood" and the tiredness also associated with menstruation, arguing, "Really, if it was necessary never to be in a bad mood, never tired, discontent, crabby even, in order to fill the duties of an intern, what man would merit being one?" He particularly condemned a whole class of surly chiefs of staff.

What is more, he suggested that if one were to be equally careful about the "interns of the stronger sex," one would have to be concerned whether "certain material appetites were not impressing too much power over their brains, and whether they are ceding too much to the continual instances of the little god, whether they are not too often complaining of Venus. How many morning and evening visits, how many hours on watch have been disrupted by these preoccupations of an inferior origin, upon which I will keep myself from further insisting!" In any case, he concluded, the work gets done, and when one is suffering it simply gets done with more courage. On the insistence of women's modesty, Manouvrier countered with the modesty of countless female patients and suggested that, especially in hospitals for women and children, it would make sense for a few "interns in skirts" to join "the monopoly of young men." Still, he was careful to add that completely sex-specific doctoring was "neither necessary nor desirable." He invited his readers to compare their rules of "sex" with the rules of "class or caste." Once, "one had to be a 'noble' to become an officer. Many disqualifications have disappeared this century. Many more will disappear in the time to come" (594–595).

In the second part of the article, Manouvrier discussed the societal aspects of the matter. "It's not the first time that people have said, apropos of women who believe they can use their intelligence in other ways than making soup . . . Great gods! What will become of the family if the wife takes the place of the husband! The husband will then have to take the place of the wife and this all over the globe." But, Manouvrier soothed, that is not what is going on. Rather, "a few intelligent women" (595) are trying to do honorable work, a comment he offset with the observation that there "probably will never be enough female interns to take care of all of the sick of their sex" (596). And of this small group, some were primarily working in order to catch a husband. "I should say buy a husband, because husbands are expensive these days." His point was that men from the working class could improve their lot, and make better matches, by passing the internship exam and becoming doctors, but women from the same class were stuck. They would become old maids without any utility, and society would lose an enormous wealth of faculties. "That's what our masculine egoism brings us to." Breaking away from ridiculous "theological doctrines," modern society had begun to allow women to learn and to work. This infuriated the priests, but when the women wanted equal schooling or equal work, the secular element became worried as well. "The priest laughs" and says I told you so, but "it will not be the priest who laughs last." Manouvrier gave considerable time to comforting his reader: he "preferred a gradual evolution to a radical revolution" and attested that "more than one doctoresse . . . has said 'yes' in front of M. the Mayor; more than one has given her breast to an infant" (595). (Note the delighted reference to civil marriage.) Several passages of this 1889 speech deserve extended quotation, for they went far beyond many claims made even by nominal feminists in this period. Furthermore, Manouvrier hid neither his anticlericalism nor his socialist concern for the downtrodden.

> How many social miseries, how many plagues considered indestructible would be ameliorated, maybe even suppressed, the day when all of a half of

human intelligence receives the whole culture with the attributes that go with it. The logic, they say, if one pushes it to the extreme will be women electors and women deputies. Not yet, because logic doesn't suffice: one has to have good reasons. There are numerous bad male electors! Electors who vote knowing neither who nor why; others who don't want to vote . . . (this said without attacking the principle of universal suffrage). (596)

———

IF OUR CONSCIENCE COULD ILLUMINATE all the dark corners of our brain, all the obscurities of our *sentiment*, we would find there, perhaps, in that which concerns the present question, traces of the "male arrogance" that grew out of a muscular superiority of which a rooster can glorify itself but that a civilized man would hardly dare to assert. We would discover, perhaps, also in ourselves traces of other instincts not less egotistical, which render us so jealous of our most unjust privileges, so resistant to all novelty that disrupts our habits, indeed, all the inferior sentiments bequeathed to the present from the past.

(596–597)

———

I SAY TO MESSIEURS THE INTERNS: accept with a good heart into your ranks these young women who ask nothing more than to prove their merit and justify their aptitudes. (597)

Augusta Marie Klumpke was the first woman intern in France. She had entered the school despite the resistance of the dean of the faculty; in 1882 she won an externship but could not go further. In 1887 the anthropologists' friend Paul Bert stepped in. He was then minister of public instruction, and Augusta became the first woman in France to be appointed *interne des hôpitaux*. She earned her degree in 1889 and did important work: in World War I and after she was a pioneer in the treatment of wounds of the nervous system—horribly plentiful among the soldiers—and especially of the spinal cord; the syndrome Dejerine-Klumpke's Paralysis still bears her (married) name.

Manouvrier did not write his defense of women interns for an anthropological audience and did not choose to use his anthropological status in his argument. As time went on, though, he came to combine his outspoken feminism with his anthropological findings. In 1889 the second French Conference on the Rights of Women was held at the Universal Exposition in Paris. The year was the centennial anniversary of the French Revolution, and innumerable groups chose the moment to define themselves in relation to the event. The Conference on the Rights of Women was largely made up of women, though many male politicians were in attendance. Among the few men actively participating was Manouvrier, who had been invited to speak on the anthropological side of the question of women's rights and abilities.[16] His essay, "Indications anatomiques et physiologiques relatives aux attributions naturelles de la femme," seemed a bit out of place to him, stuck as it was in the section on "history." The other options, however, were "economics," "morality," and "legislation" and seemed no more appropriate. While appreciative that he had been invited to speak on questions anthropological in nature, Manouvrier bemoaned the absence of a section specifically dedicated to this theme. He was certainly not the only conference participant who focused on the question. Indeed, ideas about the biological basis for equality took a more or less predominant place in many of the more than fifty speeches delivered. It was as if people wanted to talk about other things but could not get past this issue.

In his speech, Manouvrier addressed some of the anthropological conceptions of women that had been proposed at the conference. Most specifically, Manouvrier addressed the thesis of Madame Conta, *doctoresse en médecine*. Women, Conta had argued, were intellectually inferior because they had been forced into a social role that did not challenge and foster their intellects. Therefore, while men's intellect evolved and progressed, women's stagnated and even regressed.[17] Manouvrier answered this by protesting, first, that there was no evidence that women were less intelligent than were men; second, by arguing that there was no reason to believe, in the vast majority of cases, that women's work demanded less intellectual rigor than did men's (in fact, he recognized that "women's work" was entirely culturally defined, differing in every people); and, third, that even in those cases—in the upper classes—where women did receive a considerably less intensive education and public role than their male counterparts, it was absurd to think that acquired intelligence would be inherited in a sexually specific way, that is, that men would bequeath their intellects to their sons while women would define the intellects of daughters.

The larger aim of Manouvrier's essay was to report the findings of several years of studying the differences between men and women. His conclusions were that the comparison of the sexes must be made according to three criteria: nutrition, movement, and intelligence. Women, he concluded, were superior in their nutritive capabilities, which meant that they were capable of feeding unborn and then born children. Men were superior in movement— they had more advanced musculature, which enabled them to provide for and protect the women while the latter were exercising their nutritive abilities. Neither of these superiorities, Manouvrier stressed, was a cause for pride. Finally, he addressed the question of intelligence, reporting that here, in this most important of criteria, men and women were precisely equal. "Woman," he wrote, "has a smaller brain in an absolute comparison, but she has a larger brain relative to body mass. This is just what it would have to be if one were to suppose that the intelligence of men and women was equal or, in other words, that intelligence does not have a sex."[18]

Manouvrier received hearty applause for such conclusions. Although he was, to some degree, preaching to the choir, his discourse was punctuated with angry references to those "conférenciers" who endorsed a more equivocal, conservative interpretation of equality.[19] The deputy Beauquier may well have inspired Manouvrier's ire when he assured women that the legislature, by providing schooling for women, had done what they could. It was for women to do the rest, he explained, adding: "We will be incapable of maintaining serious obstacles to your emancipation if you continue to succeed in letters, arts, and sciences" (226). This essay was one of the few offered at the conference that spoke of women's weakness, ignorance, and "supposed lack of rationality." The deputy was not alone in asking that women be patient in their demands for equality. M. le docteur Verrier wrote (in a classic argument to which he grafted the suggestion of biological evolution) that "eventually, through the slow and natural evolution of humanity, when, thanks to a more complete and rational education, religious myths have less control over them," women will have political rights (40). For now, however, they ought at least to be given full civil and commercial rights, asserted Verrier. In his speech, Jules Allix demanded total legal and political equality but repeatedly insisted that the justification for men's rights lay in their superior intelligence, while the justification for women's rights lay in their superior spirituality (146–163). These positions were relatively progressive and egalitarian for the day, but all fell short of the egalitarian position held by Manouvrier. In his own words: "There is thus nothing to do but laugh at these classic citations—said to be so damning for women—that certain conférenciers draw from their old books with the confidence and serenity that hides incompetence." He added that the so-called scientific opinion that women were less intelligent than men should be placed beside the opinion of "certain theologians who attribute to women a soul intermediately between that of man and that of the animals."[20]

A few years later, Manouvrier would articulate an even more fully developed position on the question of the anthropology of sex, demonstrating an astute understanding of the relationship between preconceived notions and scientific observation. In an article of 1903, he wrote that the first scientific investigators of the differences between the sexes "were no less impregnated with prejudice than were their contemporaries," and when attempting to make objective scientific observations, these scientists were unable to see past their cultural and social assumptions. Furthermore, Manouvrier recognized that this led to much more than a mere repetition of beliefs about inferiority and superiority; these beliefs and assumptions had been transformed by their expression in numbers and in the scientific idiom in general. "The rejuvenation of old prejudices," he wrote, "through numbers and technical expressions is, in effect, what is revealed by the dissection of these theories."[21] He also recognized that the translation of old prejudices into scientific terminology was often unconscious, and he posited a wide range of social criteria that might influence an observer's vision: "There are prejudices of sex, race, caste, social class, or of profession that unconsciously make up a part of someone's mentality, even a man of science, and it is not easy to erase their influence on one's thoughts, even when positive facts have demonstrated that they are there. The prejudice of sex is without a doubt largely present in certain scientific formulas on the subject of the intellectual inferiority of women and also present in the success that these studies have attained" (405–406).

Manouvrier recognized the extraordinary social power of scientific language, and he sympathized deeply with its victims: "Women have exhibited their drawings and their diplomas. They have also invoked the philosophical authorities. But opposing them were *numbers* that neither Condorcet, nor Stuart Mill, nor Emile de Girardin had known. These numbers fell on the poor women like the blows of a sledgehammer, accompanied with commentaries and sarcasms more ferocious than the most misogynist imprecations of certain church fathers" (406).[22] He recognized that even beyond the unfair numbers derived by anthropometry, the language of evolution and of anthropology in general had provided so-called experts with a new compendium of insults.

> The theologians had asked whether woman had a soul. The scientist went further, several centuries later, and refused her a human intelligence. One can read in the works of the most esteemed scientists the opinion that women have been left behind in human evolution in the most important ways (notably the cerebral development and diverse morphological characters of the skull or the members), so that, relatively speaking, she has hardly advanced from the state of anthropoid. It is not an exaggeration to consider such observations as the result of a prejudice—even of an inflamed prejudice—because they were not expressed with the circumspection that their authors themselves consider as necessary for all other questions. (406)

Despite these strong concerns about sexist anthropology, Manouvrier did believe that science should help guide the social relations between the sexes. Indeed, he believed that such scientific guidance was crucial to the survival of society. Though he thought them equal in abilities, he did not think all men and women, across the board, should share duties. A society where men and women competed equally for work outside the home, while children were attended to by child-care institutions, struck Manouvrier as equally unsatisfactory for men, women, children, and society at large, and he suggested that only the pressures of "pathological social necessities" (413) could bring about such a society. It should perhaps be remembered here that Letourneau had imagined a very happy version of such a society, peopled with engaged and fulfilled men and women and children liberated from the tyranny of the nuclear family. Manouvrier thought everyone should be allowed to do what they wanted, but that what they wanted was not as radical a change as some seemed to fear. In preparation for the future, he relied on science despite his suspicion of its power and felt quite certain that if science did not actively guide a significant change in the role of women in society, the consequences would be disastrous. He hypothesized that if the "battle between the sexes" continued to escalate "there will surely be a revolution, and one might well ask oneself if this could happen without being fatal to the society that was its theater." Fearing patriarchal backlash as much as feminist violence, he believed that "a few scientific provisions" could greatly diminish the extent of the coming crisis (422). It is interesting to note how often early social scientists hypothesized about coming revolutions of various types. Apparently, late-nineteenth-century Europeans had seen so much social, political, economic, and technological change that they could not help projecting similar upheavals into the future (they were not far off, but that does not mean we may take the fact of their predictions for granted). These changes were specifically imagined as the result of scientific modernity and the end of the soul, both of which seemed to promise revolutionary new information, new responsibilities, and new freedoms.

By the early 1900s Manouvrier's pronouncements on the capacities of the female mind were quoted by feminists with approval and appreciation. The relationship between biology and intelligence was at the center of a range of questions about women's place in society and was frequently handled through the invocation of his studies. For example, Maria Montessori relied heavily on his work. Famous as an activist for the reform of primary education, Montessori also lectured on anthropology at the University of Rome and wrote the influential *Pedagogical Anthropology*.[23] She also seems to have been known as a freethinker in some circles: her name graced a day in the *Almanach de la libre pensée*.[24] In *Pedagogical Anthropology*, Montessori extensively cited Manouvrier against Broca's thesis of women's intellectual inferiority. She interpreted Manouvrier's work in her own way, however. Having found that women's brains were proportionately heavier than were men's, Manouvrier argued that brain weight was inconsequential and that there was no evidence that women were intellectually inferior to men. Montessori, citing Manouvrier as "one of the most gifted anthropologists of our day," took his measurements as proof of women's "anthropological superiority."[25] Celebrating his conclusion that "the cerebral volume of woman is superior to that of man!" Montessori added her own notion that woman's cranium was "more perfected . . . inasmuch as woman has an absolutely erect forehead" (258). It was her contention that men were considered "socially superior" because strength had long reigned in human affairs. Now that technology privileged intelligence over physical might, she posited, "perhaps . . . the reign of women is approaching, when the enigma of her anthropological superiority will be deciphered" (259).

There were more faithful interpretations of Manouvrier's work. Charles Turgeon, for example, devoted the entire chapter on intelligence in his important book Le féminisme français to a summary of Manouvrier's work on the subject.[26] Turgeon, a professor in the law faculty of the Université de Rennes, strongly objected to biological determinism, believing that women's minds should be judged not on their size but on their abilities. Still, he felt it necessary to counter accusations that woman's intellect was biologically inferior to man's, and he did so by reporting Manouvrier's findings that brain size and weight must be considered in relation to the overall size and weight of the subject. Paraphrasing one of Manouvrier's metaphors, Turgeon asked, "The largest brain is that of the whale: must we suspect the great beast of genius? No: the size of the brain is not, in and of itself, a sign of intellectual superiority" (131). Turgeon would likely have had little trouble convincing his colleagues at the law faculty that intelligence should not be established through biology: such establishments tended to promulgate the classicist approach to questions of intelligence, free will, and moral responsibility. But recognizing the attention that such arguments could command, Turgeon took care to introduce Manouvrier as "a true scholar" and to quote some of his more technical findings. Most important, Turgeon cited Manouvrier's explanation that any fair study of the human brain would demonstrate the equality of the sexes and that it was purely anthropologists' "prejudice of sex" that had led to other conclusions in the past (134).

The fact that a late-nineteenth-century male scientist could come to such a conclusion has been treated, by modern historians, as profoundly exceptional. In Cynthia Eagle Russett's 1989 study, Sexual Science, Manouvrier is mentioned as "the one member of Broca's school of French anthropology who rejected belief in the inferiority of women."[27] Of course, Russett's sense of the exceptional nature of Manouvrier's work is well founded, and it was by no means Russett's project to study Manouvrier; neither, for that matter, was she particularly concerned with the French. In order, however, to understand Manouvrier's critique and, indeed, French anthropology of the period, it is crucial to know that Manouvrier was associated with a profoundly left-wing, egalitarian scientific community in which proclamations of gender equality were frequent, if occasionally ambivalent. It is also crucial to understand that Manouvrier's critique of biological determinism with regard to women was part of a broader attack that he waged against biological determinism in general.

MANOUVRIER AND CRIMINAL ANTHROPOLOGY

Criminal anthropology, the study of the biology of criminals, was founded on the principle that this biology is significantly different than that of other members of society. The most influential proponent of this school was the Italian physician Cesare Lombroso, who in 1870 experienced a "flash of inspiration" that led him to believe that modern-day criminals were in fact atavistic throwbacks to prehistoric or even prehuman hominids. Lombroso saw such atavism in a host of attributes, ranging from slight anatomical anomalies (the most important of which were the short brow and jutting jaw) to a penchant for tattoos. Some details were said to correlate with particular crimes. For example, tattoos of two clasped hands indicated pederasty. A large space separating a woman's big toe from the next toe indicated a tendency toward prostitution. An aquiline face, pointed ears, bushy eyebrows, and a sharp nose were all signs of criminality and were used as such in popular depictions of malefactors—including that of Bram Stoker's *Dracula* (1897). Beyond almost verbatim Lombrosian descriptions, Stoker's text was quite specific: "The Count is a criminal and of criminal type. [Max] Nordau and Lombroso would so classify him, and qua criminal he is of imperfectly formed mind."[28] Pointed ears and other secondary attributes, suggested Lombroso, might help society identify criminals at an early age and keep them from doing harm. Unlike "classical or metaphysical criminologists," to use the Lombrosian characterization, criminal anthropologists claimed to represent the applied version of a pure science.[29]

Lombroso's theory and practical suggestions were extremely attractive, both because they offered the possibility of stopping crime before it happened and because they took the notion of sin out of criminology: criminal anthropology was the deconsecration of evil. For several decades after the first publication of these theories, they met with much acceptance. As is well demonstrated in Stephen J. Gould's *The Mismeasure of Man* and Robert Nye's *Crime, Madness, and Politics in Modern France*, Lombroso's theory had considerable effect on legal proceedings.[30] Contemporary discussions of hereditary criminality were quite common; what Lombroso had done was to add a specific scientific theory to a generally accepted—or, at least, a widely entertained—notion. Lombroso's theory profoundly influenced thought on crime and punishment, provoked numerous penal reforms, and was used to convict many suspects. The theory helped to establish criminal anthropology as an independent discipline. At the center of this discipline were international conferences, held every four years for several decades (up to World War I), where criminal anthropologists, general anthropologists, jurists, judges, and government officials had an opportunity to share and debate the discipline's latest developments. Over time, however, these conferences also provided the scene for the demise of Lombroso's criminal anthropology and, to some degree, criminal anthropology in general. The argument against Lombroso's determinism was formulated and championed in France, and, to a great extent, Manouvrier originated the secular biological indeterminism that came to be known as the "French School."[31]

Since jurists and philosophers had an intellectual and economic stake in the notion of free will, it is not surprising to find that they generated a critique of criminal anthropology. In modern histories of the period, however, it is difficult to perceive a context for an antideterminist anthropologist. As a result, Manouvrier has been seen as a strange and virtuous anomaly. Gould describes him as "the non-deterministic black sheep of Broca's fold, and a fine statistician," and Nye refers to him as "the unrelenting Manouvrier, who . . . was emerging as the most pugnaciously articulate spokesman for the French," and yet neither says much more about him.[32] His political attitude makes more sense once we know he was part of a group that, from Broca on, hand-picked fiercely egalitarian and secular people to be the officers of their society. The indeterminism that he joined to this political and religious position was more his own but reflected changes in French philosophy of which we know him to have been aware (and which I will discuss in depth in the next chapter).

Manouvrier's first major work devoted to discrediting Lombrosian criminal anthropology was his report to the Congress of Criminal Anthropology of 1889, which, like the Conference on the Rights of Women, was held as a section of the Universal Exposition of Paris of that year. The congress was arranged around thirteen essays written especially for the event by some of the most prominent specialists in questions of crime, law, and social theory in France and Italy.[33] There were representatives from many other nations present, but the congress was without question an informal debate between the Italian school of criminology (which held that criminals were biologically determined) and the French school (which countered strict biological determinism with a variety of propositions centering on either free will or social determinism). The conference was opened with discussions of the first two essays, which were written by the energetic "leaders" of the two camps: Lombroso and Manouvrier.

Lombroso's essay briefly reiterated his general position and added new, rather wonderful developments in criminal typing, ranging from the discovery of a distinct system of hand gestures innately employed by criminals to findings of a higher level of acidity in criminal urine (25–27). Manouvrier's reply pulled no punches. He referred to Lombrosian criminal anthropology as no more than a rejuvenated phrenology that, though it had been thoroughly discredited by rigorous scientific critique, was now being revived because of the addition of numerical measurements and the scientific language of evolution. Manouvrier conceded that a significant relationship between mind and anatomy was denied by no one but "ignorant metaphysicians," but he protested that this relationship in no way entailed specific correspondence between certain acts and certain physical anomalies (28). This is significant because Manouvrier wanted to keep criminology secular, and he did not want anthropology to be locked out by his own antideterminist argument. Nevertheless, his enemy of the moment was a criminal anthropology that he saw as painfully simplistic, unscientific, and dangerous. "Crime," he wrote, "is a sociological matter; it is not a physiological matter" (29).

In a metaphor that he would employ many times in his future writings, Manouvrier expressed his understanding of the inheritance of moral characteristics in the following way: A human being, he asserted, is like a musical instrument. It has qualities of its own, but the music that emanates from it is greatly dependent on the player, to which he likened the environment.[34] According to Manouvrier, many people have trouble understanding the effect of the environment because they interpret it too narrowly.[35] The environmental forces that influence us are not confined to such large factors as poverty—which could not explain why one brother turns to crime while the other stays honest—but to minutia on the order of passing acquaintances, the fact of being older or younger, and even the effect of being handsome or ugly. Manouvrier pointed out that one could make no definite value judgment on a trait that assists someone in becoming a criminal. He was not speaking here of atavistic signs but rather of qualities such as strength and courage, which, though generally considered to be good, are often necessary for a life of crime. Manouvrier also argued that Lombroso's criminal group represented those people who broke the law and got caught. What then, he asked, of all those comfortable members of the bourgeoisie who commit small illegal or merely immoral acts? Are they, too, atavistic throwbacks to a preevolutionary state?

Manouvrier's ideas received attention and respect because of his status as a physical scientist, but it was these kinds of thought-provoking, conceptual arguments that dominated his critique of criminal anthropology. He challenged

Lombroso's scientific methods in similar ways. This critique was perhaps most severe in his statement that if one is to define an anatomical characteristic correctly as criminal, it must be a characteristic that is found exclusively in criminals. He asserted that "there probably is not even one anthropologist who believes in the existence of such a characteristic" (32). He also berated Lombroso for not testing a control group of honest people. Perhaps most damning, he asserted that Lombroso's huge collection of criminal anomalies simply did not constitute a scientific group: "One criminal is plagocephalic, another's arms are too long, another has a *fossette vermienne*; this is not the way one establishes a scientific type" (34).

Most French participants of the congress sided with Manouvrier. Indeed, despite the wide range of concerns represented by the French jurists, anthropologists, doctors, and philosophers present, the general debate over biological versus social explanations of criminality was often envisioned and described as the "Lombroso-Manouvrier duel" or "the tournament between Manouvrier and Lombroso" (417, 419). The congress was also discussed in these terms in English-language anthropological journals.[36]

Still, other French positions were raised. Among the more prominent French thinkers in attendance at the congress was Gabriel Tarde, and his position differed from Manouvrier's in important ways. A jurist by profession, Tarde had written several works on criminology and statistics and would, one year after this congress, publish his masterwork, *Les lois d'imitation*, which would constitute the greatest French rival to the Durkheimian school of sociology. Tarde's contribution to the Second Congress on Criminal Anthropology was significant, though not well appreciated. This was largely because his overall analysis was seen by many of the members as too metaphysical, and, when he did apply himself to the specific question of heritable criminal characteristics, he often expressed views that contained more biological determinism than those of Manouvrier.[37] In response to Manouvrier's essay, in which the anthropologist firmly denied any evidence of a relationship between criminal tendencies and specific morphological characteristics, Tarde took a rather equivocal stance, agreeing there were no incontestable, exact traits that mark criminals, "but that should not stop us from affirming that there are organic predispositions to crime. . . . No trait has an absolute meaning, but collectively certain traits could have significance. For the rest I lean toward the preponderance of social forces that push someone into crime" (199).[38] Manouvrier's rebuttal was prefaced with a show of respect, because Tarde was "one of the too rare French magistrates who understand the necessity of basing the principles of law on things other than metaphysics" (368). Beyond this, however, it was extremely critical, systematically dismissing each of Tarde's "positive" proofs. That Manouvrier went further than Tarde in his critique of biological criminology seems to have been generally accepted. Indeed, the congress's secretary general commented on Manouvrier's reaction to Tarde by noting that it was not that the anthropologist objected to the intrusion of a philosopher but rather because according to Manouvrier "the views of M. Tarde [were] not justifiable scientifically" (433).

Another important French voice at the congress was that of Clémence Royer. Again, Royer's presence was surprising because she was a woman in an almost entirely male domain, but as usual she was equally shocking because of her unconventional views. Her belief in transformism allowed for mutability in the species, but she generally saw individuals as captives of their hereditary needs and abilities. There had been an element of Social Darwinism in her preface to the French translation of *Origin of the Species*, and this strain was evident in much of her other work. In contrast to both Tarde and Manouvrier, Royer's theories of criminality were extremely determinist. All human actions, she asserted, were entirely determined by biology and milieu. Indeed, according to Royer, "Human beings are no more responsible for their virtues than they are for their vices" (357). Royer did not, however, agree with the Italian school because, though she believed in total determinism, she did not believe in predeterminism, that is, she did not believe that the influences that would create a person's actions could be known or correctly understood in advance. A human being, she wrote, is not analogous to a clock, the position of whose hands one could predict for any given moment, but rather to a small ship being tossed on the ocean, that is, totally without responsibility for its movements and entirely unpredictable (358).

Royer was also notable for being the only member of the congress to suggest that crime was a result of the heritable effects of racial mixing, a theory that provided an equally deterministic alternative to Lombroso's idea of atavism (170–172). With this in mind, Royer suggested, a general study of the genealogy of criminals should be performed alongside the current studies on atavistic traits. The suggestion was received by Lombroso with great appreciation (195). Few of her compatriots, however, paid much attention to Royer's stark materialism. Rather, Manouvrier's conceptual and technical critique dominated the French response to Lombroso at the Paris congress.[39] The defeat of determinist criminal anthropology was so marked that Lombroso and the members of his school angrily boycotted the following Congress on Criminal Anthropology that was held in Brussels in 1892. Manouvrier, however, was there. Once again, he took a central role in the conference, reiterating and clarifying his belief in the almost limitless power of the environment to shape individuals and his steady faith that anthropology would and should help unravel the mysteries of human life.[40]

Manouvrier's opinion dominated the French scene enough to surprise an American onlooker. It may be recalled that the journalist Ida Tarbell mentioned only three authors whose works she saw on Bertillon's bookshelf, and Lombroso was one of them (the English anthropologist John Lubbock and Francis Galton, founder of eugenics, were the others). Why did Alphonse Bertillon never try to become predictive with his measurements? The answer is that Manouvrier had won this battle in France in the very years that Bertillon's studio was getting under way. It is worth returning to Tarbell's 1893 interview with Bertillon, because she wondered the same thing:

> "But your archives, M. Bertillon?" I asked. "Are you not going to use your observations for purely scientific deductions, for anthropological conclusions, as, for instance, to establish a criminal type?"
>
> "Undoubtedly," he responded, "the statistics of the service will be used more and more for ethnographical and anthropological statistics. I have already done something with them. Here is a chart showing the color of the eyes in the different parts of France, from the maroon of the Spanish border to the blue of the Channel; and there is another, giving the relative length and breadth of the head. As for the criminal type, that is a delicate question."
>
> "Then you have never sought to confirm the doctrine of Lombroso's school, that certain anatomical characteristics indicate the criminal?"
>
> "No; I do not feel convinced that it is the lack of symmetry in the visage, or the size of the orbit, or the shape of the jaw that make a man an evildoer. A certain characteristic may incapacitate him for fulfilling his duties, thus thrusting him down in the struggle for life, and he becomes a criminal because he is down. Lombroso, for example, might say that, since there is a spot on the eye of the majority of criminals, therefore the spot on the eye indicates a tendency to crime; not at all. The spot is a sign of defective vision, and the man who does not see well is a poorer workman than he who has a strong, keen eyesight. He falls behind in his trade, loses heart, takes to bad ways, and turns up in the criminal ranks. It was not the spot on his eye that made him a criminal; it only prevented his having an equal chance with his comrades. The same thing is true of other so-called criminal signs. One needs to exercise great discretion in making anthropological deductions. Nevertheless, there is no doubt but that our archives have much to tell on all questions of criminal anthropology."[41]

This is Manouvrier's argument. He was very cautious, but he still wanted anthropology to explain the human experience. Despite his outrage at "pseudo-

science" practiced in the name of anthropology, Manouvrier consistently rejected "metaphysics" and defended anthropology as one of the primary disciplines through which human society should be studied.

On this issue, Manouvrier's chief opponent was Paul Topinard. Having
failed to attend the meetings of the Second Congress on Criminal Anthropology, Topinard made an appearance on the last day of the congress and delivered
a paper denying the existence of any relationship between anthropology and
the study of criminals.[42] Topinard had struggled against the rising leadership
of the freethinking anthropologists at the Société d'anthropologie. He lost and
was ousted from his chair at the Ecole d'anthropologie, but he did not leave
the society, settling in instead to war against their wildly political version of
the science. This guided his position at the conference, where Topinard argued
that anthropology should be strictly confined to the study of the human races
and completely devoid of a political agenda or, indeed, any application to the
workings of society. In his estimation, anthropology was "the zoology of man,"
and he asserted that according to Broca "anthropology . . . studies men as
a naturalist studies animals" (490). He claimed that Broca allowed no infiltration of ethnography, sociology, or psychology into the science of anthropology
(491). "In anthropology," wrote Topinard, "one must separate pure truth from
its applications to medicine, social economy, politics, and religion" (492).

Manouvrier reacted strongly to such delineations. In his opinion, neither
biology nor sociology, taken in isolation, was sufficient to understand and
guide human beings. Rather, he wrote, "what is required is an education that
is both biological and sociological, which is to say, anthropological."[43] Anthropology had to guide society because so many errors were being committed in
its name. Though he wrote comparatively little about it, Manouvrier did believe that anthropologists would eventually be able to identify physiological
characteristics that denoted specific temperaments and aptitudes. These somatic characteristics, "totally innocent in and of themselves," might, when
mixed with wretched conditions, make someone prone to crime, "for it is exclusively the conditions of the milieu that dictate the mode of utilization and,
as such, the moral or social value of organic aptitudes" (457). Not to recognize
the importance of the milieu as the decisive factor was, in Manouvrier's eyes,
"a flagrant abuse of the notion of heredity" and an "abusive use of the theory of
transformism" (458). As to the relationship between anthropology and law,
Manouvrier considered it to be extremely important. He consistently reminded his readers of the "Lombrosian mistake" made in the name of criminal
anthropology but held that this error had at least brought the concept of a socially active anthropology to the attention of the general public. His idea was
to change the name of criminal anthropology to "judicial anthropology" in
order more clearly to define its goals, which were, in essence, to advise mag-

istrates on such matters as "general information on crime, its mode of perpetration, the condition of the victim, the existence of mental trouble in the accused, the influence that could have created this trouble in the criminal's character, and on his responsibility"(264). That is, Manouvrier wanted anthropology to do the work that psychology and sociology successfully appropriated in the twentieth century.

Manouvrier frequently found it necessary to negotiate a delicate balance between rejecting particular ideas within anthropology and yet defending anthropology as a whole. In the opening of his essay "La genèse normale du crime" (the keynote speech of an important conference on evolution), he carefully explained that a negative critique of evolutionary-based criminal anthropology was by no means a negative critique of evolution.[44] Along similar lines, in this same essay, Manouvrier rejected a compliment he had received from a religious journal: he had been congratulated for disproving the theory of atavism (422–423). In truth, Manouvrier had only attacked atavism as an explanation for criminality, and in his response he insisted that atavism might indeed be a useful theory in certain instances, even if it were totally without value in certain other kinds of analyses. We are reminded of how difficult it was to reject a particular scientific theory without allowing the argument to be co-opted by opponents on the religious side of the debate. In each case where someone suggested a forceful and pragmatic way for anthropology to intervene in society, Manouvrier could not help noticing some major fallacy. He thus had to fight against the fame garnered by his own burgeoning science, while supplying erudite ammunition for rival ideologies. To make matters worse, he was trying to argue a negative—trying to prove that biology was not the root of crime. In so doing, he gradually stopped referencing numbers, charts, and measurements, backing away from the whole notion of understanding human beings by the centimeter. Instead, he labored to disclose the contradictions and absurdities of Lombrosianism (and similar theories) while championing an argument for the immense power of the milieu (see, e.g., 422 and 454).

Manouvrier bisected crime into two components: the ability to be violent and the decision to break the law. He then showed that violence was by no means confined to the criminal—as demonstrated, he wrote, by any meat butcher—and that breaking the law was by no means confined to violence. What then was in the special biology of a criminal? Infants may suckle by instinct, but, Manouvrier asserted, the notion that they might commit forgery because of their inborn makeup is an entirely different thesis (427). If we are to invoke atavism to explain violent crime, we should also do so to explain the Parisian gentlemen's passion for the hunt (439). Thus would be born the science of "L'homme chasseur" alongside Lombroso's "L'uomo delinquente."

Manouvrier pushed the absurdity of this to its limits with suggestions of the "born bourgeois," "born bicyclist," "born landlord," and the "born rotisserie chef.""Intellectual errors," wrote Manouvrier, "are, in effect, like crimes: they do not need atavism, or immediate heredity, or even tradition for them to be repeated. Causes of error or causes of crime, their sources are far from being exhausted. They are always abundant" (456).

MANOUVRIER AND SCIENTIFIC RACISM

The third of Manouvrier's great battles was waged against the theories put forward by Georges Vacher de Lapouge. As I discussed in the previous chapter, Lapouge described the history of civilization in terms of two unequal races that he had essentially invented: long heads and round heads. Lapouge believed that the proper balance of power between these races had been disrupted by democracy and capitalism, and he devised a eugenical scheme for shaping the future of humankind. His belief that Jews were a decadent, venal strain of the superior race was often spoken of as the first proponent of "scientific" anti-Semitism. Removed from their proper station, the dolichos could not hold out against the swarms of brachies and were intermixing and losing their biological distinction. Meanwhile, the brachies were dragging the country into decline through their mismanagement and ignorance of racial laws. Eventually, the Jews would take over.

While Lapouge's proposed remedies were generally based on guiding reproduction rather than on purposefully murdering the inferior, this latter idea was also represented in his writing. Yet despite the extreme nature of his vision, Lapouge found considerable support in France during the first half of his career (roughly until 1900). Even figures strongly committed to fostering social morality and cohesion gave him a hearing. As noted, Louis Liard, the director of higher education and rector of the Sorbonne, secured for Lapouge a salary for his courses at the University of Montpellier. Emile Durkheim included a section on anthroposociology in his journal *L'année sociologique*, and in 1897 the journal ran an extremely positive review of Lapouge's first book, *Sélections sociales*. Durkheim did not like anthroposociology but reasoned that his journal ought to put forward all points of view regarding sociology. He also argued that while the main goal of an endeavor might be futile, interesting "secondary points" might come of its practice.[45] For these reasons, he wrote, he included a section on anthroposociology in his journal and, in order to give a fair representation of this school of thought, he assigned its coverage to one of its partisans: Henri Muffang, the lycée German teacher who was an energetic disciple of Lapouge and the French translator of Otto Ammon. Aside from his review of *Sélections sociales*, in 1897 Muffang wrote several articles that drew on the work of Lapouge. All were more paraphrases than analyses, and he treated their conclusions as established fact. This is not surprising, as Muffang sent all his articles to Lapouge for "correction" and suggestions before they were sent to Durkheim for publication. Durkheim's close collaborator, Célestin Bouglé, was well aware of this, and Durkheim certainly knew as well.[46] For both professional and ideological reasons, Durkheim was against the reading of sociology as a mere epiphenomenon of biology, but the authority of science and its numbers was difficult to discount entirely without the testament of another scientist. Indeed, widespread critique of Lapouge was not to be found until his theories were attacked from within the anthropological community—by Léonce Manouvrier.

According to a letter to Lapouge written by his disciple Henri Muffang, Gustave Rouanet, the socialist deputy for the Seine (from 1893 to 1914), had approached Manouvrier asking him to write a definitive, scientific rejection of the theories of Vacher de Lapouge. Muffang reported to Lapouge, "I know that it was requested by a deputy, Rouanet, who was worried about the spread of your ideas and of the arguments they furnish to anti-Semites and nationalists and thought that a definitive refutation was indispensable."[47] Manouvrier was already well known as an antideterminist, so it makes sense that Rouanet approached him for this. He may have even heard about Manouvrier's anger over Lapouge: as early as 1887, just after Lapouge's first anthroposociological article was published, Topinard had reported to Lapouge that this work had been met with great hostility among the Paris anthropologists and that "Manouvrier, in particular, is fuming."[48] Manouvrier's response to Rouanet was "L'indice céphalique et la pseudo-sociologie" (1899), a pointed attack on Lapouge which defended anthropology and sociology against anthroposociology.[49] Manouvrier began by saying that he found the preoccupations of anthroposociologists to be bordering on the comical. And yet, wrote Manouvrier, ignorance concerning anthropology among otherwise well-educated people had brought anthroposociology considerable "literary success . . . at least thirty books devoted to anthroposociology were published in 1896 alone" (249). People had also been seeking him out in the Laboratory of Anthropology, anxiously requesting that he measure their skulls—as if the secret of their lives lay in the number.

The cephalic index, Manouvrier repeatedly insisted, has absolutely no relationship to moral or intellectual characteristics. His argument was that a person's cephalic index is comparable in meaning to his or her hat size; that blondes and brunettes are fools and sages, cowards and heroes, villains and saints; and that there is sufficient variation in any given group or race to explain the appearance of any trait. What angered him in Lapouge's racial theories was of a piece with what angered him about claims of women's intellectual inferiority or the notion of the born criminal: that in each case an old myth—generally one that was particularly attractive to a specific group of people—was rejuvenated and fortified through its expression in the authoritative style and language of science. Manouvrier was both defending his science, anthropology, from being associated with what he saw as an extremely simplistic theory and practice and attempting to stay what he perceived as a growing threat. The index was, in his terms, nothing more than the latest lay anthropological craze—following the tradition of Gall's phrenology and Camper's facial index. But he felt that the anthroposociological craze was growing so steadily that "soon it will be impossible to convince the public that blond dolichocephalics are not of a superior essence, destined to govern the brachycephalics" (252).

Manouvrier believed that the stakes were high, that what he saw as a rather silly idea was gaining ground, and that its proponents were intent on restructuring society according to racial categories. Recognizing that numbers were the convincing and authoritative aspect of most "pseudoscientific" exercises, Manouvrier nevertheless chose not to fight numbers with numbers but to demonstrate the arbitrary nature of the categories. He sought to show that the numbers assigned to dolicho- and brachycephalism were "pure conventions" and that there was an infinite selection of physiological indexes that might be calculated. Some of these indexes might be informative to the student of the human body, he quipped, but none of them would be informative to the student of the human mind (254–255). Having requested that Manouvrier write "L'indice céphalique," Rouanet then wrote a review of it for *La petite république* in which he no more than briefly introduced the ideas of *L'Aryen* before turning to quote and paraphrase Manouvrier for the rest of the article. "M. Manouvrier," he wrote, "does not leave one stone of this grotesque edifice in place. . . . He proclaims in a loud voice, and from his mouth the argument is peremptory, that there has not been established any direct correlation between the shape of the skull and a corresponding mentality."[50]

Future reviewers of *L'Aryen* took Manouvrier's criticism to heart. The archaeologist Salomon Reinach's review in *Revue critique d'histoire et de littérature* is exemplary in this regard, especially because he, like so many others, gave Lapouge a lengthy hearing before dismissing him with a quotation from Manouvrier.[51] Reinach wrote that Lapouge had demonstrated a surprisingly sophisticated education and erudition and stressed that "not only has he read a lot, and read very good things—written in French and in other languages—

but he is remarkably informed on contemporary discoveries and hypothesis; he knows how to go back to Greek and Latin sources and to cite them pertinently and very correctly" (124). However, Reinach concluded, Lapouge was still wrong. How to prove it? The best Reinach could do on his own was to ridicule Lapouge's style ("trenchant affirmations, framing prophecies") and politics, strenuously pointing out that anthroposociology was a "fatalist and materialist thesis where no place is left for education, assimilation, or all of the invisible agents of moral progress." Here, materialism was placed in opposition to "invisible agents of moral progress." God was kept out of it, but indeterminism was being revived. Still, Godless indeterminism was not yet fully capable of the job at hand: for a solid refutation of Lapouge, Reinach cited Manouvrier. "If some historian," he wrote, "were tempted to take seriously the fundamental thesis of M. Lapouge, it would be necessary to recommend to him the excellent articles wherein M. Manouvrier has demonstrated their futility." Following this statement, Reinach quoted Manouvrier extensively.

It is extraordinary that Reinach, and so many other learned people, did not feel fully justified in rejecting Lapouge's work on their own. After citing several choice statements by Manouvrier that accused anthroposociology of being a pseudoscience, Reinach admitted, "I prefer to have this said to an anthropologist by an anthropologist; those who have never measured skulls must abstain from making a personal judgment." The comment reveals a lot about scientific belief. Reinach returned to his literary critique, writing that while great thinkers are modest, "M. Lapouge does not have their scruples; he asserts, he vaticinates, he suspects the honesty of his adversaries." All this reminded Reinach, interestingly enough, of the politically slanted scholarship of Gabriel de Mortillet, the left-wing egalitarian anthropologist—a fact that gives witness to Reinach's evenhandedness (125).

As for Durkheim, when Manouvrier's critique appeared, he used it as justification to cut the anthroposociology section from his L'année sociologique, and its erstwhile author, Muffang, was no longer included in the production of the journal. The first issue that did not include the section instead ran back-to-back reviews of Manouvrier's article and Lapouge's latest book, L'Aryen. Not surprisingly, Manouvrier was seen as defending sociology against impostor sociologists, and his work was lauded. L'Aryen, on the other hand, was strongly censured. Even here, Lapouge was given credit for an "incontestable erudition," but his theories were summarily dismissed through the frequent evocation of "the magisterial critique by M. Manouvrier" (145). In a 1900 essay "La sociologie en France," Durkheim dismissed anthroposociology from the history of the progress of sociology because "its goal is to submerge this science in anthropology." But again Durkheim was more comfortable with a scientific refutation. "Furthermore," he added, "the scientific bases on which this system rests are very suspect, as M. Manouvrier has recently demonstrated."[52]

Muffang knew about the critique in *L'année sociologique* before it came out and wrote to warn Lapouge.

> I must tell you that there will be mention of *L'Aryen* in *L'année sociologique*. I believe that I've already told you that Durkheim informed me that the anthroposociology portion of the journal has been discontinued because the publisher does not consider it to be sociology. But, in reality, it has been transferred to another editor, H. Huberd. This guy wrote to me that he delivered a review of *L'Aryen* "giving very clearly his opinion," which announced a thorough, unmerciful thrashing. He said that it was only because I had met him in Paris that he took the trouble to warn me before I read the thing in *L'année*. In a sense, I think that it will be better for you to be attacked by the Durkheimists than praised by them.[53]

Lapouge was largely discredited in France during his lifetime, but as I have discussed, his work still had considerable influence in the twentieth century, especially in Nazi Germany and Vichy France.

That anthroposociology resurfaced in this way lends further significance to the fact that it was scientifically dismissed earlier in the century—by one anthropologist. As Muffang wrote to Lapouge in 1900: "Look at the two articles by Durkheim in the *Revue bleue* that discuss sociology in the nineteenth century. You will see the execution, in one paragraph, of anthroposociology. As always, it is to the article by Manouvrier that they refer."[54]

MANOUVRIER'S REDEFINITION OF ANTHROPOLOGY

Manouvrier was more significant in his role as critic of anthropological theories than he was as creator of anthropological theories. He did, however, have a strong conception of what the science of anthropology ought to be, and it was from this vantage point that he was able to limit the influence of the reductionist, determinist anthropological theories discussed above. Manouvrier's notion of anthropology stood somewhat precariously between two extremes. Objecting to false anthropological panaceas, Manouvrier fought against claims that one or another anthropological process would revolutionize the manipulation of human behavior. But he fought with equal conviction against notions of anthropology that refused to include the study of social interactions and denied the possibility that anthropology might create for itself an involved, advisory role in society. Significantly, he did not keep his negative remarks within the community of anthropologists. In fact, he directly addressed sociologists (a group that can be seen as competing with anthropology for the role of social expert) in René Worms's prominent *Revue internationale de sociologie*. In one article, he went so far as to declare that "about three quarters of current enterprising anthropological research serves no purpose other than to augment the already enormous stockpile of useless numbers."[55] Yet in this same article he argued that legal theory and practice required a scientific orientation and that anthropology was the science that was "precisely appropriate" to the task (241).

To make law without science was to abandon the fate of human beings to haphazard traditions, accidents of history, and the metaphysics of religion. Manouvrier cited French civil law on the rights of women: a woman's salary was not her own in the name of the law, it was the right of the husband to sell his wife's property against her will if he so desired, and so on. "Science," concluded Manouvrier, "has already disproved certain errors and certain prejudices relative to the indignity of the female sex. It has promoted the idea that the amelioration of the situation of women and the improvement of her social condition could have the happiest consequences from the point of view of the interests of children, of men, and of society." There was, however, much more to be done, and some of it, he believed, could be done by anthropology. "A little more science," continued Manouvrier, "might contribute to the enlightenment of the law on this point and might make the sentiments of legislators conform a little closer to morality, that is, to progress and to happiness" (359–360). Clearly, when Manouvrier spoke of science in such a way he was referring to *his* science, and perhaps to the science practiced by his freethinking colleagues at the Society, School, and Laboratory of Anthropology. He was referring to a notion of science that was defined in opposition to dogma, imposed hierarchies, and unjustified authority. As such, he could only believe that more science would lead to more enlightenment.

Despite numerous explanations of what lay within the purview of anthropology, Manouvrier encountered more protestations of confusion than of dissension. In the first decade of the twentieth century, he was still straining to guard against science being seen as either incapable of filling a function in society or capable of curing (or at least totally redefining) the world's ills. One of the best examples of this effort to find a happy mean is his speech delivered to the Congress of Arts and Sciences at the 1904 Saint Louis World's Fair. The speech, entitled "L'individualité de l'anthropologie," stressed that what distinguished anthropology from other sciences—social or biological—was the effort of anthropologists to understand every facet of the experience and nature of human beings.[56] That is, the very individuality of anthropology was its inclusiveness. Manouvrier argued that whether one studied individuals or the whole species, the study of human beings always consisted of inquiry into diverse sorts of phenomena, "from a quadruple anatomo-physio-psycho-sociological point of view. Voilà the individuality of anthropology" (408). To illustrate the point further, Manouvrier reminded his audience that mineralogy was nothing more than geometry, mechanics, physics, and chemistry, but that minerals must be studied through all of these and no one of them could replace mineralogy. Though no one would ever say that mineralogy, being a composite science, was not, in itself, worth studying, anthropology was thus assailed. The error, explained Manouvrier, was that people kept studying only one aspect of a complicated process, "as if there were no real relations between physical conformations and intellectual and moral characters or, indeed, as if intellectual and moral characters were without relation to exterior conditions, social or otherwise, in which the human being or the category of human beings being studied had to live and evolve." It is a valid point, and one could agree that twentieth-century development of the social and biological sciences has not been paralleled by any broad, coherent study of these various disciplines. According to Manouvrier, anthropology could have served to link the somatic, mental, and sociological points of view. These aspects of humanity are connected, he argued, and even create each other. As such, scientific study of them must also be connected.

Manouvrier's message was that anthropology was the science through which human beings would be increasingly understood. But he consistently warned against its abuses: anthropology would shed light on human beings, but it "is still waiting for that light" (409). Without "this scientific light," wrote Manouvrier, society's attempts to progress end up sidetracked by "sterile agitation and dangerous experiments." At the base of all social questions, he warned, "there are anthropological questions; but we know the worth of the hasty responses of an anthropology still incompletely organized and, above all, incompletely conceived" (410).

The Bohemia-born American anthropologist Aleš Hrdlička came to Paris to study with Manouvrier for six months in 1896, and Hrdlička's biographers agree that those six months determined a great deal for American anthropology.[57] The two men remained friends throughout their lives. Hrdlička founded the American Association of Physical Anthropologists, created the *American Journal of Physical Anthropology* in 1918, and "wrote the classic American work on the subject," *The Skeletal Remains of Early Man*.[58] One of the consequences of his time in France was that, like Mortillet and Manouvrier, he always argued for a unilinear evolutionary progression, though many contemporary paleon-tologists were beginning to recognize that some hominid lines died out and did not lead to modern humans. That was perhaps an unfortunate conse-quence. Another one of questionable scientific value was that Hrdlička spent his life trying to attain what Manouvrier happened to have: an anthropological laboratory, unattached to a university, in possession of thousands of skulls. The reason the Smithsonian has a collection of 7,500 "non-white crania" is that Hrdlička convinced the United States National Museum to support expedi-tions to Alaska, which he led, for the purpose of gathering skeletons and skulls.[59] Yet the greatest influence seems to have been unquestionably positive, for Hrdlička championed an egalitarian, nonracist anthropology at a time when racist anthropology had many supporters indeed.

THERE WAS A GREAT WIT TO MANOUVRIER, and yet his manner was relaxed and unadorned. People seem to have liked him, though they smiled at his pref-erence for scientific solitude. The archives show him several times taking up subscriptions for men and women who were somehow connected to the So-ciété d'anthropologie and were in sudden financial difficulty or some other dire straits. He was often described as generous and kind but a bit distant. When Clémence Royer received a letter of congratulations from Manouvrier, on the appearance of one of her books, she described it to a friend as "warmer than his normal temperature."[60] Maybe he did not like her or her work—she was a biological determinist in terms of criminals, races, and the sexes—but it also seems he was a quiet man with everyone outside the lab or the society hall. I will show in this book's Coda that Manouvrier added two important in-stitutions to that roster in the first part of the twentieth century. In all, it was enough for a rather rich social world, and there are indications that his late marriage (he was close to sixty) was a very happy one. They had a son. Manou-vrier came to be among the most successful anthropologists of the period, lauded in France and abroad. In 1903 he was awarded the prestigious Legion of Honor, and for many years he reigned as secretary general of the famous So-ciété d'anthropologie de Paris.

THE EXAMPLE OF CLINICAL PSYCHOLOGY

In 1885 a group of scientists founded the Société de psychologie physiologique under the leadership of Jean-Martin Charcot, the eminent neurologist and psychologist. Manouvrier was one of the important figures in this group. I examine it here in order to contextualize further Manouvrier's move away from numbers, toward more discursive arguments, and finally toward a secular indeterminism. More than that, studying Charcot's clinical psychology offers an opportunity to examine another late-nineteenth-century French science and its relationship to atheism. Clinical psychology had its own impassioned project of deconsecration.

Charcot is a crucial figure in modern medicine. His work utterly transformed our understanding about a whole range of neurological diseases. For many of these, including multiple sclerosis and Parkinson's, present-day knowledge is still based on his research and insights. Yet, to his consternation, he was best known in his own time for his work on hysteria and hypnosis. He is best known today for his influence on Sigmund Freud. Charcot was a secular republican. Throughout his life, however, he was interested in a few subjects that seemed rather spiritual to many people, particularly hypnosis and faith healing. A posthumous biographical sketch offered by one of his students, Gilles de La Tourette (of Tourette's syndrome fame), was clear on this point.[61] It was also more than a little defensive regarding the late master's rigorous scientism— apparently, he had come to be seen as a bit of a "miracle worker" (608).

Charcot was born in 1825, trained in medicine under Pierre Rayer (he who had silenced Broca's hybridization paper at the Société de biologie), and in 1862 was given a post running Salpêtrière, a hospital intended for the mad but mostly peopled with sane or senile old women with no resources. A contemporary commentator's description of the place was vivid: "We ignore on purpose, perhaps it pleases us to pretend we do not know, that in the great city of Paris there exists another city, a city of old women and of madwomen, and which counts close to five thousand inhabitants."[62] There, apparently, "pandemonium" reigned, and the consensus was that nothing could be done about it. Charcot, however, "understood . . . what a mine was at his disposition" and turned the place into a "scientific station."[63] He set up a lab, and through microscopic study, some brilliant experiments and deductions, and the reclassification of a number of diseases, he was able to make extraordinary contributions. But the work that would bring him popular acclaim had to do with hysteria, a category of disease largely applied to women in the late nineteenth century; the word itself derives from the Greek for "wandering womb," which many thought to be the cause of certain forms of female distress. Madness itself was female in this period. Women were seen as having a more tenuous hold on themselves, and many madhouses were constructed with twice as many dorms for women as for men. Laws made it relatively easy to put away an unruly or unwanted woman, and this accounts for some of the dorms, but many of the houses also served as old-age homes—mostly for redundant older women. Then there were the truly disturbed, and these were mostly hysterics. Hysteria was understood as a disease of greatly varying degrees: a pubescent girl, suddenly set off from her brothers' activities, silenced and constrained, might find that shouting or even pouting led to a diagnosis of mild hysteria. Serious hysteria was something else again, though it likely derived from a similar cause in many cases. In serious hysteria, the woman's arm or leg muscles would contract, so that the hands and feet were like claws; she would suddenly flail about and then return to calm; sometimes she could not walk or speak; and sometimes a whole range of physical ailments were understood as symptomatic of the disease. While some investigation of male hysteria took place before World War I (notably by Charcot), the war produced too many male hysterics to be ignored. Soon they were called "shell-shocked" to save them the embarrassment of a label with female connotations, but Freud came to understand these boys in terms of hysteria: when society asked them to be brave men and run toward machine-gun fire, some could neither obey nor disobey: something snapped, and they took a "flight into illness." The complex and constrained role of women in many Victorian households seems to have been producing a similar effect for decades.

According to Gilles de la Tourette, it was while traveling that Charcot made the observation that would bring him popular acclaim: "his penetrating eye, which nothing could escape, discovered an ex-voto, a painting that permitted him to write his *Démoniaques* and his *Difformes dans l'art*, which are both evocations of nervous pathology in the past" (611). While using the asylum as a source of experimental patients and lab facilities, Charcot grew familiar with the symptoms and gestures of the small portion of his citizenry that was hysterical. When he saw a painting of a possessed woman, he was struck by her resemblance to the hysterical women in his hospital. He began to use this epiphany in his teaching, bringing in medieval paintings and drawings of the demonically possessed and comparing them to living women, twisted and agonized by their condition. He even posed them to match the images. Especially with this help, the comparison was remarkable, and Charcot believed that he had figured out something important about the history of religion: the miraculous, the demonic, and the ecstatic were all, in fact, various manifestations of hysteria. He was soon hanging medieval possession imagery in the halls of his asylum and describing everything from religious piety to sorcery as superstitious, backward conceptions of the scientific category of hysteria. As Jan Goldstein has shown in *Console and Classify*, her study of French psychiatry in the nineteenth century, there were deep divisions here based on religion and irreligion.[64] Catholic leaders had long mistrusted Charcot's profession, which was widely seen as republican and materialist; in fact, in 1874 the monarchical, religious leaders of the "republic of dukes" banned all clinical instruction in psychiatry in all municipal asylums (the ban was lifted in 1876; 360). Charcot's work on hysteria, however, took place in an era better suited for it: Gambetta's scientistic republic.

In his lectures, Charcot used hypnotism on women who had been diagnosed with hysteria. Hypnotism was generally called "magnetizing," "mesmerism," or "somnambulism"—sleepwalking. The German doctor Anton Mesmer had brought "magnetic sleep" to France in 1778. He was a practitioner, and he had written a book explaining that a universal fluid courses through all nature and can pass through the human body. The reason this was called "animal magnetism" was that Mesmer used to put magnets on the subjects while inducing the trance; he thought it communicated the universal fluid to the body. Mesmer found a number of students, but he and his medical art were soon condemned by the Paris Faculty of Medicine. "Meanwhile, clients arrived in crowds."[65] Soon his house was too small for them all, and he bought a hotel; over the next five years he magnetized eight thousand people, before the tide turned against him, he was abandoned by his disciples, and he left France in 1785. After that, animal magnetism had been used as a carnival trick.

Knowing the connection Freud would make, it is notable that Charcot used hypnotism as part of a theatrical pedagogy but not as a therapy: he put hysterical women in trances so that he could demonstrate their symptoms in a controlled manner. This made sense because, having removed the magnets from the equation, Charcot concluded that only hysterics could be hypnotized, so the event of hypnosis was itself a symptom to be demonstrated. Tourette repeatedly clarified that the whole affair had nothing spiritual or supernatural about it: "Hypnotism is a morbid state that, as with the hysteria from which it derives, has its determinism and its laws." Charcot's disciple Paul Richer doubled as the asylum's artist, and Charcot had him draw up detailed charts cataloging the minute shifts and dramatic stages of a hysterical fit. Charcot insisted these were perfectly predictable, part of a "physiological law." But outside Salpêtrière, Charcot was teased for having stepped outside real science. In reaction, he "threw the doors of his amphitheater wide open and put the question to the great public." These were his famous Tuesday lessons, and it was here that he earned the nickname "the Napoleon of hysteria," because he had a commanding presence, a "cold, severe mask," and because he posed with his hand in his jacket. He paced a lot and "split the air with trenchant and quasi-sacerdotal gestures."[66] He won the day, as people saw hypnotism with their own eyes and the elite of all Europe and the curious of Paris flocked to witness the spectacle. Charcot also published two books—mentioned by Tourette in the quotation above—full of drawings by Richer flanked by images of possessed women and explanations by Charcot. He also set up a photographic laboratory in the hospital and photographed his hysterics in a variety of beatific and demonic positions. Again, sometimes they were posed, but this did not seem like bad science, since the poses were so reminiscent of authentic hysterical gestures. At least one of the freethinking anthropologists attended these lessons: a painting by André Brouillet called "Un leçon clinique à la Salpêtrière," shows Charcot standing near an attractive hypnotized woman.[67] She is collapsed backward a bit so that her frock has fallen open, revealing much of her well-lit neck and chest, her arms contorted behind her. An older, female nurse stands at the ready to catch her should she fall. The rest of the room is packed with men, and one wonders if she was not putting on a show for them, since aggressive sexual candor was one of the acknowledged symptoms of hysteria and thus not only tolerated as female behavior but rewarded with attention as an edifying medical case. In the background is an illustration of a woman in the throes of a "demonic" hysterical episode; she is in the same pose as the patient here, Blanche Whitman, but on the ground, having entirely collapsed backward. Everyone in the painting has been identified; Tourette was there, as was Désiré-Magliore Bourneville (to whom I will soon return), Richer (sketching), Alfred Joseph Naquet (whose divorce law of 1884 was considered a major victory for secularization), several novelists, political friends of Gambetta's, Charcot's son, and the freethinking anthropologist Mathias Duval.

Along with the lectures, the books were a big success and vindicated hypnotism as scientific. As Tourette complained, however, from this point forward, Charcot's other accomplishments were forgotten by the great mass of people who saw him only as a "hypnotizer." He came to regret the victory because of its effect on his reputation. Worse, he was "profoundly saddened" to have opened the way for imitators and charlatans: "Science had nothing to see in these experiments that were dominated by money or a morbid curiosity—we saw the reflowering of magic [and] occultism" (618). Nevertheless, for many his name came to stand for the deconsecration of demonic possession, religious ecstasy, and the faith cure. One figure who brought attention to Charcot's scientific translation of religious phenomena was Charles Richet, the editor of *Revue scientifique* who would so favorably review the works of the freethinking anthropologists and who later still would defend science against the accusations of Brunetière. In 1880, Richet wrote two important articles on Charcot for Brunetière's widely read *Revue des deux mondes*: "Les démoniaques d'aujourd'hui" and "Les démoniaques d'autrefois."[68] The first was subtitled "hysteria and somnambulism," the second "witches and the possessed." He presented them in this reverse chronological order because, "when one is more familiar with the positive facts elucidated by contemporary scientists, one reads descriptions of the superstitions that misled our ancestors with more interest."[69] The two articles thus formed a lesson in this translation or deconsecration process. Richet put forward a thesis that was defined by his historical moment and then proceeded to take medieval people to task for not seeing outside their own paradigm. What Richet did was to begin his discussion by talking about the difference between "light" and "serious" hysteria and dwelling for some pages on light hysteria, which he described as extraordinarily common. "I imagine everyone knows more or less the bizarre characteristics of nervous women," and nervous women, he asserted, were clinically hysterical. It was a hereditary disorder that could appear even in a family with no history of the disease: like facial features that shift from generation to generation, any nerve disorder could transmute to hysteria. Still, he explained, it was generally made manifest because of frustration with "life's obstacles."

Richet noted that it was most frequent in women who had been given some education and, separately, that the women who experienced hysteria tended to be very intelligent. Instead of guessing that smart, educated human beings broke down when refused all personal authority, he took the women's intelligence as a symptom of the disease. In all his descriptions, he wrote of how easily the women, especially the adolescents, became short tempered or weepy, "the lightest joke often becomes a cruel offense." They "are all, more or less, liars," as well (a trait, he mused, that they share with inferior races and children). They "are not masters of themselves," wrote Richet. The combined description is enough to make anyone crazy: intelligent persons given some education but nothing to do with it, never allowed to be truly "master" of their own selves, and made subject to jokes. Meanwhile they saw their every expression of distress or independence dismissed and pathologized.

At one point Richet even wrote that "this light hysteria is not a real disease. It is one of the varieties of the character of woman. One could even say that hysterics are more woman than other women" (346). He then used the descriptions of women in recent novels to illustrate his point—Madame Bovary was foremost among them. When he moved on to serious hysteria, he continued to describe the position of these women without seeing it as a causal agent. "The hysterics desire only one thing, that one pay attention to them, that one be interested in their little passions, that one take part in their affections or in their angers, that one admire their intelligence or their attire." What's more, "no sense of shame or false modesty stops them . . . and they talk with men as if they were the same sex. Nothing embarrasses these female Diogeneses: they have a response for everything. . . . Self-love is not missing from them, however, and if one seems not to pay attention to them, they grow indignant" (360). He reported with scientific calm that it was easy to make them cry, too (368). One has to wonder what he went around saying to them.

After all this, Richet made it clear that he was scandalized that these "sick women would have been burned in the past. . . . Three centuries ago their illness would have been taken as a crime" (340). Here, by contrast, their misery was taken as illness, but Richet could not be expected to have seen things that way. For his readers, he described the hysterical condition in some detail. Richet then told of the unfortunate career of Mesmer's "magnetic sleep" and his "bizarre, almost mystical book" and then announced that "today, all enlightened doctors recognize that somnambulism exists" (364). He had visited Salpêtrière to see the hypnotism and had been overwhelmed by the experience. Now he had the very difficult task of arguing that this mysterious thing actually did exist but was not supernatural. He compared it to dreams and to sleepwalking (also mysterious but real) and explained that "magnetizers" were mostly fakes, especially those who claimed that the entranced subject could tell the future. "The famous Lucile," he explained, was, in fact, hysterically anesthetized, but "she knew very well that it was her job to tell the future." The women's intelligence was "overexcited" by the disease, so their responses were often ingenious. "In a word, the somnambulists of the fair and the theater are really sleeping: they are not future tellers but sick women, and their true place is in the mental hospital" (369).

Richet made two connections that foreshadowed Freudian psychology: First, he recognized that a given hysterical attack was usually set off by something that reminded the victim of her original "disappointment" or trauma (359). Second, hypnotism suggested to him "that what makes the *moi*, we can call the collection of our memories, and when it turns out that there are memories reserved to a special physical state, one is almost able to say that the person is doubled, because she remembers in sleep a whole series of acts of which she is ignorant while awake" (369). This was 1880; Freud would not publish his first paper until 1895. Richet ended his essay insisting that anyone who did not believe in hypnotism had to see it for themselves. The empirical approach also worked in description, where Richet insisted, with Charcot, that the phenomenon had a physiological basis:

Thus hypnotism may be considered as a real disease, a disease of which the symptoms are as well described as those of hysteria or epilepsy. The only strange and obscure side of its study is that this nervous disease can be provoked by exterior maneuvers of which the mode of action escapes our understanding. But just because we are ignorant of the cause of phenomena does not give us a reason to deny them. Soon, in a few years maybe, we will arrive at an exact knowledge not only of its symptoms, which are pretty well known today, but of its physiological causes. It is permitted to hope that the empirical procedures that we employ today will be replaced by scientific methods that no one will be able to doubt and of which everyone will be able to agree on the efficacy. In sum, we have seen that without producing actual insanity, there are maladies that profoundly trouble the functions of intelligence. Certainly, these troubles are strange and surprising, but one can affirm that they are controlled by natural laws and not the fantasy of seven million four hundred five thousand nine hundred twenty-six devils of hell. That was not the opinion of the judges of the seventeenth century, and it is not one of the smaller benefits of science that it has affirmed and proven the innocence of these poor women who were once made to climb onto the stake. (372)

In "Les démoniaques d'autrefois," Richet introduced a range of descriptions of demonic possession, making a point of their similarity to the hospitalized hysterics of his day. He got some of this material from a book on witches by Jules Michelet and some from *A History of the Devil*, by Jean Réville, the historian of religion who reviewed Véron's *La religion*. Most of the material, however, came from Alexandre Axenfeld's study of Jean Weir. Axenfeld was a member of the Faculty of Medicine, and one of those singled out as "materialist," along with Broca and Charles Robin. Weir was a sixteenth-century doctor who wrote a book arguing that the people being burned as witches were either ill and deserving of medical attention or harassed poverty-stricken women deserving of charity and pity. In arguing this, Weir offered extensive portraits of the women's symptoms and behavior. Axenfeld had found the manuscript and given a series of lectures on it; the book Richet consulted was a collection of these lectures. As Axenfeld, Richet, and several others would agree, Weir was a fantastic author to advance: he detailed the occult and the darker side of the religious imagination, he was morally disgusted and enraged, and he demonstrated that, rare though it may have been, it was possible to have conceived an almost scientific, medical model even in the still dark world of the sixteenth century.[70] This made the church all the more responsible for its cruelty.

As Goldstein has shown, there was a political and ideological alliance between Charcot and Gambetta (they socialized as well), and when the latter became prime minister, a chair in diseases of the nervous system, funded by the government, was created for Charcot at the Faculty of Medicine.[71] The republican, anticlerical scientist Paul Bert—whom we have seen assisting at the Society of Mutual Autopsy—had been made minister of public instruction and religion and had facilitated this "consecration universitaire."[72] Bert's articulated goal was to reform the "university of France, inheritor of Jesuitical methods [that] have reduced scientific instruction to sterility."[73] Like secular anthropology and sociology, Charcot's neurology/psychology was supported by a government eager to carry out a general program of deconsecration.

Freud was one of the many young physicians who came to Paris to attend Charcot's lectures, and he was always clear about the debt he owed to this experience. Along with his philosophical materialism, Charcot's brilliant neurological work had led him to see both hypnotism and hysteria as somatic disturbances. Freud would bring them to the center of cultural discourse in a very different way, which nevertheless continued the deconsecration of madness. Charcot died in 1893, while on a guided archaeological excursion with a few of his disciples—very likely run by Gabriel de Mortillet and his son, Adrien. Had Charcot lived only a little longer, he would have seen the study of both hypnotism and hysteria advanced by this other violently antireligious, scientistic researcher. Interestingly, Freud saw Charcot as a classifier, "not of the reflective type" but one who could spot patterns and organize them.[74] Charcot himself, according to Freud and others, recommended nosography as a deeply satisfying experience; he had arranged his hospital patients to form a great nosological chart, and he immensely enjoyed walking people through it and admiring its form. It is reasonable to see this as ritual. Wrote Freud, it was like "the myth of Adam, who must have experienced in its most perfect form that intellectual delight so highly praised by Charcot, when the Lord led before him the creatures of Paradise to be named and grouped" (11). Grouping things according to the similarities that pertain to some particular problem can, of course, be the primary factor in finding a solution. Still, the history of ideas, belief, and behavior is as tricky as animal evolution: developments sometimes prove fabulously useful for purposes other than their original function. Charcot's naming and regrouping was pragmatic and functional in many cases; it was also an act of deconsecration that changed the meaning of people, objects, and human conditions.

Many students and doctors gathered around Charcot at Salpêtrière, but Désiré-Magloire Bourneville was the disciple most prominently linked with Charcot. He was also the most famously anticlerical. Bourneville rivaled the freethinking anthropologists in his deconsecrating zeal. He served on the Paris Municipal Council from 1876 to 1883 and as a socialist deputy to the National Assembly, during which time he led a crusade to secularize Parisian hospitals. Nuns made up about half the nursing staff in these public institutions, and Bourneville worked to have them kicked out. It is worth noting here that in these years all sorts of religious figures were dismissed from all sorts of activities and many religious orders were expelled from France, but nuns were somewhat exempt from this because they did so much at such a small expense. The secular republic took the nuns out of the prisons but let them continue their work running hospitals, poorhouses, asylums of every kind, orphanages, and, despite the illegality at some points, schools for girls and adult education classes.[75] In the case of the Paris hospitals, however, fervid anticlericalism won. I have already reported how well the freethinkers held the Paris Municipal Council in the 1880s. In 1883, in response to Bourneville's plea, that body threatened to withhold funding unless the sisters were expelled and the hospital chaplains were dismissed. The hospitals complied. Bourneville was less successful in obtaining government funds for a secular nursing school but managed to create one on private funds (Charcot donated) at Salpêtrière. If all this sounds remarkably like the behavior of the freethinking anthropologists, consider this: in an article on the Society of Mutual Autopsy, Le petit bleu listed Bourneville as a member.[76] Further, in 1880 Bourneville helped found the Society for the Propagation of Cremation and became its president in 1884.[77]

Bourneville's literary attack on religion was equally pronounced: by 1885 he had published a flurry of books indicting the church for having tortured and burned thousands of innocent women who really belonged in asylums. The women were not possessed, and they were not witches. They were hysterical. Further, he explicitly argued, with a pointed, almost liturgical repetition, that all sorts of religious states, including religious ecstasy, were, in fact, insanity. Religious cures were also translated into the language of science. The "cured" illnesses had simply been hysterical: he found ample evidence of "contracted" limbs (the most common symptom of hysteria) in several shrine records. The cures were no more than a result of the power of suggestion. Bourneville called these books his "Bibliothèque diabolique"; the rubric was printed on the title pages and mentioned frequently, with the brash theatrics of the Mutual Autopsy crowd. While his "Bibliothèque diabolique"—there were nine volumes in the end—was full of demonic descriptions, the title also jokingly referred to the way the church would surely view the little library.

In 1872 Bourneville coauthored De la contracture hystérique permanente; ou, Appréciation scientifique des miracles de Saint Louis et de Saint Médarde (On permanent hysterical contraction; or, A scientific interpretation of the miracles of Saint Louis and Saint Medard).[78] In 1875 he came out with Louise Lateau; ou, La stigmatisée belge, which insisted that blood blisters were another symptom of hysteria and that recent claims regarding a young Belgian's sanctity were based on a mistaken interpretation of her so-called stigmata.[79] Again religion had offered young women a powerful if crushingly burdensome role, and science was stealing it from them with derision and yet liberating them from it as well.[80]

His *Iconographie photographique de la Salpêtrière* was a multivolume compendium of photographs of hysterical women in religious poses: trances, possessions, ecstasies, and prayer.[81] In 1885 Bourneville republished Jean Weir's *Histoires disputes et discours des illusions et impostures des diables* and supplied a preface.[82] He described the book as follows: "The point of this work is to make evident that the crimes imputed to witches are imaginary; that these women are not criminals but sick people, damaged in their mental faculties; and that they are not rightly under the jurisdiction of priests, monks, and judges. In consequence, they must not be imprisoned, tortured, and led to the flames of the stake but should be entrusted to the care of doctors" (iv). Bourneville discussed Charcot's work in the preface, setting the context for his claim that "anyone interested in the grand struggles of the scientific spirit against barbarism [would] find ample satisfaction in reading the book of Jean Weir" (vi). As Goldstein puts it, redefining the once supernatural as natural was a purposefully secularizing act, but "the redefinition of the supernatural as the natural-pathological went further and had the effect of debunking religion; it was consonant with the frenetic crusade for laicization that marked republican politics in this era." Here, as Goldstein says, Charcot was a strong force of moderation and did not share Bourneville's "much more strident" tone.[83] But though he was a bit of a loner as a strident atheist in his profession, Bourneville otherwise stands as another example of the freethinking anthropologists' crusade.

One last article rounds out this portrait of clinical psychology and its particular deconsecration project. The piece was called "La foi qui guérit" (The Faith Cure), and it was the last thing Charcot published before he died. It came about in this way: the editor of the British *New Review* had read of Zola's disgusted reaction to the supplicants at Lourdes and asked Charcot for his opinion on the matter. Charcot's response caused such a stir that two French journals translated and reprinted it (retaining a number of English phrases).[84] It deserves attention because Charcot here demonstrated that he was still actively antireligious but had come to believe that materialism could not explain everything. That he still enjoyed translating religious experience into scientific language was clear in his claim that "St. Francis of Assisi and St. Theresa, whose shrines are both in the first rank of those renowned for miracles of healing, were themselves undeniably hysterical." According to Charcot—and, once again, foreshadowing Freud—miracle cures were real but not miraculous: the diseased were hysterical, as were the healers, and the process; "throughout all the ages, among the most diverse civilizations," it was always the same, "its laws of evolution immutable" (21). Charcot's explanation noted the "fatiguing journey" that brought the patient to the shrine "in a state of mind eminently receptive of suggestion" and showed that the eight-day stay most shrines prescribed further

helped the "mind to obtain mastery over the body" (23). Having well demonstrated the real nature of paralyzed limbs as "hysterical contraction" (24), he
turned his attention to a more prickly question: "the water of the sacred spring
is of avail nowadays against tumors and sores; it cures the most stubborn ulcers
in a moment. Is it to be supposed that these complaints have a neurotic origin?"
(25). In answer he offered a lengthy and rather gory description of a Madamoi-
selle Coirin's battle with breast cancer. Among her many symptoms she had a
huge tumor, a shriveled paralyzed leg, and a running sore around her breast so
severe that "the nipple fell off bodily" (26). No doctor could cure her, but when
she was given a cloth that had touched the tomb of Francis of Paris, her every
symptom began to heal. Charcot made a point of the progress each made and
used this to prove that the body was acting according to natural law: a miracle
could have cured everything at once, but here the paralysis disappeared long before the muscular atrophy began to fade. Of the cancer Charcot wrote: "It must
be understood, of course, that the term 'cancer' is not to be read literally" (28).
A Dr. Fowler had recently taken on eight cases of tumors of the breast, "some
as large as a hen's egg," and, "with better judgment" than those who would am-
putate, he "subjected his patients, who were all hysterical like Coirin, to a
course of treatment in which, so to speak, the psychical element was made the
chief point, and tumors which had been pronounced reducible by the knife
alone vanished as if by magic." Charcot concluded that had the eight women
"gone to a shrine, it is impossible to avoid the conclusion that many of them
might have been healed." To make the analogy more complete, he quoted
Fowler saying, "Like all women of a similar temperament she had a fetishlike
faith in her regular medical attendants" (29).

How it all worked Charcot did not pretend to know, but he repeatedly invoked "facts" and insisted that the cure followed "natural laws" (29) and "physiological laws" (30). Charcot then offered a surprising confession: "I have seen
patients return from the shrines now in vogue who have been sent thither with
my consent, owing to my own inability to inspire the operation of the faith-
cure." He then had occasion to chart the very natural, lawful procedure of the
physiological healing. But the point here was that he did condone the faith cure
and even tried to manage it himself. The biggest surprise came at the end of
the article: "Can we then affirm that we can explain everything which claims
to be of supernatural origin in the faith-cure, and that the frontiers of the
miraculous are visibly shrinking day by day before the march of scientific at-
tainments? Certainly not. In all investigation we have to learn the lesson of patience. I am among the first to recognize that Shakespeare's words hold good
to-day—'There are more things in heaven and earth, Horatio, Than are
dreamt of in thy philosophy'" (31). That "certainly not" was a smack at the

hubris of his own materialism and a shift toward secular indeterminism. There would be many more.

Throughout the 1880s and into the 1890s, Manouvrier met with this circle, bringing them his findings from the Society of Mutual Autopsy, including, for instance, the detailed comparison of the brains of Louis-Adolphe Bertillon and Gambetta; a discussion of the brain of a man deaf in the left ear (Bertillon); and an essay musing on mental images. This work was presented to Charcot's Société de psychologie physiologique and subsequently published under the heading "Société de psychologie physiologique" in Ribot's *Revue philosophique*.[84] My intent in detailing the beliefs and behaviors of Charcot's clinical psychology has been to demonstrate their conceptual similarity to the freethinking anthropologists and to suggest that a similar deconsecrating project guided a host of endeavors at the dawn of professional science. This discussion also helps us to understand Manouvrier's increasing comfort with discursive proofs, despite his discipline's passion for numbers. Charcot and Bourneville offered another model. The freethinking anthropologists and the clinical psychologists both used classificatory models more than equations; the clinical psychologists also used symptoms instead of numbers. All were evangelical in their atheism and intent on transforming the sacred—here, religious ecstasy, demonic possession, saintliness, stigmata, and faith cures, as well as the Paris hospital staff—into the profane. In Charcot, Manouvrier also had the model of a secular scientist for whom scientific determinism could no longer suffice. Manouvrier was thus part of a dynamic cultural process that was shifting, in a variety of ways, toward naturalist indeterminism.

———

THERE WAS AN AWFUL LOT OF FEAR hidden behind the hierarchical anthropological doctrines of the late nineteenth and early twentieth centuries. Men were afraid that women were becoming emancipated and would no longer need them and no longer serve them. Women were afraid of having to prove themselves in the public world after having been trained only for the private world. They were afraid of the loss of masculine protection, and they were afraid that their emancipation might destroy the family. Bourgeois society was afraid of crime and criminals, afraid of its own capacity for criminality, afraid of convicting the innocent, and afraid of releasing the guilty. Many were afraid that a secular society might not be capable of morality, and the spate of anarchist violence at the turn of the century heightened the sense of decline. Much of society was also worried that the young nation-state, defined only by the unity of its populace, was being corrupted from within. Some feared a corruption that was more of the bumbling variety: the working class, and perhaps the middle class as well, had too much power and was making a mess of things. Some were specifically worried that the Jewish people among them were the cause of French decline. Anthropological theories addressed all these fears. There were serious attempts to revive Catholicism at the end of the century, with the specific intention of handing these problems back to the church. For those who did not consider this to be an acceptable option, other arrangements had to be made. Materialism had won out over mysticism, and the material world now had to be held responsible for somehow providing answers.

In the first decades of the twentieth century, such figures as the psychiatrist, the statistician, and the sociologist were taking their place alongside the judge and the doctor as secular authorities on human behavior and potential.[86] But in France, though a considerable battle had been waged, physical anthropology did not succeed in its struggle to be counted among society's primary counselors. Advocates of psychiatry and sociology positioned themselves to help established authorities develop and promote the new ideologies necessary to a modern, capitalist democracy. While sociology provided a secular morality and psychiatry provided a theory of behavioral normality, anthropology no longer had anything of this sort to offer. The discipline's most aggressive, vibrant, proactive theories were deeply antiegalitarian, logically untenable, and at least vaguely antidemocratic. These theories were exciting and compelling, however, and they received a great deal of attention. At first, they seemed to offer much more than what might be offered by psychiatry or sociology because they were so profoundly physical, which seemed to suggest empiricism, quantification, and certainty. The visible signs and numbers that anthropology could provide were a perfect antidote to the religious world of invisible players, assumed priestly authority, and coded biblical guides. It was convenient and persuasive to assign to the body what had been the purview of the soul. But the switch from soul to body tended to promote doctrines of determinism and human limitation. Manouvrier struggled to maintain a secular position and yet create a new indeterminacy; he wanted a materialist replacement for the soul broad enough to include the unlimited possibilities once suggested by elusive spirit. In the early years of his struggle against reductive biological determinism, Manouvrier dismissed a scientific study partially because it "rested on no numbers whatsoever." By the turn of the century, however, he had issued as harsh a critique of quantification as it would ever receive, publicly lamenting the "enormous stockpile of useless numbers" still being augmented. In a sense, Manouvrier's shift from the science of numbers to a discursive, conceptual science was, in fact, a movement away from science—

as it was then envisioned—and toward social philosophy. That he understood this on some level will become clear in this book's Coda, as I examine Manouvrier's pursuit of one of the most celebrated chairs of philosophy in France. For now, I turn to some late-century French philosophers and sociologists and their strained return to indeterminism.

CHAPTER SEVEN

∞

The Leftist Critique of Determinist Science

The freethinking anthropologists of Paris had managed to earn the cultural authority to weigh in on the question of the human soul. They yearned to get the scent of the church off everything, public and private: to remove its claims to their own bodies, to the city in which they lived, and to the conceptual notions that structured their civilization. Their students and fellow travelers—the young Bertillons, Lapouge, and Manouvrier—continued to use the techniques and ideas of physical anthropology to struggle over the end of the soul and to search for new ways to understand human individuality, free will, accountability, personal meaning, and national identity. This chapter is not about the politics, irreligion, or materialist anthropology of Paul and Augustine Broca, Gabriel Mortillet and his son Adrien, Clémence Royer, Louis-Adolphe and Zoé Bertillon and their family, Eugène and Jeanne Véron, André Lefèvre, Charles Letourneau, and the rest, or of their students and fellow travelers. Instead, it follows some of the cultural reverberations of their theories and their claims about the end of the soul. I will first examine the three major contemporary responses to Lapouge and then show that two of the most important figures in France at the turn of the century, the philosopher Henri Bergson and the sociologist Emile Durkheim, were responding directly to the freethinking anthropologists' idea of the soul in some of their most celebrated works.

In many ways, scientism was besieged in the 1890's.[1] There was the Catholic revival, best marked by Brunetière's "Après une visite," Pope Leo XIII's Ralliement, and the rise of the miracle cult. This same period saw a revival of metaphysics. Paul Bourget published a famously antiscientistic novel, *Le disciple*, in 1889, and a small flood of metaphysical Russian novels arrived in

France throughout the 1880s.[2] The great scientistic philosophers Renan and Taine both died in the early 1890s, and though their late work was not fully consistent with their reputations, their deaths added to the sense that the ideology of scientism was passing.

In the academy, the metaphysical revival was heralded by a new journal, the *Revue de métaphysique et de morale* (or *RMM*) that was founded in 1893 by the philosophers Xavier Léon and Elie Halévy, with Léon serving as editor until his death in 1935. It was designed to take its place alongside the two other philosophical revues that had served France for twenty years: *Critique philosophique* (Renouvier's neo-Kantian journal) and *Revue philosophique* (Théodule Ribot's science, philosophy, and psychology journal), both of which are discussed in chapter 3. In the introduction to the first issue of *RMM*, the editors dismissed *Critique philosophique* by damning it with faint praise: "Whatever one thinks of the characteristic thesis of neocriticism—the foundation of a morality of imperatives on the basis of phenomenalism—one must conclude that the achievement of Renouvier . . . is considerable." The *Revue philosophique*, for its part, was said to have offered the great service of publishing the works of both scientists and philosophers and thus introducing them to one another's work.[3]

The *Revue de métaphysique et de morale* was created as a forum for all those who were not interested in either neo-Kantian philosophy or the intermingling of scientific, philosophical, and psychological ideas. It was intended to present doctrines of "philosophie proprement dite" (true philosophy), to put aside science and bring public attention to general theories of thought and of action, "from which the public has turned away of late and which have meanwhile always been under the currently discredited name metaphysics, the only source of rational beliefs" (2). In contrast to the stereotypical image of science as dynamic and progressive, the new journal charged that scientific morality was static and conservative, able only to describe truths found in nature. The contention was that French society was in crisis owing to an imbalance between intellectual and moral thought. In the absence of a strong, credible moral code, some members of society were returning to "a very simple, very sweet, very sad Christianity," while others buried themselves in "specialized scientific projects." Meanwhile, society was falling prey to "blind and terrible forces."[4] The light of reason, explained *RMM*, was "as weak and shaky as ever," stranded in the middle of all these worries and lost "between positivism that stops at facts and mysticism that drives one to superstitions" (4).

As suggested by the reference to a "a very simple, very sweet, very sad Christianity," the philosophers at the *Revue de métaphysique et de morale* were particularly frustrated by the debate between Brunetière and the scientific materialists. The philosopher Alphonse Darlu, the influential professor of Xavier Léon (and, famously, Marcel Proust), was a frequent contributor to *RMM*. He reviewed Brunetière's "Après une visite" (1895) and made the journal's position clear: "We would like to remind him that philosophy exists."According to Darlu, Brunetière's problem was that his intellectual education had been excessively shaped by positivism.[5] He had been raised on Renan and Taine, and, following that, he had "read and reread Darwin, at that young age when one lives for intellect and at that moment in the century when M. France and M. Bourget were reading it at the same time, with drunken passion." Royer's translation and preface to the work were not mentioned but may well have contributed to the force of their conversion experiences. Darlu argued that Brunetière's change of heart was by no means a unique experience. Rather, it represented a widespread phenomenon affecting men and women who had embraced positivism and evolutionism with too much faith and then, pulled by their strong senses of morality, eventually swung back to Catholicism with great force. "To stop midway," explained Darlu, "requires a philosophical frame of mind and very deep moral beliefs" (248). Yet what would this "midway" look like?

RESPONDING TO LAPOUGE:
BREAKING THE "NATURALIST OBSESSION"

The rejection of Lapouge among nonscientists took place within the political doctrine of solidarism, and, in fact, the reaction against Lapouge helped to create the doctrine. Solidarism has come to be known, in the historian J. E. S. Hayward's terms, as "the ideology of the Third Republic."[6] The philosopher Alfred Fouillée formulated the doctrine beginning in the 1870s, and by the 1890s it dominated French political discourse. Léon Bourgeois was its most dedicated political champion, and after he became prime minister in 1895, he continued to articulate and popularize solidarist goals.[7] The appeal of solidarism was its concerned moderation: it was a reaction against laissez-faire individualism, but it stopped short of socialism. Liberalism had once stood for a government that removed artificial economic barriers, such as guilds and noble privileges, contending that without these arbitrary rules the marketplace would become a just and fair field of competition. By the late nineteenth century it was clear that the new field had developed a new kind of viciousness. Very quickly, novel privileges and cruelties had appeared, based on family connections, wealth, and education, and these were keenly exacerbated by the excesses of capitalism, urbanization, and industrialization. This is how liberalism went from meaning government with its hands off (i.e., with its guild-and-privilege-ordaining hands off the market so that the average person would not be constantly blocked by preexisting networks of power) to meaning government with its hands on (i.e., using regulatory control over the market so that the average person is not constantly blocked by preexisting networks of power).[8] But as much as proponents of the new kind of liberalism wanted to help out in a new, active way, their policies were more a stopgap against socialism than they were a path toward it. Solidarism marked the emergence of this new liberalism, equally anxious about the possible abuses of socialism and capitalism.

It would be difficult to exaggerate the extent to which solidarism's theorists discussed political ideas in terms of Darwinian evolution. The penchant for using sociobiological rhetoric must not be taken as a mere borrowing of scientific jargon in order to increase the doctrine's cultural authority. Rather, Léon Bourgeois and the theorists of solidarism plainly felt both burdened and blessed to be the first thinkers with access to anthropology's stunning new information. It was with a sense of duty and resolve (and a sense that misinterpretation could be calamitous) that they brought natural science to the old questions of social contract, human character, general will, and a prepolitical "state of nature." Solidarism's concern about the political implications of evolutionary theory concentrated firmly on Darwinian, *not* neo-Lamarckian, theory. The current historical analysis of this period is that neo-Lamarckianism crowded Darwinism out of French political discourse. It is certainly true that the French revived, reformulated, and celebrated Lamarck even long after Darwinian evolution arrived on the scene. They did so partially in order to commemorate the origin of evolutionary theory in France and partially because with a few adjustments Lamarck's "inheritance of acquired characteristics" served to support French republican theories of social amelioration. Historians do not argue that the French utterly ignored Darwin, but the way in which the French used neo-Lamarckianism has captured our attention, obfuscating the importance of Darwinian "struggle" in French political theory and debate.[9]

In fact, the social significance of Darwinian evolution was a central topic in many cultural and political debates. As it was understood, civilization's moral goal of taking care of the "unfit" was preventing, even reversing, the work of evolution (which was supposed to function by killing off or at least limiting the reproduction of the unfit). And yet that goal could not be abandoned: to return to the political, social, and economic equivalent of the "state of nature" would lead to a brutal world. Solidarism was partially conceived of as a defense of civilization: a humane call for society to remain above nature's base struggle. But it was also born of the notion that the natural world was more just (impartial, uncorrupted) than the world of human society, because human society creates artificial barriers (unequal wealth) and artificial hazards (war, machinery, and voluntary celibacy) to the survival and propagation of the most fit. Because of these conflicting interpretations, solidarism was, at first, sometimes described as the policy of a just society working to ensure that natural, cruel competition was tempered by human reason—I will call this "civil solidarism"—and sometimes described as an effort to return to a natural condition in which the "fittest" have the opportunity to succeed, regardless of their original social station—I will call this "natural-competition solidarism." It further confuses matters that some solidarist theorists saw the natural world as more cooperative than competitive. This gave rise to a "natural-cooperation solidarism" that held that mutualism was a natural fact, either because animals were seen to be interdependent or because society was conceived of as a single organism with individuals and classes acting as cells and organs, respectively. All three forms had their champions from the beginning, but it is possible to discern a clear shift in emphasis over time: from the natural solidarism of the early years, wherein modern society was generally held to be the villain (indicted either for promoting artificial rather than natural inequality or for replacing cooperative nature with artificial competition), to the later civil solidarism wherein it was assumed that whatever evils could be found in society, nature was worse. In Hayward's words, by 1908 Bourgeois "had (following Fouillée) recognized that natural solidarity, the fact of interdependence, was amoral and that it was only through the rational intervention of men that it could be made the foundation of social justice."[10] But to understand why Fouillée had come to rest, after much vacillation, on the idea of civil solidarism, one has to consider the battle that he was waging against Lapouge. The brutality of nature had once meant a lion killing a zebra and, further, a less apt lion dying of starvation. Now the brutality of nature might mean the erasure of morality from public life, European races dominating or even slaughtering one another, state-determined laboratory pregnancies, the end of the family, the end of democracy, and the elevation of race and the state above all.

By the time Lapouge published his first book, Fouillée had long been engaged in a fight against the naturalist politics articulated by Herbert Spencer. Fouillée held that an understanding of biological facts was necessary for the creation of an ideal state. Unlike Spencer, however, he believed that these biological facts argued for mutualism as strongly as for individualism and that human reason must, in any case, mitigate the harsh interpretation of "survival of the fittest" prescribed by Spencer and other Social Darwinists. From his doctoral thesis, "La liberté et le déterminisme," of 1872 through his *Humanitaires et libertaires au point de vue sociologique*, which appeared posthumously in

1914, Fouillée published twenty-seven books that grappled with the relationship between biology and politics. He deliberated extensively on the morality of redirecting or accelerating Darwinian evolution and devoted considerable attention to anthroposociology.

Fouillée's acceptance of some degree of Social Darwinism made him particularly sensitive to Lapouge's claims. In one of the earliest articles to take up the discussion of Lapouge, Fouillée complained that "the 'struggle for life' among the whites, blacks, and yellows was not enough; some anthropologists have also imagined a struggle for life between blonds and brunettes, longheads and short-heads." He argued that Lapouge had grossly exaggerated the biological aspect of psychological differences between the Germans and the French, and he cited Léonce Manouvrier as an anthropologist who denied the significance of the cephalic index.[11] Though Fouillée objected strongly to Lapouge's "fanaticism," he still quoted him consistently and, in general, with approval. Fouillée maintained that, if used prudently, inquiries into the national physiological differences among Europeans could help to establish their psychological differences. He would devote much of his large oeuvre to defining the characteristics of various nationalities.

Fouillée is remembered as one of the most important late-nineteenth-century French philosophers, but he should also be known as a central figure in what might be termed "fin-de-siècle national character studies." This phenomenon seems best explained by the coincidence of discussions of evolutionary heredity with the moment at which Western Europe became a solid bloc of nation-states. After 1871 character studies of these states proliferated, fetishistically describing the natural likes and dislikes, virtues and failings, and friends and enemies of each national group. French works of this sort generally claimed several high virtues as inherently French, but the national character studies were also sites of anxiety and self-doubt. As authors attempted to reimagine the nations of Europe in the light of shifting political balances and new anthropological data, they considered not only the strangeness of others but also how strange (or decadent) their own nation seemed in others' eyes. They even exported this imagined criticism.[12]

Fouillée was certain that nationalities had biologically determined intellectual and social characteristics, but he firmly objected to many of Lapouge's larger claims. For example, in his study of the psychology of the French people, Fouillée cited Lapouge's national characterizations but found his pragmatic suggestions distasteful. Wrote Fouillée, "This ethics of breeding studs founded on naturalist hypotheses and on the dreams of utopians is not really human morality."[13] In any case, Fouillée doubted Lapouge's belief that "one could obtain any desired psychic type, on an intellectually uniform level 'the same as that of the highest minds of today's society'" (281). He also ridiculed Lapouge's suggestion that one could create races of naturalists, fishermen, farmers, and blacksmiths. He found this last notion particularly amusing: "A race of naturalists! As if the quality of naturalist follows a cerebral formation distinct from that of a fisherman or a farmer! What audacity it would take to want to intervene in the creation of men, on the basis of information as vague as that of the forms of skulls and of their problematic relationship to mental superiority!" (281–282). Perhaps most important, he noted that "we have no idea of the real cerebral causes of intellectual superiority or inferiority; we do not know if, in suppressing this or that individual carrying some vice, we would be also suppressing, in the same stroke, the seeds of beautiful and important qualities" (282).

It is surprising that Fouillée was so amused by the notion of a race of naturalists, because the gradations he did endorse were almost as precise. In fact, such characterizations were the whole point of his book, and though he intended them to be used to help the nationalities understand one another, he offered scientific explanations as to why any given group was more or less nervous, imaginative, prone to dreaming, sexually energetic, and so on. For this reason, Fouillée was sometimes referred to as an anthroposociologist, though he himself strictly rejected the appellation within his published works, in his correspondence with Lapouge, and in reported conversations with Lapouge's disciples.[14] Fouillée never rejected the idea that national character types were based in heritable biological traits, but the years he spent arguing against Lapouge's antimoralist naturalism shifted his thinking toward civil solidarism. He found himself codifying solidarism as civil protection against natural law precisely because Lapouge's natural laws were so convincingly nasty.

In 1903 Fouillée published a lengthy attack on anthroposociology. The work was, as he described it, a study of the various psychological profiles of European nations, but while the central chapters of the book kept to a sociobiological agenda, the introduction and conclusion were devoted to combating anthroposociological ideas. "The real law of human societies," asserted Fouillée, "is not natural selection and the struggle for life but rational choice and cooperation for life."[15] He had argued in the past both that mutualism in human society was scientifically based in natural models and that the competition that did exist in nature was preferable to the corrupted competition of human society. Now, he characterized solidarism as distinctly human. Modern society might be brutal and amoral, but, if Lapouge was even partially correct, natural forces were even less humane. Human beings must then create a world based neither on religious dogma nor on natural science. Only "rational choice" could serve as the "real law of human societies" (529).

In fashioning his notion of solidarism, Léon Bourgeois drew heavily on the work of Fouillée, increasingly promoting the idea that human logic and morality dictated mutual assistance. As he explained at the Ecole des hautes études conference on solidarity (1902), Bourgeois believed that "humanity, according to the ingenious image by M. Fouillée, is not comparable to an archipelago of small islands of which each has a Robinson. Every group of men . . . is, voluntarily or involuntarily, a solidarist ensemble, the equilibrium, conservation, and progress of which is obedient to the general law of universal evolution."[16] Though anthroposociology was herein rejected, anthropological ideas such as "the general law of universal evolution" were accepted as significant, indeed paramount, to the proper formation of the state. Though civil solidarism triumphed in the last decade of the century, notions of natural-cooperation solidarity and natural-competition solidarity never entirely disappeared from the arguments of Fouillée and Bourgeois. Solidarism's other central theorist, Célestin Bouglé, substantially altered the debate by explicitly rejecting science as a viable means of arriving at sociopolitical truths.

Célestin Bouglé was one of Durkheim's primary disciples and closest collaborators. Having written a doctoral thesis entitled "Les doctrines égalitaires," he went on to teach at the Faculté des lettres de Toulouse, and in 1901 he began teaching social philosophy at the Sorbonne. In 1920 he was named the director of the Centre de documentation sociale at the Ecole normale. In his many works, Bouglé expressed a position on egalitarianism and solidarism that respected the validity of scientific information on humanity but increasingly considered it to be inconsequential to society. In his 1897 article "Anthropologie et démocratie," Bouglé argued that whether or not science could prove the existence of biologically based differences in the capabilities of races or individuals, these differences should have no effect on the philosophical decision to maintain political equality.[17] He directly attacked anthroposociology and Lapouge, whom he recognized as the French founder and leader of this movement.[18] He would later use the same arguments to refute a wider range of anthropological, racist doctrines, but in "Anthropologie et démocratie," Bouglé's argument aimed squarely at Lapouge, claiming that his descriptions of inequality might be factual but they should be functionally insignificant to the republic. Indeed, he suggested that the republican attachment to notions of natural equality may have been no more than a necessary but transient stage. Thus "a morality suffused with the idea of solidarity may not need to consider the idea of equality as anything more than provisional. . . . If it is true that, in declaring men to be equal, we deliver a judgment not on the way nature made them but on the way society must treat them, well then, the most precise craniometry could not prove us right or wrong" (461). Although Bouglé questioned whether human capabilities could be deduced from physiological measurements (citing Léonce Manouvrier's scientific argument), his invective against Lapouge was largely based on the assumptions that anthroposociology made concerning the *significance* of the biologically based inequality of human beings.

Even after Manouvrier had dismissed the scientific validity of Lapouge's claims, Bouglé continued to write refutations of anthroposociology. For whether or not brachycephalics constituted a distinct race, Lapouge had put forth a profoundly disturbing challenge to liberal democracy and laissez-faire capitalism. A crucial aspect of that challenge was that it provided a way of referring to the nation's less capable, less intelligent members as a distinct group. Bouglé believed that the essence of Lapouge's questions (if not their particular formulation) was, in fact, extremely important. When anthropologists claimed, through the erroneous method of comparing cephalic indexes, that human beings differed in their capabilities, they were, explained Bouglé, pronouncing "a truth as old as the world"; people's abilities differed whether or not they were commensurate with cephalic indices (457). However, Bouglé concluded, that fact should have no bearing on their political, judicial, or economic rights. Yet even amid his plea for a revival of political philosophy, Bouglé demonstrated the importance of the issue of Darwinian struggle in this period. "Darwinian anthropologists," he wrote, insisted that the facts of nature condemn democracy, "but . . . the suppression of struggles is not one of democracy's goals; democracy only wants—and this is totally different—to regulate the struggles. By opening the same field to all individuals without distinction, does democracy annihilate competition in any way? One could not even say that it attenuates it" (459). If indeed there is a superior race, he argued, "its natural superiority will triumph just as well in a fair fight" (458).[19]

Bouglé published "Anthropologie et démocratie" in the *Revue de métaphysique et de moral*, and the journal's hostility toward scientism was echoed in the article: while he did consistently point to Vacher de Lapouge's anthroposociology as the primary offense, he seemed to include all of anthropology, and many anthropologists, in his indictment.[20] Indeed, despite the fact that Bouglé depended on Manouvrier for his dismissal of anthroposociology, he wrote of Manouvrier's contribution as if it were an almost accidental betrayal of all anthropology. "The anthropologists themselves," wrote Bouglé, with a footnote citing several of Manouvrier's antireductionist articles, "have observed that, if the anatomical constitution of an individual implies certain very general aptitudes, it is the social milieu that determines them. The anthropologists have also observed that, because of this, it is a chimera to try to deduce from an examination of purely biological characteristics, the necessity of sociologically defined acts" (450).[21] In future works, Bouglé would give much more credit to Manouvrier as a critic of anthropology's claims, but it is important to see that Bouglé and many other writers tended to issue strong critiques of all science and all anthropology (and not sociology) when they began to critique anthroposociology.

The *Revue de métaphysique et de morale* was a natural home for Bouglé's more theoretical work, but he worried that those who were awed by Lapouge's scientific data would be moved only by a scientific, data-laden rebuttal. He thus straddled several different academic fields as he progressed from his early critique of bad science to a later, more general indictment of the natural-science-as-politics enterprise as a whole, always remaining true to an antireligious secular republicanism. Despite his strict rejection of the pragmatic "return to religion," he gingerly approached Brunetière on the subject of combining their resources in a struggle against anthroposociology. In a private letter now in the archival collection of the Bibliotèque nationale, Bouglé candidly admitted to Brunetière that he had "combated several of your ideas and methods with all possible vigor." But he went on to say that, in light of an earlier conversation, he was sure Brunetière would be eager to fight the "pretensions of anthroposociology." Bouglé proposed to discredit anthroposociology by writing a study of the caste system in India and apparently hoped that Brunetière would publish sections of it in his *Revue des deux mondes*. He asserted that the work would show that Indian marriage rules had not given the results predicted by anthroposociology and that "it is impossible to find, even in this land, a true parallel among social differences, physical differences, and mental differences."[22] We do not have Brunetière's response, but it seems that he turned down the proposal. Bouglé went ahead with his study nonetheless; it was published in sections in *L'année sociologique* and *La grande revue* and as a book entitled *Essais sur le régime des castes*.[23] This work was explicitly aimed against anthroposociology (especially the section on race), and it came to be one of Bouglé's most influential sociological studies. In the words of historian of sociology Don Martindale: "More than any other single study, this essay laid the basis for the modern theory of caste."[24]

In 1904 Bouglé, who was now a professor of social philosophy at the Sorbonne, sharply criticized all attempts to describe history through natural history. In *La démocratie devant la science*, he divided such attempts into Social Darwinism, organicism, and anthroposociology. To the extent that he could, he countered each of these notions with anthropological, "scientific" arguments, resisting the idea that Enlightenment ideals might have to be divorced from Enlightenment methodology. He was no longer naming Lapouge as his pri-

mary opponent, but he still referred to him when articulating the philosophical problem of science and political equality: "Anthropology, according to M.
Vacher de Lapouge, victoriously refutes the errors of the eighteenth century,
'the most fantasy-believing, the most antiscientific of centuries,' and demonstrates that a democratic regime is 'the worst condition in which to make good
[hereditary] selection.' "[25] Denying the scientific validity of antidemocratic anthropology was not a sufficient reaction to Lapouge's critique. Bouglé made it
clear that if he were forced to choose between Enlightenment political ideals
and Enlightenment trust in science, he would choose the ideals: "Even when it
is established that solidarity exists within the best organized animal societies,
this animal solidarity does not seem to approach the human ideal: respect for
the equal dignity of each of society's members. Democratic societies recognize
from this that they are attempting to go above and beyond nature. . . . At
times acquiescing and at times resisting nature, society seems to say to natural
science both: 'I will apply your laws' and 'Your laws do not apply to me'"
(288). It was a deft solution to the problem of humanity's place in nature. The
disavowal of the authority of science inherent in this idea clearly went beyond
the mere negation of unpleasant scientific findings. Bouglé insisted that neither
naturalism, nor logic, nor rationality would ever manage to make society "lift
its smallest finger" toward equality and social cohesion unless they were joined
by sentiment. "In other words, the indispensable condition of moral efficacy of
these sociological inferences [of solidarity] is the preliminary existence of a
'social spirit'" (301).

Though Bouglé called for a return to a philosophical justification of democratic ideals that lay beyond scientific discovery, he did occasionally argue that
a truly objective science would demonstrate that natural laws dictated a solidarist society. He was arguing not that science would discover equality but that
it would discover that human societies should be run on principles of equality.
He acknowledged that if science were ever able to do that, there would be no
need for philosophy. Indeed, he hoped that in the distant future science would
"relegate all moral philosophy to the frontiers of society as totally useless." But
until then, he mused, France should concentrate on the revival of moral philosophy. Even those who are most dedicated to science should stop assaulting
philosophy as an unempirical and thus unnecessary discipline, because until
science was able to fulfill its promises, philosophy would be needed. Bouglé
warned that

> if it is true that the most objective scientific observation cannot yet suffice to
> demonstrate to human beings that they must work for the coming of a just
> city, of which the members aid each other to rise; if right up until the new

order it will be necessary to come to this by a sort of rational choice, then
maybe it would be imprudent (and in a democracy more than in any other society) to denigrate moral philosophy, which is the art of rational choice and
of methodically ordering the purpose of a human life in terms of a universal
purpose.

Here again we see a late-nineteenth-century scholar in the explicit expecta-
tion of a revolutionary "new order" that will be based on unfathomable new information and forever end the need for moral philosophy. In the meantime,
the ineluctable mystery of human interaction was referred to consistently,
sometimes in somewhat spiritualist language, and the act of becoming civilized
was explicitly contrasted to the natural world. "In a democracy, more than in
any other society, it is important that the culture is widely spread out so that
a communal consciousness becomes the point of the spiritual life and, learning to surpass nature, literally humanizes itself" (302).

Bouglé hoped that someday there would be an objective, egalitarian, "scientific morality," but he was quite sure that contemporary scientific morality
was unacceptable. Anthroposociological doctrines were wrong, he argued, because of the society they imagined. "Against these we can propose, according
to experience, our firm conclusions. Henceforth, we will know them by their
fruit." Bouglé knew it was very unscientific to dismiss a methodology because
it drew unpleasant results. He was uncomfortable with the position and worried that others would disagree with him and argue that egalitarian principles
are impossible to employ and that "it would be dangerous to try; it would be
much better to listen to the lessons of nature." According to Bouglé, this was
an "adroit effort to put into conflict the two great contemporary ideas; to exploit the prestige of science against the attraction of democracy," but he felt
content that this effort had been paralyzed by his analysis. Admitting that he
offered no positive proofs, Bouglé was able to declare that he was correct, anyway. "Our conclusions," wrote Bouglé, "if not imperative, are at least emancipatory. They liberate our society from its naturalist obsession. They remind it
that no one has the right to discourage the ambitions of the spirit in the name
of a so-called scientific morality. The way is clear" (303).

Several years later, Bouglé came back to these ideas in his *Qu'est-ce que la
sociologie?* no longer sure that science, even sociological science, would ever
furnish a true moral code. "Sociology," he wrote, "does not seem to us to be
ready—if, in fact, it will ever be ready—to substitute itself for morality."[27]
Further on in the same work Bouglé emphasized this idea more broadly, writing that "so far as morality is concerned we have recognized that sociology is
in no way ready to supplant it, and we have denounced the error of those who

propose the example of organisms when dictating the laws of societies" (160). Like Fouillée, Bouglé had moved from an investigation of natural law to a rejection of naturalist political science. Neither "ran back into the citadel of clericalism," to quote Lapouge,[28] but both came to reject the terms of scientific materialism and explicitly to welcome feeling and sentiment back into the discourse.

Despite the bold calls for equality among Bouglé's statements, he issued no direct denunciation of racism. In a political sense, Bouglé had gone beyond the question, because he insisted that society should be blind to natural differences among groups. Again, in a sense, this is the ultimate answer to sociobiological claims, whatever they may be: each human being is assumed to be different from all others and is responded to on the basis of his or her particular characteristics (again, this is not a call for total equality; it is a call for a truly fair meritocracy). No differences between human groups are to be recognized. In a sense, the central issue in *La démocratie devant la science* was that "the attentive study of the laws of heredity do not at all prove that professional qualities are transmitted from father to son."[28] But this is a rather limited claim for biological indeterminism. That is why Bouglé had to have recourse, in the end, to the position that he summed up as *noli me tangere*: natural science cannot touch human values. Even if racialist science were correct, human dignity and the resulting claim to equal treatment must be set above scientific pronouncements of inequality between human groups (288). Had he been able to discredit not *bad* race science but *all* race science—that is, had he been able to announce that the search for objective natural racial laws was inherently fallacious—he would not have needed this elegant intellectual device.

Jean Finot's work demonstrates that such a critique of race science was, in fact, conceivable. Finot, born Finklehaus, was a Polish journalist who became a French citizen in 1897. In his adopted country he founded a journal, *La revue des revues*, which he edited and to which he contributed numerous articles.[29] This work put him in close contact with many of his most illustrious contemporaries. His correspondents included writers as diverse as Zola, Tolstoy, Brunetière, and Lombroso, who all published essays and other writings in *La revue des revues*.[30] Finot was himself very well known in his time. When he died in 1922, the sociologist René Worms eulogized him as an eminent "philosopher, philanthropist, patriot, hygienist, feminist, and sociologist."[31] Above all, however, Worms praised Finot for having fought against the whole "school" whose doctrine is "generally known under the name anthroposociology" (229).[32]

Finot's greatest fame came from his *Le préjugé des races* and from his later *Le préjugé et problème des sexes*.[33] Both these works put forward lively, witty arguments against the existence of innate, biological character traits and intellectual abilities. Finot, too, drew on Manouvrier's work for its scientific clout.[34] His own arguments against racism concentrated in part on debunking scientific race theory and in part on considering the sociopolitical origins of perceived racial differences. His indictment of "race prejudice" was broadly conceived, including discussions of American racism, animosity between the English and the French, and the idea of Aryan supremacy. He often cited obsolete racial categories as evidence of the transience and historical specificity of such delineations. In the foreword to the 1906 English translation of *Le préjugé des races*, Finot pleaded for the "indulgence" of his readers by reminding them that he had claimed, as early as 1901, that, contrary to popular belief, there was no innate, immutable hatred between the English and the French races.

> When my first works appeared in 1901 on that subject, mocking voices were raised to show the impossibility of an *entente* between *two races* which were so inherently different and, presumably, antagonistic. . . . The *Times*, in a remarkable article on my efforts in this direction (November 1st, 1902), was right in maintaining that it is often sufficient to breathe on the subjects of our discord to see them vanish. The union of a few men of goodwill has succeeded in overcoming the stupidity of the theory of races and of age-long prejudices![35]

Finot held that the salient differences among human beings were only individual, and though he criticized Fouillée for his racialist thinking he made use of the philosopher's notion of solidarism on behalf of international peace.[36] *Le préjugé des races* began with a discussion of English and American eugenic theorists, and, though Finot dismissed them, he did so without anger, explaining that they were simply trying to ameliorate the public health. "In France and Germany," he added, "the gospel of human inequality has taken on even stranger aspects. It is Vacher de Lapouge who is the most authoritative representative of the new doctrine. Loyal to his principles, convinced of their truth, he defends them in his work with a keenness and a talent worthy of esteem." Finot had so much respect for Lapouge's scholarship that he cited him as the quintessential opponent, writing that "in M. Vacher de Lapouge, the new doctrine finds a defender of the greatest eloquence and it suffices to examine his books for one to know all the weapons that are taken up by his coreligionists, adepts, and students" (27). Finot bemoaned the uncritical acceptance that anthroposociology had found among journalists, politicians, literary writers, artists, and the greater public and noted with disgust that the doctrine was finding its way into manuals of history and pedagogy. He assured his reader that "without doubt, this doctrine will some day take a place of honor in the history of human errors," but he lamented that, for now, "out of any one thousand educated Europeans, nine hundred and ninety-nine are persuaded of the authenticity of their Aryan origins" (356–357).

Finot's critique of racist anthropology held that the science began with certain assumptions about racial difference; he described it as "teleological" and thus without value (491). The anthroposociologists, he claimed, were "hypnotized by their primordial idea," which they supported by bringing together, "without examination, everything that seems propitious to their theory, a theory that is more political than scientific" (312). Finot's conclusions were far-reaching. He argued that character traits were specific to individuals and that even if, indeed, a trait could be found in one human group more than in another, that was due to environment and culture. Beauty, he argued, was a purely social convention, and no single standard could be set for all human beings. Neither a language type nor any system of government could be established as a native capacity of any single racial group. Finot was remarkably suspicious of racial reasoning, going so far as to assert that "the term 'race' is but a product of our mental gymnastics, the workings of our intellect, and outside all reality. Science had need of races as hypothetical groupings, and these products of art . . . have become concrete realities for the vulgar. Races as irreducible categories exist only as fictions of our brains" (501).

His analysis of the work of Manouvrier confirmed his own conclusion that "craniological measurements teach us almost nothing concerning the mental capacity and the moral value of peoples" (109). Ridiculing anthropology's "instruments of precision," he declared their data "fantastical" and meaningless. Having dismissed the scientific validity of classing people according to their cephalic indexes, Finot asked, "What is left to the anthroposociologist once the cephalic index is gone?" and answered, "Only analogy" (110). As he stated elsewhere: "Analogy does not constitute identity" (74). Still, Finot was not entirely free of the prejudices of his age, and once in a while, very rarely, came out with some bizarrely racialist descriptions. For example, though he argued equality of intelligence, he agreed that the psychologies of "primitive" peoples, "especially of Negroes," resembled the lesser classes of Europe and cited the following behaviors as common to them both: narrow-mindedness, a love for noisy knickknacks, and a penchant for gossiping among the women (456). Even here, however, in his least impressive moment, he argued that the existence of these traits was due, in both cases, to a lack of exposure to civilized culture. As he elsewhere wrote:

> The science of inequality is emphatically a science of white people. It is they who have invented it and set it going, who have maintained, cherished, and propagated it, on the basis of their observations and their deductions. Consid-

ering themselves above human beings of other colors, they have elevated into superior qualities all the traits that are peculiar to themselves, commencing with the whiteness of their skin and the pliancy of their hair. But nothing proves that these vaunted traits are real traits of superiority. (490)

Finot may not have avoided all the conceptual traps of his era, but what he did was extraordinary. He certainly recognized the folly of racial typing even when accompanied by the most left-wing politics. Consider his analysis of Alfred Fouillée: "Optimist by nature, leaning toward skepticism in regard to the exaggerations made by anthropology, he brings reserve and scruples there where his coreligionists have only global condemnations or benedictions to pronounce." So he was one of the better ones. "However, it suffices to examine his *Psychologie du peuple français* . . . or his *L'esquisse psychologie des peuples européens* to realize just how far the aberrations of this new science can go. Carried along by his subject, he, too, distributed honors and reproach, hereditary and innate virtues and vices, onto the mysterious aspirations of peoples" (298).[37] In this and many other statements, Finot was able to speak directly to an issue that his contemporaries generally dealt with by blustering avoidance. There was something mysterious in the aspirations of peoples. For Finot, it was okay not to know exactly what that was, without feeling compelled to give it a religious name or define it out of existence.

Consider the last words of *Le préjugé des races*: "As the differences among men are thus only individual, theoretically there will be no more room for internal and external hatreds, as there will be no more room for the social and political inferiorities of classes. On the ruins of the lie of race, solidarity and true equality will be born, both based on the rational sentiment of respect for the dignity of human beings" (505). One strand of materialist anthropology had led to anti-morality and, separately, to an argument for a racialist state marked by controlled breeding and compulsory sterilizations. In defense of republican ideals, Finot rejected the republican dedication to science and championed "rational sentiment" instead. For Finot, progress was not aiming toward an endpoint of utopian stasis but rather toward a kind of plateau on which human beings could endlessly change in an environment of peace and equality. "The character of a people," wrote Finot, "is thus nothing but an eternal becoming. The qualities of our soul and its aspirations remain as mobile as clouds chased by the wind" (345). In this nice formulation, the "character of a people" is indeterminate because it is ever-changing. Here, what is mysterious and inimitable in human identity, the "soul," is seen neither as an objective supernatural entity nor as a bodily secretion nor as the product of advanced animal instinct. Instead, it is described with the naturalist metaphor of a cloud chased by the wind: ever changing, unpredictable, delightful, and patently real.

Like Bouglé, Finot struggled his entire life to strike a balance among science, philosophy, and the religious needs of the masses. Later in life he wrote several books on longevity, a theme that grows in significance in an atheist context. Finot's longevity theories all carried a wistful optimism that he maintained in the explicit absence of God. It was this balanced optimism that the philosopher Henri Bergson celebrated when he presented Finot's *Progrès et bonheur* to the Académie française in 1914.[38] Two months before the beginning of the war, Bergson stood before the Académie praising Finot for discussing morality without erring either on the side of the "abstract deductions of the old metaphysics" or on the side of "pure empiricism" and, instead, balancing between the two, "for a knowledge that is clearly of the philosophical order but, without pretending to embrace the totality of the real, concentrates its attention on human activity" (1093).

Finot's other major work on longevity, *La science du bonheur*, also struggled against the dogmas of science and of religion.[39] In dealing with the issue of morality, Finot insisted on the priority of rational expectations over religious ideals, refusing to be "intoxicated by the religion of self-sacrifice . . . and especially by that of future existence. . . . We have wrapped it carefully in a purple shroud, there where the dead gods rest" (20–21). Yet following such comments, issued always with a combination of spite and sadness reserved for lost deities, Finot similarly lamented the lost prestige of science: "Nature, we are told, knows only the species. She neglects and dooms the individual. Nature is calumniated. Science is libeled in the same way" (21). The gods are really dead, while science is merely libeled. In any case, there was still philosophical drama in this matter of religion and sciences and, to varying degrees, the loss of faith in both.

Finot proceeded to devote a large portion of this book on happiness to the rehabilitation of science, albeit of a modest ilk. Certainly, science is fallible, he concluded, and much of what we now believe will be overturned, but "science endures, like the famous session of the Chamber of Deputies, which did not cease for an instant after the anarchist outrage." We may take this to mean science endures and so does democracy and civil behavior. Proclaiming that the eternally reoccurring conflict between scientific conclusion and revision is itself "beautiful, fertile, and profitable," Finot chided the pessimists, especially Brunetière and the like-minded novelist Paul Bourget (97). Finot chose a quotation from Bourget to indict scientific pessimism (and simultaneously point out that believers were in a panic): "In the presence of the final bankruptcy of scientific knowledge," wrote Bourget, "many souls will fall into a state of despair akin to that which would have seized Pascal, if he had been deprived of faith. Tragic rebellions whose equal no age has ever known will then burst forth."[40] Finot wanted his readers to note the existence of this frightened element of culture. He never abandoned science but constantly reminded his readership of its fallibility. Science and religion both preached that human beings were regressing, but, according to Finot, "the religions and the sciences are equally mistaken."[41] To the extent that both were pessimistic, both were wrong. In the struggle between science and religion, or "free-thought against the dogmas," he believed that science would and should prevail, but not to the exclusion of a devotion to human solidarity and love (234–240). Finot rebuked scientists who pretentiously dismissed metaphysics: "Science does not cease to progress, but the paths through which it leads us are not always infallible. If in every truth there is a portion of falsehood, in every falsehood there is a fragment of truth. From the scientific standpoint, nothing authorizes the logic of the sectarian mind violently rejecting everything that is not in harmony with its comprehension" (242). Conversely, he warned spiritualists to stop mocking secular moralists. "Dogmatic religions are also wrong in seeking to struggle against lay morality. The latter takes the place of religious morality when the other weakens or disappears. Social harmony requires their mutual respect. Mankind can exist only upon moral foundations. Why discredit those of science and of experience, if a portion of the nation must live by these latter?" (244). Despite such calls for tolerance, Finot tended to treat the religious as rather backward. He believed that many people could not manage to be pure materialist atheists, but to argue this he invoked the human need for spiritualism rather than its truth (242, 246). He believed that human beings long for some participation in eternity and crave a connection to something absolute: "The most positive rationalists," he wrote, "now admit the existence of spiritual needs and eternal aspirations toward the infinite" (248). He also wrote that in the future human beings would experience in a more useful way what will forever be the "same awe of and the same longing for the Infinite" (257).

In *The Science of Happiness*, Finot devoted some three hundred pages to the demise of two major models for human happiness—scientific materialism and religion—and then outlined his proposals for the well-being of humanity. They included such things as avoiding anger and envy, believing in human dignity, and respecting one's physical health. These were modest, realistic ideas that he thought could have a big impact on individuals and society. Ultimately, Finot argued that happiness would "transform the moral universe" (331). Indeed, it was already at work, perhaps most notably in the strides that women were making toward equality. In his conclusion, Finot outlined the reasons for his optimism about the progress of the world:

The Infinite, subjected to rigorous laws, seems to be more friendly. At any rate, it is less threatening. . . . Discounting, in advance, the duration of our stay on earth, we desire it to be equitable. . . . We are daily more respectful toward one another. Our dignity is ascending step by step, as well as our sentiments of justice and of truth. . . . Someday mankind will shelter in its bosom, with the same love, the children of every color and of every creed. Meanwhile, half the human race, namely the women, are profiting by more equity. From the ranks of the slaves of man, or of inferior beings, we behold them elevated to the level of his equals. The State is multiplying its duties and performing them in a more satisfactory manner. It is becoming reconciled to the principle of equality. It is more attentive to the voice of Justice. It is urging, in any case, a more and more equitable distribution of burdens and duties.

(331–332)

Finot, too, like Fouillée and Bouglé, stayed out of the citadel of clericalism, but he managed to reject scientific materialism without rejecting science as a basic worldview. This science, however, had to be guided by the very unscientific concepts of sentiment. What had so thrown Lapouge and other attentive contemporaries was that the worldview of science had its own terrors. These scared Finot, too, but he got over it: the Infinite had become easier to think about and less harrowing: "more friendly" and "less threatening." Life, he seemed to sigh, was brief and limited, but, let us do a decent job of it anyway.

Finot was a powerful adversary. When Lapouge complained, publicly or privately, about his detractors, he blamed Finot more than any other nonscientist for the general repudiation of his theories. In his final work, *Race et milieu social*, Lapouge attacked the intelligence, honesty, and education of Manouvrier, Bouglé, and others, with Finot bearing the worst of Lapouge's vitriol: he revealed Finot's real name, Finklehaus, with much anti-Semitic drama.[42] Lapouge was so angry because Finot was so effective. He made this explicit in a letter to Madison Grant, the famed American racist and author of *The Passing of the Great Race*.[43] The year was 1919, and Lapouge was explaining that he had not written much on race in the past few years: "Jews like Finklehaus (called Jean Finot) have so excited public opinion against the theory of races that it would be as dangerous as it would be useless to try to do anything."[44] Lapouge continued his campaign until his death in 1936, but not in France.

In the late nineteenth century, French theorists had to revise their understanding of science and republicanism as innately joined in the struggle against authoritarianism and dogma. Lapouge's vision of a scientifically engineered society jolted republicans into a realization that science had the potential to be extraordinarily antirepublican. Fouillée, Bouglé, and Finot each struggled to maintain the connection between republicanism and naturalist scientism. They devoted large sections of their many works to arguing that anthroposociology was bad science. What is striking, however, is that they each managed to step outside the scientific argument and question its relevance. Brunetière had revived religion, declaring that we must willfully remove ourselves from animality. Fouillée, Bouglé, and Finot made similar claims without turning to religious dogma. For them, there was no soul, but there was something— something in the paradox of human consciousness and community that justified the elevation of our ideals. Each of their positions required a certain philosophical bravery. For a long time, republicans had based their political ideologies and their public rhetoric on the conviction that scientistic empiricism was the sole road to truth. In the battle against religious and political dogma, they had used this empirical conviction as both weapon and shield: it gave mettle to their public polemics and supported them in their private existential malaise. Abandoning a commitment to empiricism without returning to religion meant a nerve-racking submission to relativism and uncertainty. This experience defined a generation of theorists and deserves our attention.

The common idea that it took Nazi eugenics to silence racialist genetic science may have to be reconsidered. That interpretation harbors the notion that, were it not for those cataclysmic excesses, today's genetic science would be unencumbered by the burdens of politics. The ability of these late-nineteenth-century thinkers to dethrone science and insist that it be treated as a mere servant of moral philosophy is deeply significant to this question. Tensions between present-day scientists and the academic left are also illuminated by this history. Given the tenor of late-twentieth-century debates on the nature of scientific truth, it is useful to note that the relativism at which these earlier theorists arrived was not an abandonment of the pursuit of objective truth in favor of a valueless universe. It was, instead, an appreciation that scientific theories of humanity are, inherently, in eternal flux. These theorists argued that some intuitive moral wisdom must be held above science. They gave themselves license simply to "know" what is right and to assert that some scientific proclamations are wrong, despite indices, bell curves, Latin names, and calibrated tools. Henceforth, they hoped, we should know them by their fruit.

BERGSON AND DURKHEIM: PHILOSOPHY AND SOCIOLOGY REJECT SCIENTIFIC MATERIALISM

When Fouillée, Bouglé, and Finot rejected Georges Vacher de Lapouge's materialist anthropology, they did so for essentially pragmatic reasons: their goal was to defend the republic and its ideals. They persisted in connecting democracy with scientism but insisted that scientific authority had to be monitored by humanist feeling. But in referring to this humanist feeling, which was not at all scientific, they bent over backward to keep it from sounding religious. At the turn of the century, two left-wing theorists came to propose grand visions of humanity that were based on the same quest for scientific indeterminism: Henri Bergson and Emile Durkheim. Both of them explicitly formulated their theories in a response to the scientific materialism of late-century anthropology.

Henri Bergson was born in 1859, the son of Jewish parents, his father from Poland and his mother from England. He studied philosophy at the Ecole normale from 1878 to 1881, taught at a series of lycées, and eventually found a place for himself teaching philosophy at the illustrious Collège de France. From early on, his lectures at the college were a standing-room-only sensation. His philosophy was vitalist. It proposed that life and consciousness consisted of an élan vital, or life force, something beyond the material world and beyond the ken of traditional science. The study of life, Bergson insisted, required a profoundly different kind of science than did the study of the material world. In one of his many conceptual illustrations, Bergson asked his readership to imagine that Western science had been created in order to study life—paying no attention at all to material technology or theory—and that it had been doing so these several millennia. Further, he asked, imagine that sometime in the nineteenth century a great steamship were to approach Europe from an unknown place. Would not the mechanics of the ship be utterly inexplicable to these hypothetical life scientists? The real Western world had so dedicated itself to the study of material stuff that any attempt it made to understand the nature of the life force would require an equally alien methodology.

The basic idea of vitalism is that the phenomena of life and consciousness are not explicable through physics, chemistry, and biology: something sort of spiritual is going on. Vitalism instead posits the idea that some natural force is responsible for all life and suggests that we speculate about its nature from the facts at hand. All living things tend to be understood by vitalists as meaningfully united in this force, so our individual solitude in this life is either a mistaken impression or a temporary exile from our place within the universal life force. Generally, vitalism holds that no separate consciousness will survive death as a self-aware entity but that we will nevertheless continue to exist as part of the unified force. The notion of a life force had been in discussion since the ancient world, yet a significant reason for the success of Bergsonian vitalism was that it was profoundly unexpected, appearing as it did at the end of a century of increasing materialism. Philosophical vitalism was not antiscientific in its stance, stating only that a new science was going to have to be created in order to deal with the questions of life, consciousness, and free will. Vitalism could thus be entertained by people who considered themselves rationalist republicans. Without having to reintroduce the notion of the individual soul, republicans could regain the sense that they were not entirely alone and that life did not really end with death.

Bergson's ideas clearly had something in common with religious spiritualism, but they were seen as threatening to the ideology of the Catholic Church. In fact, in 1914, the same year that Bergson was elected to the Académie française, his writings were put on the Catholic index of prohibited material. Any attempt to pinpoint the nature of Bergson's followers, however, founders against the enormity of the category. According to Charles Péguy, an author on whom Bergson had a profound influence, Bergson's classroom auditors—so numerous that they regularly spilled out into the street—included "elderly men, women, young girls, young men, . . . Frenchmen, foreigners, mathematicians, naturalists, . . . students in letters, students in science, medical students, . . . engineers, economists, lawyers, laymen, priests, . . . poets, artists, . . . well-known bourgeois types, socialists, [and] anarchists."[45]

Bergson's early works had concentrated on vitalist-materialist questions, as his titles *Time and Free Will* and *Matter and Memory* suggest. In 1907 the publication of Bergson's great work *L'évolution creatrice* (*Creative Evolution*) brought the philosopher terrific fame, earning him a Nobel Prize in 1927. Bergson's work is well known today for its attack on science and for the tremendously enthusiastic popular response with which that challenge was met, yet the attack was not on all science or on the scientific method itself. Rather, it was leveled quite specifically at anthropology. *Creative Evolution* was fundamentally, as its name implies, a critique of the pure materialism of Darwinian evolution. It sparked the imaginations of his generation because of its rationalist insertion of a non-Catholic, creative, purposeful life force into the discussion of the biological progress of human beings. Bergson argued that evolution was not accidental; it was guided by the life force, it was creative. It is worth mentioning that such a self-directed version of evolution does not need human help, hence Bergson's élan vital functions as an argument against eugenics. Yet its great attraction was that it combined into one doctrine the most comforting aspects of religion and the most emancipatory aspects of empirical science. His philosophy allowed for the questioning of dogma and the increasing manipulation of the environment (that is, the power of science) and the belief in progress (rather than the fall of humanity), while at the same time the élan vital provided generational continuity, a sense of partaking in eternity, and even a possibility of overcoming the finality of death. Consider, for example, the final paragraph in Bergson's chapter "The Meaning of Evolution" in *Creative Evolution*:

> But such a doctrine does not only facilitate speculation; it gives us also more power to act and to live. For, with it, we feel ourselves no longer isolated in humanity, humanity no longer seems isolated in the natural world that it dominates. As the smallest grain of dust is bound up with our entire solar system, . . . so all organized beings, from the humblest to the highest, from the first origins of life to the time in which we are, and in all places as in all times, do but evidence a single impulsion, . . . itself indivisible. All the living hold together, and all yield to the same tremendous push. The animal takes its stand on the plant, man bestrides animality, and the whole of humanity, in space and in time, is one immense army galloping beside and before and behind each of us in an overwhelming charge able to beat down every resistance and clear the most formidable obstacles, perhaps even death.[46]

Bergson got the attention he did because his work spoke to the crisis engendered by materialist, atheist anthropology. Indeed, all of this emanated not only from a critique of the accidentalism of Darwinian evolution but also from a direct critique of the established interpretation of Broca's aphasia. Bergson devoted two chapters of his *Matter and Memory* to the argument that damage to the brain inhibits physical action but not mind. The brain merely translates spirit or mind into action. He claimed to have reached this conclusion in a way remarkably similar to the way Broca reached the opposite conclusion. Broca found that damage to a certain part of the brain coincided with certain inabilities of speech. Bergson studied Broca's work and concluded that damage to the brain could disable a body's physical action, which could inhibit the communication of thought and memory, but that there was no evidence that thought or memory themselves had been damaged. As long as an absolute parallel between mind and brain did not exist, Bergson saw room for any amount of dualism, and it was this that allowed the possibility of immortality. Perhaps consciousness and ego were wholly independent of the body and could thus exist long after corporeal demise.

As late as 1911, Bergson was still using Broca's work as his central foil, as is clear from a *Times* summary of one of Bergson's extended conferences. The *Times* reported that a primary section of Bergson's lecture was entitled "Lessons from Pathology" and that Bergson used the observations of Broca to contend "that the doctrine of parallelism [of mind and brain] was contradicted by the facts."[47] This connection between Broca and Bergson is essentially absent from histories of the period, and yet the intellectual contributions of these two men are vastly more meaningful when seen in relation to one another. Broca's anthropology seemed republican and avant-garde at midcentury, because the enemy was the church, but by the time Bergson's philosophy dominated the scene, many saw biological politics and mechanistic determinism as the republic's chief foe. Broca stood for democracy in the decades after 1860, and Bergson stood for democracy in the decades after 1900. But into the twentieth century, even though Bergsonianism depended on a rejection of Broca's central work, the two names coexisted as champions of emancipatory, progressive, republican science. The opposition noted the republican significance of Bergson. For example, the ultra-right-wing Action française tried to keep him out of the Académie française in 1913. Yet old issues die hard: later in the twentieth century, antirepublicans would also use Bergson as a bludgeon against Broca's republican scientific materialism. In his 1928 study of Bergson, Jacques Chevalier presented an interesting version of this. Wrote Chevalier:

> Forty-five years after the famous "observation" made by Broca in 1861, someone took it into his head to re-examine the two brains of the aphasics in the Dupuytren Museum upon which he had "demonstrated" the lesion of the third frontal convolution of the left cervical lobe, and it was found—a thing which seems scarcely credible—first of all that these two brains had never been dissected, and then that their frontal lobes bore the marks of many other lesions besides that of the third frontal one. . . . Until 1906 nobody had ever thought of getting at the facts . . . no one dared call [the theory] into question. When Bergson first laid a hand upon it in 1897, physicians treated his action as a nonsensical move, even as "pure madness."[48]

Chevalier was the son of an army general, Marshal Pétain's godson, and a professor of philosophy at Grenoble. In 1940, twelve years after his study of Bergson was published, the Vichy regime made him minister of education; in December of that year, he reinstated religious instruction in all state schools.[49] After the war he was put on trial as a collaborator and sentenced to prison, losing his property and his civil rights, including the right to vote. The above version of the Broca-Bergson debate was thus fashioned by a voice of the far right. Despite Bergson's left-wing convictions, he could still be called in as an antidote to the powerful republican icon that Broca remained well into the twentieth century. Bergson was a republican, but he was not also a materialist.

By the end of her life, Clémence Royer had shifted toward monism, a doctrine related to vitalism: both saw life and consciousness as other than physics, and both were free from the baggage of ordinary metaphysics and religion.

Still, most of the freethinking anthropologists dramatically challenged themselves to live without a whole range of religious comforts and certainly without hope of an afterlife. Bergson re-created these comforts. His lectures and texts had a rousing, impassioned quality that spoke directly to the atheist crisis of the era, and he wrote and lectured about what he considered to be a philosophical foundation for real immortality. Emile Durkheim's new antideterminism did not offer immortality, but it did translate and revive a range of religious comforts and ideals. Durkheim, too, labored to find a midway between scientific materialism and religious dogma but concentrated his investigation on human behavior. Bergson returned indeterminacy to the questions of our existence, our thought, and our disappearance at death. Durkheim concentrated on a different aspect of religious work: community, belonging, and an educated devotion to a shared moral field.

Durkheim was born in 1858 at Epinal in Lorraine. His father was the chief rabbi of the region, and his grandfather and great-grandfather had also been rabbis. He entered the Ecole normale a year after Bergson and Jean Jaurès, the future socialist leader and historian, and he was much in their company. These two seem to have influenced him in his rejection of religion and dedication to science.[50] He passed his *agrégation* in philosophy in 1882 and went on to teach philosophy in two provincial lycées. Late in 1885 (possibly 1886) his career plan was changed by an important meeting with Louis Liard, then director of higher education. Liard held that the clergy-bound educational system of the Second Empire was holding France back and that it partially explained the French defeat by the Prussians in 1870. By the end of the meeting it was decided that Durkheim would be sent on a fellowship to Germany in order to study how philosophy and moral science were taught there. After taking the fellowship, Durkheim came back convinced that the role of the philosophy teacher was to arouse "in the minds entrusted to his care the concept of law; of making them understand that mental and social phenomena are like any other phenomena subject to laws that human volition cannot upset simply by willing and therefore that revolutions, taking the word literally, are as impossible as miracles."[51] It was an appealing message for the republican government, both as a comment on the ideological work of indoctrinating youth in science and as a caution against proposals of excessively swift social change. In 1887 Liard created a post for Durkheim at the Bordeaux Faculty of Letters: *chargé de cours* of social science and pedagogy. He became *chargé de cours* for a chair in the science of education in 1902, winning full professorship for that chair in 1906.

All these titular references to pedagogy were meaningful. One of Durkheim's main points was that an individual's moral sense and personal wellbeing were products of social cohesion—of the numerous social ties, celebrations, and interactions in which the individual is enmeshed. In his famous study *Suicide*, he concluded that there were not enough social groups beyond the family and the state in France; people were lost and atomized in the interim space and the rise in suicide was a result of this individualization. As W. Paul Vogt has pointed out, the murder rate had declined in the same period; Durkheim knew it and never said the suicide rate was more meaningful than the murder rate, but he preferred to concentrate on the bad news.[52] The cure for this social atomization was to be created in the republic's secular primary and secondary schools, as teachers learned to inculcate republican values in the new generation. In the 1880s Ferry had secularized the schools in an articulated attempt to instill "the scientific spirit" in French youth; in the decades that followed, this goal was advanced by further curricular changes, particularly those of 1902 relating to secondary education. For the students taking the teaching degree at the Ecole normale supérieure (almost all were), there was only one required course: Durkheim's pedagogy. It began in 1904, was made mandatory in 1906, and was taught by him every year thereafter. As Vogt concluded, "Obviously, his message was considered an important one" (65). The message he had for the new schoolteachers was that moral education had one central goal: to instill "respect for reason, for science, for the ideas and sentiments which are at the basis of democratic morality."[53] The position was secular and scientistic, but it referenced sentiment and ideas.

Durkheim was not the only sociologist on the scene. There were a few ideological camps, and the one that rivaled Durkheim's for a while was led by Gabriel Tarde.[54] The two had a lot in common. Both scorned any deviation from scientific materialism. Yet though Tarde and Durkheim both declared that sociology was an empirical science, they both went on to suggest ideas that their strictly materialist contemporaries found extremely metaphysical. Because groups of people behaved in ways very different from individuals, it seemed as if some force took over whenever people functioned collectively. The social group has moods, fads, outbreaks, and tension. When you are in action with the crowd, you do and feel things you would not ordinarily do and feel. The company of a sympathetic crowd—a crowd with whom you have an authentic connection—is a powerful thing, as peculiar as hypnotism. Colossal group efforts, personal conversions, and many atrocities are thereby explained. Sociology was billed as an expansion of scientific fact-finding into this particular moral world—the social world, the world of the nation. But right from the beginning, Durkheim and his followers found it useful to refer to the mystical weirdness of popular moods and public outbreaks as aspects of the collective soul. They used religious language for something that was not religious when they explained it, and the reasons they did so will become apparent.

Both Tarde and Durkheim had formulated systems for understanding social cohesion and social change, and both sounded quite determinist at times; Tarde had "laws of imitation," Durkheim had "social facts," "scientific morality," and a world mapped out with social and institutional "coercion" and "constraint." Against critiques that they had left no room for free will, both Tarde and Durkheim argued that individuals were significantly in control of their actions and were capable of making choices between the finite options available to them. Durkheim stated, in the conclusion to his *Règles de la méthode sociologique*, that his sociology would not come down on either side of "the metaphysician's great division." It supported neither determinism nor liberty. Individual people were determined by the mind's material conformation and by society. But society itself possessed a mind, free of any material construction and thus capable of maintaining the fundamental character of the republican public. "In joining together," wrote Durkheim, "the individual souls give birth to a being, a psychic being, if you will, but one that constitutes an individuality of a new genre."[55] The religious tone of this was not lost on contemporaries.

French philosophers and sociologists shared social space for a long time: almost all the Durkheimian sociologists trained in philosophy and taught philosophy in the lycée before getting college posts, sometimes in moral education or social science but usually in pedagogy. Furthermore, the grand new journal for "truly philosophical" philosophy, *Revue de métaphysique et de moral*, made itself into a welcoming home for the Durkheimians, and a few philosophers also published their work in *L'année sociologique*. The editor of the *RMM*, Xavier Léon, certainly found Durkheim's work exciting; so did Liard, who also trained and published in philosophy; and so did Lucien Lévy-Bruhl, a philosopher who was considerably influenced by Durkheim and argued with him on a number of points over the years. But many philosophers were less enthusiastic. In defense of their own discipline, French philosophers repeatedly accused Durkheim of both scientism and spiritualism. Philosopher Alphonse Darlu, Léon's philosophy professor and a regular contributor to *RMM*, took Durkheim to task for engaging in scientific morality and for suggesting that definitive truths were forthcoming. "Durkheim is pursuing, I am convinced," wrote Darlu, "this chimera of one day causing moral facts to emerge from the crucible, in a pure state, immune forever from the revisions of conscience and reason."[56]

Durkheim had a more nuanced description of his position that well recognized the intermediary stance he had taken. "It is thus," expressed Durkheim, "that this spirituality by which we characterize intellectual facts, and which seemed in the past to be either above or below the attentions of science, has itself become the object of a positive science and that, between the ideology of the introspectionists and biological naturalism, a psychological naturalism has been founded."[57] This meant that theorists should stop fighting over whether to discuss spirituality because the thing that we call spirituality is an actual thing that must be discussed, though it is not really spiritual. Science can never win by saying humanity has no spiritual feelings, Durkheim would say, because, obviously, it does. So theorists must make these feelings the subject of science. When Durkheim said "between the ideology of the introspectionists and biological naturalism, a psychological naturalism has been founded," he meant that he was defining sociology as secular, pragmatic, and rationalist, and yet he insisted that it think about spirit (knowing that it is not really spirit) and even try to reproduce spiritual feeling (knowing that it is not really spiritual). Durkheim was locating sociology between philosophy and anthropology. Consider the language and the tense precision of the following passage from the same article:

> Beyond the ideology of the psycho-sociologist and the materialistic naturalism of the socio-anthropologist there is room for a sociological naturalism which would see in social phenomena specific facts, and which would undertake to explain them while preserving a religious respect for their specificity. Nothing is wider of the mark than the mistaken accusation of materialism which has been leveled against us. Quite the contrary: from the point of view of our position, if one is to call the distinctive property of the individual's representational life "spirituality," one should say that social life is defined by its hyperspirituality. By this we mean that all the constituent attributes of mental life are found in it, but elevated to a very much higher power and in such a manner as to constitute something entirely new. Despite its metaphysical appearance, this word designates nothing more than a body of natural facts which are explained by natural causes. It does, however, warn us that the new world thus opened to science surpasses all others in complexity; it is not merely a lower field of study conceived in more ambitious terms, but one in which as yet unsuspected forces are at work, and of which the laws may not be discovered by the methods of interior analysis alone. (34)

Durkheim here articulated a position that was crafted as a middle ground among "science," "fact," "naturalism" and "spirituality," "religion," and "metaphysics." In his reference to the "socioanthropologist," he was also loudly distancing himself from the anthroposociologist Lapouge and his like. Again, Durkheim was claiming that he was not a materialist since he was interested in the spiritual, which, of course, was not really spiritual but does exist.

Human communities have an amorphous something that creates, invents, and acts in ways that seem alien to the material world and the materialist explanation of things. But Durkheim did not want to concoct a fictitious explanation for this quality, and neither did he want to accept anyone else's fictitious explanation. The credo was: no metaphysics and no miracles, but especially no miracles. He endorsed the term "metaphysics" only insofar as it "warns us" that in social life something is happening that surpasses everything else in its complexity, something guided by "as yet unsuspected forces" that will require a new method of analysis to discern. Truth is not to be found in materialism because it is stranger than materialism could yet allow.

In 1898 Durkheim had not yet dedicated himself to the establishment of a factual morality—his central project from about 1906. Still, the endeavor is present in his earlier works, and the factuality of moral law appears in the terms quoted above, as "natural facts" that are "manifestations of social life." Of these facts, wrote Durkheim in 1898, "all are expressly obligatory, and this obligation is the proof that these ways of acting and thinking are not the work of the individual but come from a moral power above him, that which the mystic calls God but which can be more scientifically conceived" (25). That was what he instructed a generation of teachers to teach: morality is obligatory even "without God" (pace Lapouge), because all those strange internal yet external forces that had always seemed to be the properties of God were all, really, the properties of society. Durkheim's "collective soul" and his notion of "that which the mystic calls God" were set out in such religious language because he wanted to bring attention to the spooky quality of something very real and almost mundane: "there are ways of behaving, of thinking, and of feeling that possess this remarkable quality: they exist outside of individual consciousness."[58] In a way, he was announcing that we get to keep God because we never had him, that is, we get to keep what we had all along taken to be him: the phenomena of our collectivity.

Durkheim handled such difficult negotiations with intellectual grace, but these ideas could get clunky, especially in other hands. As Durkheim transformed the collective mind into a somewhat spiritual more-than-the-sum-of-its-parts, sociologists as diverse as Gabriel Tarde, Gustave Le Bon, and Célestin Bouglé all argued that the collective mind was less than the sum of its parts. They all believed, to varying degrees, that any group of people (be it a parliament or a committee, an academy or a gathering in the street) is more impulsive, irritable, credulous, and intolerant than any of the group's individual members. "Such," asserted Bouglé, "are the hard truths that sociology delivers regarding democracy."[59] Le Bon even believed (and to this Bouglé took exception) that because of this phenomenon, a group of intelligent men would have the same intellectual abilities as would a group of fools. This notion was not easy to integrate with democratic ideals.

Materialists and metaphysicians alike balked at both Bouglé's and Tarde's negative image of the group mind and Durkheim's positive image of the group mind. The common critique was well expressed by the philosopher Charles Andler in the *Revue de métaphysique et de morale*, when he simply stated that "these words, 'popular intelligence' and 'popular will,' are not intelligible. They are metaphors." Critics agreed that democracy embodied a host of profound contradictions, but many insisted that sociology had no business approaching the question. It was too metaphysical, Andler contended, more literary than scientific and, in general, a pseudoscience.[60] When Bouglé responded to these critiques, he cited the great physiologist Claude Bernard, arguing that just as society is more than the sum of its individuals, a living being is more than the sum of its parts.[61] Bouglé turned this around and argued that just as a living being is more than the sum of its parts, a society is more than the sum of its individuals. Social ideas prove natural laws by analogy, and then the natural law proves the social idea was true.

A lot of this was rather sloppy, but it is nice to see Durkheim offer an apology for Tarde's awkward steps into spiritualism in the criminal anthropology debates of the 1880s and 1890s. Wrote Durkheim in 1915: "In order to understand its full significance, it is necessary to place it in the epoch in which it was conceived. This was the time when the Italian school of criminology exaggerated positivism to the point of making it into a kind of materialist metaphysics that had nothing scientific about it. Tarde demonstrated the inanity of these doctrines and reemphasized the essentially spiritual character of social phenomena."[62] Durkheim and Tarde had their differences, but just as the fight against anthroposociology brought Bouglé to treat with Brunetière, the biological determinism of Lombroso served as a unifying common enemy among the founders of sociology. In any case, Durkheim's turn-of-the-century and early-twentieth-century work came to revitalize and reorganize the conceptual field.

ELEMENTARY FORMS OF RELIGIOUS LIFE

Durkheim's magnum opus of 1912, *Elementary Forms of Religious Life*, is particularly interesting because, in a way, Durkheim was speaking directly to the freethinking anthropologists and their followers.[63] One of the main points of *Elementary Forms* is that "there are no religions that are false. All are true after their own fashion: all fulfill given conditions of human existence though in different ways" (2). By this Durkheim did not mean to imply that all were equal: some religions "bring higher mental faculties into play" or are "richer in ideas and feelings" (2), and we might add, without betraying Durkheim's type of value judgment, that some religions are more egalitarian and generous as well. The fact that they all do a certain kind of work—and Durkheim said they could not have survived if they did not do this work—does not mean that they are all good. I make a point of this because in the late twentieth century, arguing against what came to be known as Durkheim's functionalism, scholars have suggested that his mode of thought beckons us to respect anything that exists in society—racism, for example—because if it has managed to survive it must be fulfilling an important function. This misunderstanding can only happen if one forgets what Durkheim was arguing against. He was telling Christians that non-Christian religions were not false and telling atheists that no religions were false. Unlike most human arts, religions make truth claims and pretend to have physical power. A rival doctrine of truth and physical power, science, had grown up alongside religion and was much better at these particular tasks. Those seduced by science turned back to view the religion they had left and found it to be wrong in its knowledge of the world and in its promise to manipulate that world through prayer and ritual.

Durkheim's point was that explanations of the world and the ability to manipulate it were false claims for religion but that these claims were pretty much beside the point.[64] Instead, the point of religion was to create order in the human group and the physical world of that group, and this on the most profound level. Durkheim was offering an origin for Kant's "categories" and agreeing that human beings do live in a dreamworld: a socially defined, communally agreed-upon mirage. For Durkheim, however, this dreamworld of shared meaning first comes into human purview as religion (the primary expression of human culture) and later takes the forms of philosophy, science, and other human arts. As with Kant's categories, the common dreamworld generates such basic aspects of perception as our belief in the reality of time and space, but Durkheim localizes these phenomena in the social world. Of space, for instance, "in itself it has no right, no left, no high or low, no north or south, etc., . . . and since all men of the same civilization conceive of space in the same manner, it . . . implies almost necessarily that they are of social origin" (11). There was room here to take human relativism rather too far, but it was still an exciting and useful thesis. The epistemological explanation doubled as a description of social control: "Does a mind seek to free itself from these norms of all thought? Society no longer considers this a human mind in the full sense and treats it accordingly. This is why it is that when we try, even deep down inside, to get away from these fundamental notions, we feel that we are not fully free; something resists us, from inside and outside ourselves. . . . This is none other than the authority of society, passing into certain ways of thinking that are the indispensable conditions of all common action" (16). This may be disturbing, noted Durkheim, but there is not much that can be done about it. In a slightly different context, he explained: "Of course, the mental habits it implies prevented man from seeing reality as his senses show it to him; but as the senses show it to him, reality has the grave disadvantage of being resistant to all explanation" (239). Since Durkheim's study was about a totemic religion of the Australian aboriginal, he was easily able to show the extent of this: here, individuals of a particular tribe said they were the same as the white cockatoo. That sameness might escape us, but it is essentially just as good as insisting on the similarities of a pencil and a pen: there are no real similarities, and each society creates a system of similarity and difference that is so grounded in their order of things (it is the origin of their order of things) that it feels right to all its members. These divisions and likenesses may be totemic (complex but not equipped with intrasubjectively verifiable proofs) or scientific (complex and equipped with such proofs), but they begin as a single division. "All known religious beliefs display a common feature: They presuppose a classification of the real or ideal things that men conceive of into two classes . . . sacred and profane. . . . Such is the distinctive trait of religious thought" (34).

Durkheim added to this the notion that collective action, a group engaged in a common behavior, creates powerful feelings in its members: a "collective effervescence." The feelings are elevated, they seem to come from outside oneself, and to the extent that one participates at all, but especially if one participates noticeably, collective action creates deep feelings of power and pride. These may be utterly incommensurable with the individual's usual self-image. Durkheim calls it a fact that, "Nowhere can a collective feeling become conscious of itself without fixing upon a tangible object" (238). Thus these feelings get projected onto a totem and seem to emanate from it. Gods later derive from the totems. Our moral life is socially determined, but because we are ignorant of this, we locate the dualism of individual and society as a personal duality: "To make this duality intelligible, it is by no means necessary to imagine a mysterious and unrepresentable substance opposed to the body, under the name 'soul.' But in this case, too," cautioned Durkheim, "as in that of the sacred, the error is in the literal character of the symbol used, not in the reality of the fact symbolized. It is true that our nature is double; there truly is a parcel of divinity in us, because there is in us a parcel of the grand ideals that are the soul of collectivity" (267).

Durkheim was secularizing God and the soul in a way that not only preserved and validated them but also made their concepts meaningful and useful
in new ways. "In sum, belief in the immortality of souls is the only way man is
able to comprehend a fact that cannot fail to attract his attention: the perpe-
tuity of the group's life. . . . Since it is always the same clan with the same
totemic principle, it must also be the same souls" (271). Immortality was
equally explained through social feeling, in a way that spoke directly to the issues raised by Lapouge, and in a broader sense, by the idea of the nation-state,
with its personality, memory, and continuity over time. As Durkheim continued: "Thus, there is a mystical sort of germinative plasma that is transmitted
from generation to generation and that creates, or at least is held to create, the
spiritual unity of the clan over time" (271–272). Durkheim thus allowed himself to talk about soul (a social phenomenon) and immortality (another social
phenomenon) and even God: "gods are only the symbolic expression" of society (351). "Thus if the totem is the symbol of both the god and the society, is
this not because the god and the society are one and the same?" (208). We feel
as if we were being acted on by a force outside ourselves, "a moral being upon
which we depend." Wrote Durkheim, "Now, this being exists: It is society"
(352). There was no reason to argue against religion anymore.

> In this way, religion acquires a sense and a reasonableness that the most mili-
> tant rationalist can not fail to recognize. The main object of religion is not to
> give man a representation of the natural universe, for if that had been its es-
> sential task, how it could have held on would be incomprehensible. In this re-
> spect it is barely more than a fabric of errors. But religion is first and foremost
> a system of ideas by means of which individuals imagine the society in which
> they are members and the obscure yet intimate relations they have with it.
> Such is its paramount role. And although this representation is symbolic and
> metaphorical, it is not unfaithful. It fully translates the essence of the relations
> to be accounted for. It is true with a truth that is eternal that there exists out-
> side us something greater than we and with which we commune. (227)

These conclusions offered French republicanism an amazing package of solutions to its most pressing problems. To Lapouge's nihilist lament that the universe had no "up nor down" without God, Durkheim responded that the functions of the old conception of God were easily filled by society, because that is
who God had always been, anyway. To the crisis over mortality and Lapouge's
solution of heredity, Durkheim responded that what had been mistaken for the
immortal soul had always been, in reality, the simple fact of group continuity.
We need not worry about how to create group continuity in the absence of

soul, because group continuity is more real than the notion of soul, which was a secondary, fully dependent idea. Neither was morality troubled by the loss of God. The freethinkers had tried to demonstrate that morality was only habit strengthened by heredity, and they argued this in order to defeat the idea that our moral sense was either a metaphysical manifestation of the noumenal or an invisible law made sensible to us by God. For Durkheim, this secularizing effort was no longer necessary, because morality was a real thing that came from the place we had always thought it came from, we just had the name wrong: it was society, not God. Long before Durkheim came up with this conception, he was deeply engaged in revitalizing morality, belonging, commitment, and community through the manageable, midsized collectivity of educational institutions. This was a practical, pragmatic behavior, and now it had a rather dramatic and all-inclusive theoretical meaning. Merged with the new character studies of the nation-states and the growing pastoral power of the natalist welfare state, the result was a potent, romantic conception of the modern national community.

Durkheim was specifically saying that the freethinkers' style of studying religion in order to eradicate it was wrong; it was based on thinking that those who believed in God, or in the totems, or in their own status as cockatoos, were marked by "a kind of thoroughgoing idiocy," and this was simply not admissible (177). We hear echoes of Réville's answer to Véron, but Durkheim went further. Indeed, he specifically asked, "How could this amazing dupery have perpetuated itself through the whole course of history?" (66) and "What sort of science is it whose principal discovery is to make the very object it treats disappear?" (67). The first question is rhetorical. The answer to the second is that this "sort of science" is itself very religious. Its intent was not to understand religion but to mark it off as profane. In a way, Durkheim was speaking directly to the freethinking anthropologists and their followers and also, less consciously, describing them. The following passage begins with Durkheim chastising the freethinking anthropologists and their ilk for the anthropological dismissal of religion:

> To grant that the crude cults of Australian tribes might help us understand Christianity, for example, is to assume—is it not?—that Christianity proceeds from the same mentality, in other words, that it is made up of the same superstitions and rests on the same errors. . . . I need not go into the question here whether scholars can be found who were guilty of this and who have made history and the ethnography of religions a means of making war against religion. In any event, such could not possibly be a sociologist's point of view.

At this point, it is interesting to apply Durkheim's next comment to the institution of freethinking anthropology, taking the behaviors of science as our object of study where Durkheim takes the behaviors of religions. The text continues (with no breach):

> Indeed, it is a fundamental postulate of sociology that a human institution cannot rest upon error and falsehood. If it did, it could not endure. If it had not been grounded in the nature of things, in those very things it would have met resistance that it could not have overcome. Therefore when I approach the study of primitive religions, it is with the certainty that they are founded in and express the real. . . . No doubt, when all we do is consider the formulas literally, these religious beliefs and practices appear disconcerting, and our inclination might be to write them off to some sort of inborn aberration. But we must know how to reach beneath the symbol to grasp the reality it represents and that gives the symbol its true meaning. The most bizarre or barbarous rites and the strangest myths translate some human need and some aspect of life, whether social or individual. The reasons the faithful settle for in justifying those rites and myths may be mistaken, and most often are; but the true reasons exist nonetheless, and it is the business of science to uncover them. Fundamentally, there are no religions that are false. (2)

By contrast, historically, there are sciences that are false. But in Durkheim's terms their longevity suggests that, while they are false as sciences, they "translate some human need and some aspect of life." The idea that perfectly reasonable people did bizarre science for long periods of time is not only explained by the scientiests' social and political agendas, but also by their emotional and philosophical needs. Durkheim described religion as a function, that is, he claimed that ritual works: "The essence of the cult is the cycle of feasts that are regularly repeated at definite times" (353). His emphasis was on behavior over doctrine, experience over knowledge. When "preachers undertake to make a convert, they focus less upon directly establishing . . . the truth of some particular proposition . . . than upon awakening the sense of moral support that regular celebration of the cult provides" (364). As Durkheim continues, "In this way they create a predisposition toward believing that goes in advance of proof, influences the intellect to pass over the inadequacy of the logical arguments" (365). A further thought recalls the dinner celebrating Mathias Duval: "One is more sure in one's faith when one sees how far into the past it goes and what great things it has inspired. This is the feature of the ceremony that makes it instructive" (379). For Durkheim, in religion, "the sacred is thrown into an ideal and transcendent milieu, while the residuum is abandoned as the property of the material world" (36); the "rotting garbage" of the Society of Mutual Autopsy is again called to mind.

Durkheim read religion as centrally about the same kind of separating project (between the sacred and the profane) as enacted by the freethinkers (between science and religion). Durkheim wrote: "A society whose members are united because they imagine the sacred world and its relations with the profane world in the same way, and because they translate this common representation into identical practices, is what is called a Church" (41). And further: "There is religion as soon as the sacred is distinguished from the profane, and we have seen that totemism is a vast system of sacred things" (185). We have seen that late-nineteenth-century French anthropology was a vast system of unsacred things. They raced around the physical and intellectual landscape claiming things for the profane and converting the sacred: every concept they could think of was translated through evolution and materialism, to the point of dividing up the pieces of their own bodies.

In this great work on religion, Durkheim offered very little intentional commentary on his own country's religious experience, but there was some. In a brief paragraph in the middle of the massive *Elementary Forms* and then for a sentence or two in the book's conclusion, Durkheim referenced the way this worked in "modern" Europe, though he never brought it fully into the nineteenth century:

> Nowhere has society's ability to make itself a god or to create gods been more in evidence than during the first years of the Revolution. In the general enthusiasm of that time, things that were by nature purely secular were transformed by public opinion into sacred things: Fatherland, Liberty, Reason. A religion tended to establish itself spontaneously with its own dogma, symbols, altars, and feast days. It was to these spontaneous hopes that the cult of Reason and the Supreme Being tried to give a kind of authoritative fulfillment. Granted, this religious novelty did not last. The patriotic enthusiasm that originally stirred the masses died away and the cause having departed, the effect could not hold. But brief though it was, this experiment loses none of its sociological interest. In a specific case, we saw society and its fundamental ideas becoming the object of a genuine cult directly—and without transfiguration of any kind. (215–216)

That was in the middle of the book. Some two hundred pages later, Durkheim returned to the theme:

If today we have some difficulty imagining what the feasts and ceremonies of the future will be, it is because we are going through a period of transition and moral mediocrity. The great things of the past that excited our fathers no longer arouse the same zeal among us. . . . Meanwhile, no replacement for them has yet been created. . . . We have already seen how the Revolution instituted a whole cycle of celebrations in order to keep the principles that inspired it eternally young. . . . Everything leads us to believe that the work will sooner or later be taken up again. (429–430)

So he did not quite see it as being taken up again by himself, by the freethinking anthropologists and other scientific materialists, and by the whole team of secularizing republicans who changed the entire school system, wrote a flood of books and articles, ran conferences, founded journals, financed scientific evangelists, threw parties to celebrate Voltaire and Diderot, donated their own bodies for dissection, pulled down crosses, changed street names, turned convents into secular schools, wrote atheist psalms, kicked the nuns out of the hospitals, and banished the church from the bedroom. By the second decade of the twentieth century, Durkheim saw that eradicating all religion in order to win an old grudge match with authoritarianism and factual error did not make sense; society simply lost too much in the divorce.

Science is said to deny religion in principle. But religion exists; it is a system of given facts; in short, it is a reality. How could science deny a reality? Furthermore, insofar as religion is action and insofar as it is a means of making men live, science cannot possibly take its place. . . . Faith is above all a spur to action. . . . Science is fragmentary and incomplete; it advances but slowly and is never finished; but life—that cannot wait. Theories whose calling is to make people live and make them act, must therefore rush ahead of science and complete it prematurely. They are only possible if the demands of practicality and vital necessities, such as we feel without distinctly conceiving them, push thought beyond what science permits us to affirm. In this way, even the most rational and secularized religions cannot and can never do without a particular kind of speculation which, although having the same objects as science itself, still cannot be properly scientific. The obscure intuitions of sense and sensibility often take the place of logical reasoning. (432–433)

Durkheim never stopped scolding people like the freethinking anthropologists who argued that religion was not scientific and therefore was useless, evil, and wrong. Yet in a way they had created an example of the vital, ideal collectivity that he was describing. Indeed, Durkheim's own era was unusually marked by just what he seemed to long for: passionate, effervescent movements that sought to realign modern mentality and that, along the way, provided a rich religious community for their members. He himself may be said to have led such a movement. Durkheim was not only talking about what would be good for society; he wanted a real answer to the tremendous problem posed by the freethinking anthropologists: for many people, the idea that there was no soul was uncomfortable, counterintuitive, and silenced more questions than it answered. Durkheim found a way to open those questions again. To listen to Durkheim one last time:

> In this way, there really is a part of us that is not directly subordinate to the organic factor: That part is everything that represents society in us. The general ideas that religion or science impresses upon our minds, the mental operations that these ideas presuppose, the beliefs and feelings on which our moral life is based—all the higher forms of psychic activity that society simulates and develops in us—are not, like our sensations and bodily states, towed along by the body. . . . The determinism that reigns in that world of representations is thus far more supple than the determinism that reigns in our flesh-and-blood constitution, and leaves the agent with a justified impression of greater liberty. The milieu in which we move in this way is somehow less opaque and resistant. In it we feel, and are, more at ease. In other words, the only means we have of liberating ourselves from physical forces is to oppose them with collective forces. (274)

Such a claim is dependent on the thought-as-material-product idea that the freethinkers championed, and it provides a wonderfully creative way out of the problem. It was the collectivity, much maligned though it was, that had given us the idea of the soul. In fact, it was the soul. The problem of religion had risen with the flowering of the sovereignty of the people, as it seemed that human authority had to be recognized as the true description of reality and the final word on justice. With the republic well established and in need of its own language of values and feelings, Durkheim found the "true" soul and the "true" God emanating from the people, and not only in the republic: "the people" had always been the reality of the sensations of God, soul, and immortality.

From the 1890s to the First World War, despite persistent scientism in the political world and despite a populist revival of Catholicism, the intellectual and cultural trend was toward new theories of scientific mind-body indeterminacy: "clouds chased by the wind," bad science that you "know by its fruit," vitalist philosophy, and the sociological "collective soul." For Finot, the infinite was growing "more friendly" and mortality less threatening, and the vehicle was humanity's increasing mutual respect, dignity, and sentiments of justice and of truth. Similarly, for Durkheim, an understanding of "collective forces" justified a general impression of greater liberty and existential comfort. The emblematic ideas of the period were those that eased the strain between science and mystery, and rehabilitated "spirit" and "soul" for defenders of secularism and science. Freethinking anthropology—with its determinism, its head measuring, its materialist aesthetics, and its mutual autopsy—could not hold the same cultural space in the new century as it had in the last. In the beginning of the twentieth century, it seemed like a good idea to stop bludgeoning religion, which seemed sufficiently marginalized and toothless to be tolerated, and instead to try to speak rationally about the passion of the human experience and the variety of once-religious human needs.

CHAPTER EIGHT

∞

Coda

In November 1899 Léonce Manouvrier was nominated for the chair of modern philosophy at the Collège de France. He lost, but only to Henri Bergson, one of the most prominent and respected French philosophers of the era, and Gabriel Tarde, who rivaled Emile Durkheim as a founder of sociology. (There were two empty posts.) Léonce Manouvrier, on the other hand, was an anthropologist with no conventional philosophical training, who had spent his entire career measuring bones and skulls and weighing brains. Manouvrier got as far as he did because he pitched himself as a scientist who could police the discipline. Where others turned away from science toward religion, mysticism, or vitalism, Manouvrier persisted in his attempt to marry anthropology to philosophy. His application for the chair of modern philosophy was the culmination of that. It was given a hearing at the Collège de France not because it was positivist and scientific but because it represented a critique of the grandiose claims of scientific positivism while promising to generate a rationally based morality. He stressed that the character of social and political proposals made "in the name of the law of evolution" had deeply enhanced "the immense social importance of making it well known just up to what point the law applies to humans, biologically and socially."

"What's all this talk about the failure of science?" wrote Manouvrier. "As if it were from science that Morality has, up until now, asked for illumination!" Though he made it perfectly clear that he did not believe that science could become capable of resolving all "the extremely complicated problems of Sociology and Morality," he implied that it was capable of profoundly influencing these disciplines. Manouvrier proclaimed that in a not-too-distant future, the scientific study of human beings would be an integral part of the program in all the écoles supérieures concerned with the direction of humanity. But, until then, he wrote, "it is in the Collège de France that Anthropological Philosophy would seem to have its natural place." Manouvrier closed his address with the warning that, unless anthropology were systematically consulted and appreciated, it would continue to be used in a piecemeal and irresponsible manner and eventually lead to disaster. He wrote that a tremendous movement was under way and growing every day, that it was carried by a multitude of books, journals, learned societies, and congresses where "the lack of competence in the study of anthropology is acutely felt and the utilization of scientific facts without their being appreciated or supervised has already delivered a plethora of veritable aberrations." In a companion essay summarizing his life's work, Manouvrier cataloged his many struggles against the reductionist anthropological "moral and social movement."[1] The use of anthropology in other fields of knowledge had, conceded Manouvrier, yielded some "occasionally brilliant theories," but, he added, when they are concerned with "the direction of men, the reformation of laws or of morals, or of orienting social aspirations," they have been at least as dangerous as they have been brilliant. "It is not to be doubted that, from this point of view, the teaching proposed here responds to an urgent necessity. The movement of which I have just spoken could be fertile, but in the absence of a critique that is both scientific and philosophical, it risks becoming nothing more than a sterile agitation that is more of a retardant than a boon to the progress of morality."[2]

Manouvrier did come to work at the Collège de France, in two capacities. First, the physiologist and professor at the Collège de France, Etienne-Jules Marey, set up a photographic laboratory in the Bois de Boulogne, where he studied the movement of animals and people, using all sorts of innovative techniques. He had seen Eadweard Muybridge's famous images of running horses in 1879 in La nature and went further with the idea—with much funding by the Paris municipal council and the Ministry of Public Instruction. In 1908 he created the position of assistant director just for Manouvrier, and, at what came to be known as the Station physiologique, the two formulated and carried out a great variety of photographic experiments with horses, birds, and people. Their movement studies attracted the attention of the government and influenced methods of training French soldiers.[3] Manouvrier also came to teach at the Collège de France, as a substitute in the chair of *histoire générale des sciences*. The creation of this chair had been requested by August Comte in 1832. It became a reality for his disciple, Pierre Laffitte, but slowly: in 1882 Ferry set him up in a history of science *cours libres*, in 1888 Liard arranged another, and in the French parliament, in 1892, Léon Bourgeois championed the

creation of the chair in the history of sciences at the Collège de France especially for Laffitte. Bourgeois argued that "there is no higher education worthy
of its name that does not have a scientific philosophy at its summit."[4] When Laffitte died, applicants for the chair included Léonce Manouvrier, Paul Tannery,
and Grégoire Wyrouboff, and in the end it came down to a competition between the latter two scholars.[5] Tannery, a widely known historian of science,
was elected by a large majority. In an almost unprecedented act, however, the
minister of public instruction, Joseph Chaumié, decided to override the professors' votes and give the position to Wyrouboff.[6]

Chaumié was part of the extremely anticlerical Combes government of
1902–1905. Tannery was a devout Catholic. He belonged to the Catholic Scientific Society of Brussels, which was dedicated to the compatibility of science
and Christian faith and published *Questions scientifique*, the Catholic science
journal that covered the freethinkers' work so minutely.[7] Yet Tannery's history
of science betrayed no sign of his religious beliefs. Indeed, he was deeply influenced by Comte. Tannery's work was internationally lauded and used as a
model for years to come. In 1903, at a congress on historical sciences in
Rome, he was made president of a permanent committee on the history of science. It was in that year that he received forty of forty-seven votes for the chair
of the history of science at the Collège de France, as well as the support of the
Academy of Sciences. Wyrouboff, for his part, was arguably the foremost living representative of the positivist school. He had trained in science but was
also deeply concerned with politics and social questions. In 1867 he founded
Philosophie positive with Emile Littré, and for seventeen years he edited and
wrote for that journal. Two years after the death of Littré, Wyrouboff gave up
working on the journal and devoted himself to working in the field of crystal-
lography and physicochemistry (considered extremely progressive at the
time). He was reputed to be a competent scientist, but he was not known as a
historian of science. Indeed, as it was expressed by George Sarton, the eminent twentieth-century historian of science and founder of the journals *Isis* and
Osiris, "Wyrouboff . . . was not a trained historian of science and contributed nothing whatsoever to the subject, neither before his election nor
after."[8] Sarton was harsh because his intent was to cry foul: though no admis-
sion of privileging politics over competence was ever made by Chaumié, it was
assumed by contemporaries (and all future historians of the event) that this
very rare choice to override the professors' decision was the result of the government's affinity for positivism and hostility toward Catholicism.

The event rather dramatically demonstrates that the French government
understood its fundamental ideological standpoint to be fiercely positivist well
after a new antideterminist mood was beginning to take hold at the Collège de

France. There were other cases of republican prejudice against Catholic scientists. The geologist Albert de Lapparent was forced to choose between his part-time work as a state mining engineer and his chair of geology and minerology at the Catholic Institute of Paris.[9] The physicist, philosopher, and devout Catholic Pierre Duhem was kept from advancing from Bourdeaux to Paris by the violent opposition of Marcellin Berthelot and Louis Liard.[10] The marquis de Nadaillac (the archaeologist who sided with Paul Topinard against the strict materialism of the freethinking anthropologists) also experienced difficulties in his political and scientific career because of his opposition to republican scientism and his outspoken critique of materialist rationalism.[11] Topinard himself was ousted from his professorial chair at the School of Anthropology, twice convened a governmental hearing in his defense, and saw his plea twice rejected. As Henri Brisson, then president of the Chamber of Deputies, wrote, in defense of republican positivism: "The formula 'the bankruptcy of science' is, above all, a phrase of the political order."[12] The notion that materialist, anticlerical science was the only means to republican progress was equally political. Topinard was not publicly religious; his breach with the anticlerics came because he tried to stop anthropology from becoming actively antireligious. But Tannery, Duhem, and de Nadaillac must have presented a significant conceptual problem to republicans who had essentially defined science as "that which is in opposition to religion." The prejudice they met should be understood as having its origins not only in an ideological opposition but also in the confusion engendered by shifting ideological alliances.

It was the prejudice against Catholic scientists that allowed Manouvrier eventually to teach at the Collège de France, for when Wyrouboff was in ill health in the academic years 1909–1910 and 1912–1913, Manouvrier was asked to step in and teach his courses. This is especially ironic because, even more than in the past, Manouvrier had presented himself to the professors as the scientist who questioned science. Indeed, this time around, he actually mentioned religion, and though his remarks are those of an unbeliever, they were conciliatory in tone. In this job application, Manouvrier suggested that religious morality would be preferable to no morality at all. "The morality associated with religion," he mused, "certainly possess precepts that seem to have no need to be further perfected." He even suggested that science should support religion, because, though religion had created an almost perfect moral code, it had not sufficiently convinced people that they ought to abide by that code, partially because people no longer believed in supernatural sanctions. Also, moral decisions were not always clear, and science could help. Wrote Manouvrier: "The role of science consists in making precise what is good and what is bad in a plethora of cases wherein this is not known and thus to second or to replace faith with the positive demonstration of general precepts and their natural sanction." He was thus comfortable with the notion that science and "natural sanction" might serve to "second" faith, but he gave more attention to the idea of science replacing faith. Either way, wrote Manouvrier, "thus would appear the veritable 'positive religion,' which is none other than science herself."[13]

———

THE TURN OF THE CENTURY MOVE toward indeterminism on the left took many forms. Historian Thomas Kselman has demonstrated that spiritualism persisted throughout modern France, and he invoked Charles Richet as evidence, writing that "even in the decades when positivism was fashionable the influential scientist . . . continued to take spirit manifestations seriously."[13] As proof of this, Kselman cited only one article by Richet, from 1880.[14] In fact, this article was one of Richet's paeans to hypnotism, and it was utterly materialist: he argued that somehow the phenomenon was real and that science would eventually understand it. The arguments Richet used to prove that hypnotism existed were about the ability to repeat the experiments and to predict accurately their general course, that is, they were perfectly in line with scientific protocol. Since hypnotism was not as predictable as chemistry, he made an analogy to disease (which also follows a varying course despite scientific reality), and because people cannot be trusted as one can trust chemicals, he asked if the doubting reader could really believe that the fifty or so people he had seen hypnotized were all liars ("without exception, without a single exception"), and all in on the hoax (340). The reports he had heard from his "very own closest friends and relatives" would also have to be lies.

This style of argument may seem provincial to the twenty-first-century science reader, but it does not sound spiritualist. Richet could attest that many enlightened people believed in hypnotism, but that was not enough. "In science," he proclaimed, "one does not persuade a few people: one has to persuade everyone" (338). This article, by the way, says more about how easy it was reduce a variety of women to tears and offers a further indication that the problem was Richet. An example is of interest and will demonstrate his mood (though this was one of the more colorful moments in the piece). He reported that he often hypnotized women and then informed them that he had cut off their arms and legs or was about to do so—remarkably cruel and irresponsible behavior. All screamed and wept, some searched wildly for their lost limbs, and one went into so profound a state of shock that her heart stopped and she did not breathe: "This state lasted about half a minute, a century of anguish for me; then a deep breath announced the return of the phenomena of life. Some might say this was an act. In any case, at the risk of sounding naive, I would not repeat the experience for any price" (343). That "risk of sounding naive" was as interesting a part of this phenomenon as anything else: sounding scientific all the time was tricky. Twelve years later, in 1892, Richet reaffirmed his scientism in a book called *Dans cent ans* (In a hundred years), predicting that "metaphysics will probably be abandoned altogether. . . . Philosophy properly so-called will cease to exist, its metaphysical side will become the sphere of the astronomer, the mathematician, and the physicist, while the psychological side will be the physiologist's portion."[16] Richet's scientistic beliefs were intact.

Yet it is true that Richet became interested in spirits. In the 1890s a young medium named Eusapia convinced the Italian criminal anthropologist Lombroso that she was the real thing. As one modern scholar has put it, "Lombroso's imprimatur opened the intellectual doors of Europe to Eusapia," but "Richet soon took over."[17] Having been overwhelmed by hypnotism, Richet had concluded that the phenomenon was real and had gone on to convince France and then the world of it. In view of this, he was rather predisposed to believe and champion unusual claims about the human mind. Richet went so far as to start an Institut de Métapsychique in Paris to study Eusapia and others, and he was joined there by the Curies and a host of other European scientists.[18] There were carefully studied seances where the mediums would seem to speak with the dead, sometimes move objects without touching them, and emit from their mouths a sort of ghost goo that the scientists took as a physical proof that something real was taking place here. Richet struggled against the absurdity of the claims being made but was eventually convinced, along with many of his colleagues. Richet even coined the word "ectoplasm" for the goo (which explains why ghost movies still tend to show victims slimy after a ghostly encounter). In a period when Röntgen was discovering X rays, the Curies were making discoveries in radioactivity, and Freud was proposing the theory of the subconscious, it seemed likely that strange new discoveries were liable to follow. Richet conducted many of his studies with Sir Oliver Lodge, a physicist knighted for his work in electromagnetic radiation, who would later be an early champion of the radical theories of atomic structure advanced by Rutherford and others. In Lodge's words:

As far as the physics of the movements were concerned, they were all produced, I believe, in accordance with the ordinary laws of matter. The ectoplasmic formation which operated was not normal; but its abnormality belongs to physiology or anatomy—it is something which biologists ought to study. It was something Richet, as a physiologist, found repugnant and was very loath to admit, but the facts were too much for him. He often said, "*C'est absolument absurde, mais c'est vrai*"—or words to that effect.[19]

Richet began to use precisely the same language he had used to convince Europe of the reality of hypnotism, this time to convince Europeans of the reality of ectoplasm and related manifestations of the psychic world. His discourse remained scientistic.

I have endeavored to keep the focus of this study on France, but a few words on Freud's relation to these issues will not be out of place. First of all, Freud's first book, predating his full development of psychoanalysis, was entitled *On Aphasia*.[20] In it, Freud disagreed with Broca's one-to-one correlation of morphological location and psychic function, presaging the more complicated relationship that we hold to be true today.[21] Second, having studied at Salpêtrière, Freud returned to Vienna eagerly citing Charcot's idea that medieval demonic possession was hysteria under another name. Freud extended this to all neuroses, writing: "In the Middle Ages neuroses played a significant part in the history of civilization, they . . . were at the root of what was factual in the history of possession and witchcraft. Documents from that period prove that the symptomatology has undergone no change up to the present day. A proper assessment and a better understanding of the disease only began with the works of Charcot. . . . Up to that time hysteria had been the *bête noire* of medicine."[22] Following Charcot, Freud also preferred the hereditary concept of hysteria over the one that had to do with overexcited female sexuality. Wrote the young Freud, "As regards what is often asserted to be the preponderant influence of abnormalities in the sexual sphere upon the development of hysteria, it must be said that its importance is as a rule over-estimated" (1:50).

In the early 1890s Freud began to change his mind about the hereditary root of hysteria, a change he described as dependent on two cures by hypnotism that he had effected. But in one scholar's estimation, "Freud drew his supporting evidence from medieval religious history."[23] Wrote Freud, "It is owing to no chance coincidence that the hysterical deliria of nuns during the epidemics of the Middle Ages took the form of violent blasphemies and unbridled erotic language."[24] Freud cited Charcot for the observation but offered his own new conclusion: the nuns were repressed. This was apparently a crucial leap into his mature theory, and he republished several times the connection between convent deleria, hysteria, and repression.[25] The debate over science and religion was thus the specific context of Freud's discovery. A commitment to scientism led, in the early twentieth century, to the formulation of a new kind of invisible world: as Durkheim turned to society as the real source of religious feeling, Freud insisted that there were no religious feelings, only psychological needs. Freud argued that religion ought to be replaced by a "more mature" alternative that could help to meet the needs of the individual.[26] Durkheim was interested in shoring up the community and strengthening moral bounds, also for the sake of the individual. These are big differences, but relating to either Freud or Durkheim, the scientistically derived invisible world (neither material nor religious) has continued to be associated and most energetically supported by members of the political left.

I offer two final object lessons: First, one of Charcot's best friends, the well-known author Alphonse Daudet, had a son who studied with Charcot in the 1880s, Léon Daudet. In 1891 he married Jeanne Hugo, the granddaughter of the great republican author and a vivacious and prominent social figure in her own right. The civil ceremony "provoked cries of scandal by conservative journals and triumphal cheers by liberal anticlerics."[27] But Léon could not find a place for himself in medicine; he came to resent Charcot and his politics and began drifting to the right. Jeanne Hugo divorced him (and later married Charcot's son), and by 1905 Daudet had become a devout member of the right-wing, anti-Semitic Action française. In 1908 his new wife gave him the money to finance a journal for the group, the Revue de l'Action française. Daudet was its editor-in-chief, and his wife edited the fashion section.[28] By then, Daudet looked back on the mild Charcot and remembered a zealot: "Not only was he an agnostic, he often was overtly hostile to Catholicism, which he considered as reactionary. . . . Charcot considered Our Lord Jesus Christ, a bit like his personal enemy."[29] The issue was still hot.

Second, the ideological work accomplished by Alfred Fouillée's wife had a tremendous effect on strengthening a new republican nationalist ideal. Fouillée's concerns with the relationship of biology and politics were marked by his analyses of the works of the philosopher Jean-Marie Guyau, who became Fouillée's son-in-law when Fouillée married Augustine Guyau.[30] Yet her work was the best known of the three: under the pseudonym Bruno she was the author of the famous children's book Le tour de France par deux enfants, which sold 7.4 million copies between its publication in 1877 and 1914. As one history has explained it, the pseudonym charmingly, "paid homage to the free-thinker Giordano Bruno."[31] Subtitled "Duty and Country," the book followed two youngsters around France as they interacted with various role models and learned to take pride in their nation. In schools, republican homes, and nurseries, Le tour de France served as a primer for civic and moral virtue, extending the ideology of the newly envisioned state to its smallest citizens.

MONUMENTS OF TEXT, STONE, FLESH, AND BONE

The freethinking anthropologists did not live to see much of the new era. Most of them died in the last decade of the nineteenth century or soon after. Yet the "little church" they built held up well—long enough to support them in their final hours and, further, to carry their memories into the future. When we see that the "Hovelacque" listed among the members of the Société d'anthropologie in 1900 was "Madame the widow Abel Hovelacque" we are reminded that for all the members, participation served some combination of intellectual, sacerdotal, and emotional needs. Gabriel Mortillet died in 1898, and by 1905 a monument was erected to him. It featured a tall column with a bronze bust of Mortillet on top; around the capstone beneath him were carved the names of the four prehistoric periods Mortillet had established, each accompanied by a portrait of a homonid depicted to look less apish by degrees. Leaning against the column was a marble sculpture of a young woman reading a book; as the inauguration notes explained, "she personifies the young student of prehistory, the future scrutinizing the past."[32] At the time of the monument's inauguration, the president of the Society of Anthropological Conferences was Dr. Arthur Chervin, who had once been the energetic young disciple of Louis-Adolphe Bertillon. His address on the occasion of the monument's dedication spoke of the "scientific trinity: Broca, Bertillon, Gabriel de Mortillet."[33] Thulié was still around and was one of the few to mention Mortillet's contribution to "the evolution of free thought."[34] In 1936 the secretary general of the Société d'anthropologie welcomed a new member, "Mme Grunevald de Mortillet, ethnographer [and] niece of our once and much-missed colleague, Adrien de Mortillet, whose warm and excellent memory she revives among us."[35] She was still a member in 1965, one hundred years after her grand-uncle, Gabriel de Mortillet, had first joined.

When Manouvrier died in January 1927, he had served as the secretary general of the Société d'anthropologie for twenty-five years. He had taken over the post from Letourneau, who himself inherited it directly from Broca. The 1927 funeral addresses referred to Manouvrier as "Broca's distinguished successor." The discourse by Dr. Raoul Anthony, on behalf of the Société d'anthropologie noted that it was rare for scientific research to have so much of an effect on the social world, but Manouvrier's works seemed to do it all the time: "Sometimes they were freeing us from dangerous social errors, be they the Lombrosian theory of innate criminality or the reveries of Gobineau and Lapouge. Sometimes they led to the overthrow of opinions and tendencies that forced open the barriers of our institutions that one would have believed the most definitive: in rehabilitating the female brain, considered inferior until him, he was manifestly one of those who did the most to open scientific, artistic, liberal, and administrative careers to women."[36] Anthony also wrote that it was "beyond doubt" that Manouvrier would one day be seen as one of the most profound thinkers of the end of the nineteenth century and the beginning of the twentieth (3). Eulogies were read by representatives of the Laboratory of Anthropology, the Collège de France, and several scientific societies, all of whom agreed that "since the death of Broca, [Manouvrier] had contributed the most to establishing the worldwide reputation of the Société d'anthropologie" (2). In the name of the Washington Academy of Science, Manouvrier's former student Aleš Hrdlička and two other important American anthropologists, J. Walter Fewkes and Walter Hough, sent their condolences and spoke of Manouvrier as "unquestionably the dean of Physical Anthropology in France" (12). They also wrote of him as "a man of great talent and one of utter unselfishness, with sterling honesty and character. Men of such qualities are born but rarely" (12). A glance at the *Bulletins et mémoires de la Société d'anthropologie de Paris* from, say, 1974 yields further memorials: six monographs were advertised on the back cover, their authors including Hovelacque, Chudzinski, Papillault, and Manouvrier.[37] Several studies in that volume referenced Manouvrier's work or declared themselves to be "following his methods. And the names of the other freethinking anthropologists were peppered throughout the texts.[38] Just after the death of Louis-Adolphe Bertillon, his sons Jacques and Alphonse set up a memorial essay prize in his name at the Société d'anthropologie. It was awarded every three years, but the memorial was more constant than that: every issue after 1885 carried the announcement of the existence of the Prix Bertillon adjacent to the announcement of the Prix Broca. A further memorial is, of course, the display of skulls labeled "intellectual" at the Paris Museum of Natural History. Then there is Paris itself. While researching for this book, I lived near enough to the museum that my address was rue de Candolle, named for the Swiss naturalist, and I regularly walked rue Broca and rue Bertillon. Over in the thirteenth arrondissement there's rue Abel Hovelacque. In the medieval town of Annecy, there is a rue Gabriel de Mortillet; there's a rue Clémence Royer in Nantes and a rue du Docteur Manouvrier in his hometown of Guéret.

CONCLUSION

∞

Republicans had a vision of France as democratic, scientific, and secular, but
even if everyone had been a republican and eager for these changes, to create
this new world out of an ancient monarchy and eldest daughter of the church
would be no mean feat. Between 1880 and 1905, republicans undertook a
great number of ideological reform projects, radically transforming the educational system so it was secular and scientistic and so that more people went
to school, for a longer time. In a massive, purposeful reeducation project for
a very old culture, the new generation of students was taught to love democracy, science, and France. In 1905 the republicans also managed to separate
church from state in a land where no state had existed before the church, a
land that had remained loyal to the pope through the Reformation and, after a
brief hiatus in the Revolution, had returned to him. The republicans ousted the
Society of Jesus from France and made seminary students serve in the army—
both initiated by the same man who authorized the Society of Mutual Autopsy.
    It is in this context that we find the freethinking anthropologists purging the
ghoulishness from the dead human body. Broca's finding about the third left
frontal circumvolution of the brain gave the freethinkers their greatest "fact"
against the soul, and while they looked for more, they did not really need more.
What they were up to was more a deconsecration of the human mind than a
quest for neurobiological information. Clémence Royer's combative image of
Darwinian evolution as a weapon against the church also had a dynamic career:
the "scientific fact" of (what was taken to be) progressive evolution became the
other major authority source for the freethinking anthropologists' great translation project. This deconsecration took place in myriad ways, on myriad once-

sacred things, concepts, and persons: the convents became secular schools for boys and girls; religious words and symbols were removed from cemeteries and street signs; human "monsters" were reclassified as natural and possessed of their own natural law within the human taxonomy; marriage was given a natural history, as were economics, aesthetics, the family, infanticide, the role of women, and moral law. The freethinkers even conjured prehistoric unbelievers so as to ruin the theistic argument of proof by universal consent. They called on book editors to delete the prayers from *The Swiss Family Robinson* and other novels, and they published naturalist, laic educational books of their own. They invented secular feasts and holidays, burial ceremonies, and memorials and wrote scientific psalms to the wonder of the universe and the passing of the gods. Atheist scientists affiliated with or adjacent to the freethinkers also deconsecrated the hospitals and took over the monitoring of reproduction and the status of women. In both these latter cases, the message was mixed, but the scientism was consistent. The identification of bodies was translated into the secular; the criminal went from sinner to aberration of natural law; demonic possession and the rapture of nuns became hysteria and neurosis. Historians have observed that science did not, on its own, lead to atheism. I must agree and add that, sometimes, atheism led to science. Science was ardently embraced by people who had already lost their faith in religion, so sometimes the science was secondary. Contemporaries often mocked the freethinking anthropologists' ideas, but because the work presented a version of the world without God, editors and reviewers consistently covered the freethinker's projects, attended to their claims, and sometimes praised them. The freethinker's books sold well, their classes were full, and an appreciative government augmented their salaries. Meanwhile, throughout this period, anthropologists all over the world were putting together an argument about human origins on the basis of fossil bones and new ideas about evolutionary models. Mortillet actually helped establish the great antiquity of humanity, but, in general, the core freethinking anthropologists' influence on this important argument was merely the climate created by their tremendous enthusiasm for the idea of materialist evolution. They certainly wrote more about their own idea of cultural and social evolution than they did about the biological evolution of species.

As the freethinkers did their work of grounding the world in secular terms, they used their science to provide a meaningful community for themselves. As it developed over the end of the nineteenth century, French anthropology came to function as an atheist religion. It was communal, idealistic, and abounded with priestlike leaders, burial ceremonies, sanctified heroes, answers to existential questions, primer books for the children, and utopian goals located far in the future. The anthropologists enjoyed scientific paradox, contemplating eternity, infinity, and accident. There were moody, ritualized, relic-laden dinners; the mail brought intimate confessions from the far-flung flock. In this way, for these people, science actively assumed religious roles and took on the eschatological, sacerdotal, and soteriological tasks of modernity. These tasks became even more important as the real Third Republic failed to live up to the almost religious, mystical fantasy of republicanism cherished by many throughout the periods of monarchy and empire. Ideas and emotions were articulated in anthropological theories, which were all adamantly atheistic and antiphilosophical, despite having been invented in direct reaction to spiritual and philosophical crises. Anthropology functioned in this capacity by providing materialist explanations of human origins and characteristics, so that a secular worldview was no longer beset by ruinous unknowns. For some, anthropology also served to assuage the losses of materialism by providing a secular framework in which to experience communal ritual and imagine and work toward a real-world utopia.

Durkheim wrote that no religions are false because they are not really supposed to be generating true information about the world. To the extent that science performs religious functions, that is, to the extent that it serves emotional and philosophical needs, it, too, can survive beyond its ability to inform. Good science gets us close-up pictures of Mars and stops our dying of tuberculosis. But that is not all it is doing. It is powerful, but it is also a social endeavor: it has rules and tendencies (and fashions and patterns of celebration) that seem integral to its stated projects but are actually extraneous or counterproductive (how could it not?). Following Durkheim, when we think about the Society of Mutual Autopsy, if we overcome our tendency to dismiss as "inborn aberration" such seemingly bizarre acts as, for instance, cutting up the brain of a dead friend, measuring thousands of colonial noses, or correlating head measurments with personality types, we may still call it bad science. As I read Durkheim, the "most bizarre or barbarous" scientific rites and the strangest scientific myths translate some human need and some aspect of life, whether social or individual. The reasons the scientists and their following settle for in justifying those rites and myths may be mistaken, but there is reason in their behavior, and it is the business of history to uncover it. [1]

The amazing Bertillon family was entrenched in the emotional and ideological matter of the freethinking anthropologists, but they transformed what Manouvrier called the anthropological stockpiles of useless numbers into active artillery. They made death visible in the population: one could watch it rise and fall. They brought the secrets of sex into the numerical modern world. They made criminals memorable, made silent bodies speak, and assigned to each of us the onus and possibility of a fixed identity. They helped invent the detective. All this happened in the context of a passionate atheism: Louis-Adolphe

Bertillon was writing freethinking articles for *Libre pensée* and *Pensée nouvelle* in the 1860s, in 1866 his wife, Zoé, had a civil funeral, and by 1883 his own skeleton was hanging in the Laboratory of Anthropology. When their son, Alphonse, died in 1914, his brain went to Manouvrier at the Laboratory of Anthropology to be weighed and analyzed. It was thirty-one years after the same room, and perhaps the same scales, had held his father's brain. The cult was that vibrant over more than half a century. Georges Vacher de Lapouge was also entrenched in the freethinking anthropologist's emotional worldview and well representative of its most desperate mood. Like the Bertillons, Lapouge seems to have had a considerable influence on the world, also in the field of categorizing people with measurements. The inventors of these numerical techniques of seeing the body politic and making it visible for manipulation were all living in the framework of the end of the soul. There is something extraordinary about Manouvrier's place in all of this. An early and influential believer in evolution, he was able to make a lasting contribution to how scientists glean information from ancient skeletons, to influence innumerable students—most significantly, the American anthropologist Aleš Hrdlička—and also to be an enthusiastic fellow traveler to the freethinking anthropologists. His brilliant battles against sexism, the idea of biological criminality, and racism are not widely known today, perhaps because he so thoroughly trounced his specific opponents that the matters seemed solved and were forgotten. In his own day, his work on these subjects made him famous as a good anthropologist and a subtle student of humanity; it was this work that took him as far as the Collège de France and almost got him a chair in philosophy there. He, too, explained that his work was in direct response to the decline of religion, as theological insults about the soul were exchanged for anthropological insults about the brain.

In general, as the atheists reported it, theirs was an intellectual, "philosophical" movement. It was not always very sophisticated, but, as they understood it, they were running toward truth, not from religious guilt or the terror of hellfire. The Catholic Church could hardly be outdone in purple ritual, brimstone, and bleeding imagery, and atheists spoke of cruelty at the hands of priests and nuns. But in the atheist discourse, nineteenth-century Catholicism was about comfort, submitting to authority, and averting one's eyes from the abyss with thoughts of wingéd angels. The debate over Darwinism was frequently described in terms of fear: the church held fast to the comforting idea that humanity was special; the anthropologists bravely asserted that we are accidental, and brutes to boot. As atheists of the period saw it, science made terrifying and bizarre announcements; religion functioned as a palliative or an opium. Anthropology offered stunning truth that could hardly be reconciled with daily life; religion was mundane. Certainly, there were devout Catholics in France in the 1880s who were stunned by the difference between their daily lives and the mystery of the Trinity, but for a whole range of people the Trinity was no longer true in the same way that this other mystery was true, the mystery of accidental creation. For many people, learning of Darwinian evolution was remembered as the great epiphany of their lives. With a seriousness that is almost hard to imagine now in relation to such matters, late-nineteenth-century theorists tried to figure out humanity's place among the animals. With no souls, should we take nature as our model or define ourselves in opposition to the natural world? Should we attempt to accelerate further biological distinction from the animals, or should we essentially stop evolution by ensuring the survival of everyone? In the nineteenth century and part of the twentieth, anthropological theory was the most powerful way to approach these questions. By the new millennium, some of the most important, far-reaching, broadly popular, and intense philosophical inquiries are taking place in the arena of genetics—especially the contested biological features of gender, sexuality, and ethnicity—as well as abortion, animal rights, reproductive technology, the science of evidence and identity, and euthanasia; these are the new locations of debate about biology and truth and the many ways to imagine the dignity of human beings. What we are in relation to the animals is a constant background theme in these debates, and it is no wonder, as the discussion has its origins in nineteenth-century anthropological discourse. These doctrines respond to very distinct ideas about God, death, eternity, individual versus collective value, and the difference between human refuse and human beings. The early story lends perspective on the present.

The primary task of this study has been to re-create the experience, worldview, and context of an enclave of late-nineteenth-century French atheism, with particular attention to the real emotional purview of materialist philosophy and ideology—its joys and terrors—and how these emotions were negotiated through science. In short, I have asked the Society of Mutual Autopsy to tell us about the battle against the church in turn-of-the-century France. The other major task has been to demonstrate that the pastoral state and the social sciences—sociology, demography, anthropology, psychology, and criminology—all came into the twentieth century profoundly engaged in questions of belief and unbelief. Doctrines that still shape our lives were created in response to the experience of the end of the soul and in the hope of finding some other way to account for ourselves. There were political hostilities being expressed therein, but there were also judicious and insightful contributions to the philosophy of the human experience and useful provisos for cultivating a rich, even spiritual life in a secular state. Modern politics and much modern language and technique for understanding and manipulating populations still bear the marks of these issues. It is useful to remember the roles that atheism and religiosity played in these dramas and the varying mutability of their alliances with other causes.

It is not surprising that people who believed themselves to be the first generation of fully responsible human beings—that is, human beings who were not at all relying on Providence—would have come up with aggressive new ways for humanity to monitor and keep track of itself—to get milk to the babies, to punish the criminal, absolve the innocent, and generally to increase the suddenly visible sectors of health, peace, and prosperity. Neither is it strange that painful, atheistic, materialist nihilism raged at the source of one doctrinal source of Nazi ideology. Outside these historically specific relationships, however, I do not think atheism is innately linked to either of these impulses. Rather, this study seems to be a witness to the way that ideologies calcify in their coalitions with other ideologies and behaviors but then, to some degree, come apart again. Still, as it happened, religion and irreligion forced each other into some extraordinary positions at the end of the nineteenth century and the beginning of the twentieth, and these extraordinary positions shaped the world that came after. Even in the most specific terms, we have carried the theories of science and public policy of the late nineteenth century into the turn of the millennium somewhat divorced from the debate over spiritualism and naturalism that defined their original context. Extreme materialism and naturalism are well understood as major forces of the period, but the extent to which they shaped the conceptual armature of Léon Bourgeois, Alfred Fouillée, Célestin Bouglé, Henri Bergson, and Emile Durkheim—and Sigmund Freud, for that matter—is hard to overestimate and not sufficiently appreciated. Through Emile Zola and Hamlin Garland, literary realism was particularly influenced by the anthropological materialism of Véron and Letourneau. Materialist anthropology and the atheism of its proponents also significantly shaped the literary work of Arthur Conan Doyle and Bram Stoker and were important to the social and political battles of Maria Montessori, Jules Ferry, and Margaret Sanger. Charles Richet and Paul Bert also shared the materialist concerns and a great many of the specific dramas of the freethinking anthropologists. Valéry measured hundreds of skulls with Lapouge, Verlaine's journal published profiles of the freethinking anthropologists. Broca and Gambetta gave their very heads to the cause. Indeed, Broca and Gabetta emerge as tremendously important figures in the history of freethinking and brain science. Broca performed the first autopsy of the Society of Mutual Autopsy, and his own brain was dissected soon after, in 1880; Gambetta encouraged Constans's anticlericalism, Constans authorized the Society of Mutual Autopsy, and Gambetta's brain was dissected soon after, in 1882. Both Broca and Gambetta had already done a great deal to establish freethinking and sci-

ence in France, and the scientism of the next several decades would be fundamentally determined by the ideological positions Broca and Gambetta had
championed, the groups they had formed, and the people they had helped to
get jobs. There could be no better symbol for this grand posthumous campaign
than the image of their material brains, sketched and labeled.

More broadly, the anthropology-versus-metaphysics debate was crucial to
theorists of sociology, philosophy, clinical psychology, and politics, and all this
took place within the explanatory context of an overt and violent crisis of unbelief. The three most significant conclusions I have to offer along these lines
have to do, first, with politics, second, with the social sciences, and third, with
the idea of race.

The meaning of political right and left changed in the translation to the
secular. The rise of anthropological theories of human inequality in late-
nineteenth-century France created a schism between Enlightenment empirical
scientism and Enlightenment egalitarian ideals. The relationship between scientific authority and politics began to shift, and the resultant changes contributed to the development of a new right and new left. Broadly, the old right
can be said to have relied on monarchical tradition and revealed religion, while
the left looked to science and the concomitant vision of historical progress. Between 1880 and 1914, the right began to employ science and numbers to support its social hierarchies, and the left saw that, in some instances, the preservation of leftist political ideals required a rejection of science. A significant
aspect of this reconfiguration was precipitated by antidemocratic biological
theories, particularly Lapouge's anthroposociology. For Fouillée, Bouglé,
Finot, and even Bergson and Durkheim, the turn toward relativism and indeterminism was effectuated, in part, because the version of humanity offered by
biological determinism had turned out to be as dangerous and demeaning as
the one offered by the Catholic Church. The new right (somewhat scientistic)
and the new left (somewhat relativist, sometimes spiritual) are revisions of the
old right and old left (which, of course, still exist). Lapouge's route to a better
future welcomed an interim period of homicidal social science. Once only the
superior people were left, true liberal democracy would be both possible and
effectively indistinguishable from total egalitarianism, exactly because a "state
of nature" of equality would have been created. The "state of nature" was thus
reconfigured as the end point rather than the starting point of sociopolitical history—a trick that is only possible if history becomes the story of repopulating
the garden with creatures of an improved and consistent quality.

In these terms, Lapouge always saw himself as a republican. But people
could tell the difference. The seriousness with which Alfred Fouillée, Célestin
Bouglé, and Jean Finot took Lapouge's physical anthropology pushed them into
a difficult position: they found themselves upholding an egalitarian legal and

political system despite the possibility that inequality was a "natural fact." They had no reason to expect a material, empirical justification for political equality in the near future, and so, because they were committed to democracy, they had to drop any allegiance they might have had to pure materialism. Materialism now had to be married to secular, but sometimes metaphysical, political philosophy. The soul was back, even if just as a metaphor, "a cloud chased by the wind." Durkheim and Bergson made a similar gesture because they found they could not speak meaningfully about the human individual and community while denying that certain aspects of life feel unempirical, unpredictable, inexplicable, and wild with subjective passion. The world does not feel like rank mechanism. Bergson and Durkheim announced that this "spiritual" feeling was legitimate, deserved to be cultivated, and yet belonged to the knowable world. Durkheim, in particular, kept his study allied to the phenomenal world. They both created a conceptual space in which human consciousness was magical enough: it had to be recognized as wondrous strange, but it was knowable. There was room on the right for science now and room on the left for some kind of indeterminacy, mystical experience, and social "spirit." France had been divided by allegiances to clergy and tradition, on the one hand, and science and equality, on the other. The fight over unegalitarian science cut across that primary division, so that some of the right took on the mantle of science, and some on the left took it off. The new right cherished much that the old right stood for—the mystique of land itself, the idea of a biological social hierarchy—but it could accommodate greater numbers: racist science democratized superiority for the common European. This nineteenth-century drama may add to our understanding of the deep political divisions that became visible under the Vichy regime and that were explored as homegrown issues, rather than Nazi impositions, in the work of Robert Owen Paxton and now many others.[2]

The second observation that I would like to highlight is a general one about the nature of modern social science, based on its origins. Just like other gospels, the "good news" of atheism seemed magical and somehow suggested that once all humanity came to understand and (dis)believe, some vital change would take place. Of course, nineteenth-century atheists had examples of vast, momentous change all around them; they lived in a time of technological, industrial, democratic, and scientific revolution. Because everything seemed to be accelerating, it seemed as if the next century would be marked by even more tumultuous and transforming leaps of progress. It was in this context that the social sciences were born. In his essay "The Genesis of the 'Final Solution' from the Spirit of Science," Detlev Peukert argued that social scientists had reason to expect big changes, they had reason to trust science to deliver revolutionary results, and they were supported by the state both as promoters of republican ideology and as technocrats of the unwieldy body politic.[3] The state and the social sciences claimed once-pastoral tasks as their own, and their normative logic caused them both to identify social problems and popularize them as acute and solvable. The social sciences invented solutions for these newly identified problems—the rise of technology and the resources of the nation-state helped to make all sorts of new, vast social experiments conceivable—and the solutions worked marginally at best. When they failed, the social scientists and governors had two choices: back away from their hubristic claims or redouble their efforts using ever more radical means. Peukert demonstrated that this escalation of claims and solutions in the social sciences took place in the sphere of social welfare education in Germany, and he identified the "good and ill inherent in the human and social sciences" as the "central common factor" in "the tangle of causes leading to the Final Solution." Along these lines, he also suggested that the Nazi crimes were but "one among other possible outcomes of the crisis of modern civilization in general" (236). I am arguing that even though France produced an important progenitor of racist theory, French social theorists negotiated this crisis toward another of the "possible outcomes." For the most part, in France, the social sciences managed to scale down their claims, rein in their excesses, and dramatically defend human dignity. They forged instead a secular scientism, humanized by a political and ethical philosophy that was sometimes based more on urgent pragmatism—gut feelings of what was right—than on any logically derived argument. The social sciences, and their relationship to government policy, have a terrific potential for good and for ill, and exploring the history of these interactions seems crucial to the health of a democracy.

The third observation is about the history of scientific racism. Science is neither always emancipatory (as was suggested by positivists and materialists alike) nor always devoted to control, classification, and domination (as is often part of the contention of postmodern theory). Furthermore, science dedicated to equality and justice is not necessarily methodologically "pure." The notion that, in late-nineteenth-century France, anthropology was deeply racist and that this racism was based on the shock of confrontation with the colonial other also seems to be in need of revision. In its most prestigious center, the Société d'anthropologie de Paris, late-nineteenth-century French anthropology represented the more egalitarian tendencies of contemporary French society. This egalitarianism was primarily concerned with socialism and feminism in French society, was not always aggressively pursued in actions, and was not consistently extended to other societies or to other races. Still the anthropologists who dominated the Société d'anthropologie in the final decades of the century consistently supported the notion that progress should be understood as the measure of increasing egalitarianism.

The common narrative about racism is that in the period after Darwin published there arose a widely accepted pseudoscience of racism that sought to explain history and the future in terms of clashes between human types. This way of thinking grew into a huge eugenics movement that included endless pages of advice for those deemed to be the good people and enforced sterilization for many of those deemed to be the bad. As the narrative continues, Nazi excesses shocked the world out of this behavior, and after the war important individuals and groups, such as the United Nations Educational, Scientific, and Cultural Organization (UNESCO) declared the term "race" scientifically meaningless, and the whole misshapen drama came to an end. The problem is, of course, that it did not. Brash racialist theories are published every few years, and they do not differ appreciably from their nineteenth-century forebears. Also, population studies, the genome project, and other modern inquiries revive many preoccupations with "type" that were supposed to have been left behind.

Scholars have focused on how the new, modern racism works, now that racism is no longer publicly condoned and scientifically argued. Ann Laura Stoler has offered the insight that this commonly accepted paradigm overstates the ways that racism was publicly condoned and scientifically argued in the past, suggesting that it was always largely propagated in more domestic and more culturally fluid ways.[4] She suggests that when we take such "high-profile racists as Gobineau, Madison Grant, or Vacher de Lapouge" as the templates of racialism, we are fooled into thinking racism once had a more clear-cut meaning and a plainly acceptable place in society (195). What the present study suggests about the standard narrative is that, first of all, the most numerically conscious anthropologists in France were on the left politically. Lapouge went the other way, but the left-wing anthropologists themselves, and the philosophical and political culture around them, pulled themselves out of the tailspin. Lapouge was beaten, in his time, by the fact that his left-wing teachers dominated Paris, by the sharp wit and keen sense of justice of Léonce Manouvrier, and by the philosophical imaginations of contemporary intellectuals who were willing to shift around some notions previously deemed fundamental. As glad a story as this is, perhaps it should be understood as one of racism's most long-lived patterns in a democracy: something rating some group's intelligence or character is sent out into the marketplace of ideas; it receives a tremendous amount of negative attention; and the experience confirms a noisy kind of triumph over unjust social hierarchy, while bringing a lot of attention to the idea that equality of citizenship is always up for theoretical renegotiation. It is important to note that Finot knew that in the future we now occupy Lapouge's ideas would seem laughable. Furthermore, for the history of racism, it seems useful simply to note how contentious all these issues were, how dependent they were on the specific issues of the day, how much they were wrapped in particular social circumstances, and how affected they were by tangential ideological alliances. There was no original moment of straightforward, uncontested scientific racism. Even if one looks precisely at one of the most obvious, high-profile nineteenth-century scientific racists, one finds a diffuse pileup of very historically specific public and private concerns, the end result of which looked odd and distasteful to contemporaries but still crucially helped shape the way people thought about human potential and human difference.

There were many impulses for those who, in a manner of speaking, responded to the loss of God's gaze by learning to watch humanity in his stead. A whole range of issues once confided to religion—from the fates of unpunished sinners to the size of the average family—would now be aggressively managed by *someone*, sometimes with the best of intentions. Along with all the horror and oppressive social control that have sprung from such techniques, there has arisen a society that can see itself and that can make attempts to ameliorate its sorrows. It seems best to end on a note of optimism, for the freethinking anthropologists were surely marked by high hopes for the future, if only we all keep trying. Looking at "savages" but talking about modernity, Durkheim used the term "the soul of collectivity."[5] Michel Foucault, looking at modernity, identified the network of social power and knowledge as "the real noncorporeal soul."[6] Consider his discussion of the soul in *Discipline and Punish*. Wrote Foucault: "It would be wrong to say that the soul is an illusion, or an ideological effect. On the contrary, it exists, it has a reality, it is produced permanently around, on, within the body by the functioning of a power that is exercised on those punished—and, in a more general way, on those one supervises, trains, and corrects. . . . This real noncorporeal soul is not a substance; it is the element in which are articulated the effects of a certain type of power and the reference of a certain type of knowledge"(29). Durkheim and Foucault called this social thing "soul" because they were filling a soul-shaped hole in their understanding of human society. These definitions imply that human behaviors, the way we order ourselves and others, create feelings that account for past claims of individual souls and presently create the world of mutual meaning in which we live. In this schema, the younger Bertillons, for instance, were engaged in creating the new "soul" of late modernity. The freethinking anthropologists were involved in a transitional behavior: exorcising traces of the past and clearing out the conceptual terrain to make room for the secular assumptions of the new scientific state. The Bertillons, Lapouge, and Manouvrier were all engaged in creating those new secular assumptions. By the time they were done—though surely not by their efforts alone—it seemed perfectly reasonable to turn to science for pastoral needs and to expect traditional religion to stand as but a subset of modern public life.

Nineteenth-century atheist materialism was powerful. Its proponents were driven, proud, and intent on changing the world. Several of the major revolutions that have taken place over the past century and half have been generated by atheist men and women who specifically understood their projects in terms of the twilight of the gods, the opiate of belief, and the future of an illusion. For many atheists, all the arrangements that human beings had made throughout history suddenly demanded revision: we were on our own for the first time, beholden to no one, with recourse only to ourselves. The freethinking anthropologists shared the moment and the mood, and they let us know more about what it was to be an atheist at the turn of that century: from their romantic cult of science to their dynamic effort to translate all French public and private life from religious to secular. Through attention to their behaviors and to the emotions expressed in their public and private speech, this study has tried to describe the experience of unbelief for the freethinking anthropologists of Paris and to detail some of the uses and meanings of their work for their famous students, their philosophical, artistic, political, and scientific associates, and the wider audience that attended to their claims.

NOTES

∞

All translations are mine unless otherwise stated.

Introduction: The End of the Soul

1. Claude Blanckaert: see, for instance: "L'anthropologie au féminine: Clémence Royer (1830–1902)," *Revue de synthèse* 105 (1982): 23–38; and "Préface: 'L'anthropologie personnifiée' Paul Broca et la biologie du genre humain," in Paul Broca, *Mémoires d'anthropologie* (Paris: Jean-Michel Place, 1989), i–xliii; see also "La science de l'homme entre humanité et inhumanité," in Claude Blanckaert, ed., *Des sciences contre l'homme: Classer, hiérarchiser, exclure* (Paris: Autrement, 1993), 14–46. Nélia Dias: see *Le Musée d'ethnographie du Trocadéro (1878–1908): Anthropologie et muséologie en France* (Paris: CNRS, 1991). Michael Hammond: see "Anthropology as a Weapon of Social Combat," *Journal of the History of the Behavioral Sciences* 16 (1980): 118–132. Joy Harvey: see "Races Specified, Evolution Transformed: The Social Context of Scientific Debates Originating in the Société d'anthropologie de Paris, 1859–1902" (Ph.D. diss., Harvard University, 1983). Elizabeth Williams: see "The Science of Man: Anthropological Thought and Institutions in Nineteenth-Century France" (Ph.D. diss., Indiana University, 1983).

2. Jennifer Michael Hecht, "Anthropological Utopias and Republican Morality: Atheism and the Mind/Body Problem in France, 1876–1914" (Ph.D. diss., Columbia University, 1995).

3. Theodore Zeldin, *France, 1848–1945* (Oxford: Oxford University Press, Clarendon Press, 1977), 994.

1. The Society of Mutual Autopsy and the Liturgy of Death

1. Archives of the Société d'autopsie mutuelle, Musée de l'homme, Paris (Hereafter ASAM), procès-verbal, Commissariat de police des Sables-d'Olonne (Vendée), Commissaire de police Eugène Corneaud, October 4, 1889.

2. ASAM, Extrait du registre des actes de décès de l'an 1889, Décès de Louis Victor Eugène Véron, May 23, 1889.

3. See Michael Hammond, "Anthropology as a Weapon of Social Combat," *Journal of the His-*

tory of the Behavioral Sciences 16 (1980): 118–132; Elizabeth A. Williams, "The Science of Man: Anthropological Thought and Institutions in Nineteenth-Century France" (Ph.D. diss., Indiana University, 1983); and Joy Harvey, "Races Specified, Evolution Transformed: The Social Context of Scientific Debates Originating in the Société d'Anthropologie de Paris 1859–1902" (Ph.D. diss., Harvard University, 1983).

4. Nélia Dias, "La Société d'autopsie mutuelle; ou, Le dévouement absolu aux progrès de l'anthropologie," *Gradhiva* 10 (1991): 26–35.

5. My own work on the society includes a chapter of my dissertation and an article: "Anthropological Utopias and Republican Morality: Atheism and the Mind/Body Problem in France, 1876–1914" (Ph.D. diss., Columbia University, 1995); and "French Scientific Materialism and the Liturgy of Death: The Invention of a Secular Version of Catholic Last Rites (1876–1914)," *French Historical Studies* 20, no. 4 (fall 1997): 703–735.

6. See Karl Löwith, *Meaning in History: The Theological Presuppositions of the Philosophy of History* (Chicago: University of Chicago Press, 1949). Löwith concludes that as mere reconfigurations of Christian eschatology, these models are inherently illegitimate. For a fascinating response, see Hans Blumenberg, *The Legitimacy of the Modern Age*, trans. Robert M. Wallace (Cambridge, Mass.: MIT Press, 1983).

7. For general studies of French attitudes toward death, funerals, and cemeteries, see Philippe Ariès, *Essais sur l'histoire de la mort en occident* (Paris: Seuil, 1975); Michel Vovelle, *La mort en l'occident de 1300 à nos jours* (Paris: Gallimard, 1983); Ellen Badone, *The Appointed Hour: Death, Worldview, and Social Change in Brittany* (Berkeley, Calif.: University of California Press, 1989); and Thomas A. Kselman, *Death and the Afterlife in Modern France* (Princeton, N.J.: Princeton University Press, 1993).

8. "Statuts de la Société d'autopsie mutuelle," *Revue scientifique* 11 (November 25, 1876): 527–528; *Les droits de l'homme*, October 25, 1876; *Le bien public*, October 24, 1876; *La tribune médicale*, no. 128 (October 29, 1876): 525–526.

9. ASAM, "Société d'autopsie-Statuts," 1876, 1.

10. Cited in David Bakan, "The Influence of Phrenology on American Philosophy," *Journal of the History of the Behavioral Sciences* 2 (1966): 214.

11. Mathias Duval, "L'aphasie depuis Broca," *Bulletins de la Société d'anthropologie de Paris* (hereafter *BSAP*), 3d ser., 10 (1887): 769. See also "Société d'autopsie," *Revue de l'Ecole d'anthropologie de Paris* 3 (1893): 233–236; and Auguste Coudereau, "L'autopsie mutuelle," *La pensée libre* 17 (November 13, 1880): 1.

12. For a discussion of autopsies and the poor in nineteenth-century England, see Ruth Richardson, *Death, Dissection, and the Destitute* (London: Routledge and Kegan Paul, 1987).

13. ASAM, "Société d'autopsie-Statuts," 1.

14. Ibid.

15. This scientific exclusivity was never invoked to reject an applicant.

16. There was very little discussion of eugenics among these anthropologists. Papillault was a member of the Paris Eugenics Society.

17. ASAM, "Société d'autopsie-Statuts," 3.

18. Henri Thulié, "Sur l'autopsie de Louis Asseline, membre de la Société d'anthropologie et de la Société d'autopsie mutuelle," *BSAP* 3d ser., 10 (1878): 161–167.

19. André Lefèvre, "Louis Asseline," *Contre-poison*, 1900, 259–314.

20. Mathias Duval, Théophile Chudzinski, and Georges Hervé, "Description morphologique du cerveau de Coudereau," *BSAP* 3d ser., 6 (1883): 337–389.

21. Théophile Chudzinski and Mathias Duval, "Description morphologique du cerveau de Gambetta," *BSAP* 3d ser., 6 (1883): 129–152; and Mathias Duval, "Le poids de l'encéphale de Gambetta," *BSAP* 3d ser., 6 (1883): 399–417.

22. Gabriel Deville, editor of the republican journal *Droits de l'homme*, joined the Society of Mutual Autopsy in 1876, having read about it in the *Bien publique*. Dr. J. Bach, of the Paris Faculty of Medicine, joined in 1876, having learned of it in the *Revue scientifique*. Not surprisingly, *Droits de l'homme* ran an article on the society soon afterward, as did *Le réveil médical*, soon after its director, Dr. E. Monin, joined the society in 1880 ASAM, Dr. J. Bach to the Société d'autopsie mutuelle, September 28, 1876; "Statuts de la Société d'autopsie mutuelle," *Revue scientifique* 11, no. 22 (November 25, 1876): 527–528; ASAM, Gabriel Deville to the Société d'autopsie mutuelle, October 1876; ASAM, Dr. E. Monin to the Société d'autopsie mutuelle, June 28, 1880. The article in *Droits de l'homme* ran on October 25, 1876.

23. ASAM, Aline Ducros to the Société d'autopsie mutuelle, February 26, 1877.

24. ASAM, Stéphen (Jean Marie) Pichon, April 18, 1879. He revoked his will in 1930, three years before he died. ASAM, Stéphen (Jean Marie) Pichon, 1930.

25. See, for instance, André Demaison's *Faidherbe* (Paris, 1932).

26. ASAM, L. Faidherbe, April 12, 1878.

27. An article in *L'écho de Paris*, no. 267 (October 18, 1889), on Faidherbe's membership in the society, provoked a considerable response.

28. Henry Céard, "Sapeck l'incomparable," *Le siècle*, October 15, 1889. The *Temps* article, "Pour l'humanité" (date missing), was preserved among the papers of the ASAM. See also *Le petit bleu*, April 13, 1903.

29. Eugène Véron, *Aesthetics*, trans. W. H. Armstrong (London: Chapman and Hall, 1879), 60.

30. ASAM, n.d. This passage began to appear in testaments in the early 1880s.

31. Jean d'Echerac, "André Lefèvre," *Revue de l'Ecole d'anthropologie de Paris* 14 (1904): 386.

32. ASAM, Chaptal, April 12, 1878.

33. ASAM, Chaptal, Curriculum—Travaux littéraires et scientifique," n.d.

34. ASAM, Testament: Chaptal, May 2, 1879.

35. Kselman, *Death and the Afterlife*, 75. On the origins of this doctrine, see Caroline Walker Bynum, "Bodily Miracles and the Resurrection of the Body in the High Middle Ages," in Thomas Kselman, ed., *Belief in History* (Notre Dame: University of Notre Dame Press, 1990), 68–106.

36. ASAM, Testament: Chaptal.

37. ASAM, Testament: Georges Laguerre, February 8, 1883.

38. ASAM, Testament: Eugène Véron, April 14, 1878.

39. ASAM, Testament: Paul Robin, July 14, 1899.

40. ASAM, Paul Robin to the Société d'autopsie mutuelle, November 13, 1899.

41. On Robin, see Christiane Demeulenaere-Douyère, *Paul Robin: Un militant de la liberté et du bonheur* (Paris: Publisud, 1994). On Neo-Malthusian groups, see Francis Ronsin, *La grève des ventres: Propagande néo-malthusienne et baisse de la natalité française, 19e-20e siècles* (Paris: Aubier Montaigne, 1980), 42–74. See also Joshua Cole, *The Power of Large Numbers* (Ithaca: Cornell University Press, 2000).

42. Malthus's *Essay on Population* concluded that the poor would always exist because people reproduced food arithmetically but reproduced themselves exponentially. The later, eu-

phemistic term was a long leap but meaningful: as neo-Malthusianism, the birth-control movement was generally conceived as more social than selfish.

43. See chap. 4, below.

44. ASAM, Testament: Paul Robin.

45. ASAM, Léonce Harmignies to the Société d'autopsie mutuelle (83 and 84), January 3, 1881.

46. ASAM, Barbe Nikitine to the Société d'autopsie mutuelle, April 24, 1883. Nikitine wrote for *La justice* under the pseudonym B. Gendre.

47. Archives de la Société d'anthropologie, Musée de l'homme, Paris (hereafter ASA), box: "Correspondance: 1905," letter from Lapouge to the Ecole d'anthropologie de Paris, September 22, 1897.

48. ASA, box: "Correspondance 1905," letters and telegrams from Dr. Baudry to the Ecole d'anthropologie de Paris, September 24–30, 1897.

49. ASAM, Victor Chevalier to the Société d'autopsie mutuelle, December 1, 1889.

50. ASAM, clipped article, "Pour l'humanité! La Société d'autopsie mutuelle," no newspaper name, no page, no author, no date.

51. Cited in an obituary for Letourneau, *Revue de l'Ecole d'anthropologie* 12 (1902): 8.

52. ASAM, Testament: Jh. Joyeux, August 22, 1889.

53. ASAM, Jh. Joyeux to the Société d'autopsie mutuelle, January 1892.

54. ASAM, Manouvrier to Société d'autopsie mutuelle, no date.

55. ASAM, Paul Monnot to Société d'autopsie mutuelle, April 6, 1881.

56. ASAM, Coudereau telegram to Monnot family, April 8, 1881.

57. ASAM, Vve Véron to Société d'autopsie mutuelle, October 14, 1889.

58. ASAM, Extrait du registre des actes de décès de l'an 1889, Décès de Louis Victor Eugène Véron.

59. ASAM, procès-verbal, Commissariat de police des Sables-d'Olonne (Vendée), Commissaire de Police Eugène Corneaud.

60. ASAM, Dr. Gaudin to Vve Véron, October 13, 1889.

61. ASA, box: "Correspondance 1905"; folder (tan): "Correspondance à classer: 1906–07"; and folder (tan): "I. 1888–1902" (actually extends to 1914), series of letters from Vve Eug. Véron.

62. "Un dîner pour la vulgarisation de la dissection mutuelle," *Morning News*, February 7, 1884, cited (in translation) in *L'homme* 1 (1884): 113–118.

63. For a detailed intellectual biography of Royer, see Joy Harvey, *"Almost a Man of Genius": Clémence Royer, Feminism, and Nineteenth-Century Science* (New Brunswick: Rutgers University Press, 1997).

64. *Compte-rendu: Congrès de la libre pensée à Rome, 1904* (Ghent, 1905), 1. These congresses were held every year in various locales, and full-length books were generally published describing the events and including many of the lectures. The compte-rendu of the Congress in Paris, 1905, is particularly well organized and informative.

65. ASA, box: "Correspondance 1905"; folder (tan): "I. 1888–1902," Vve Véron, September 23. 1902 (Véron heard of Royer's death rather late).

66. ASAM, Jacques Bertillon, September 26, 1878.

67. Henry T. F. Rhodes, *Alphonse Bertillon: Father of Scientific Detection* (New York: Abelard-Schuman, 1956), 41.

68. Suzanne Bertillon, *Vie d'Alphonse Bertillon: Inventeur de l'anthropométrie* (Paris: Gallimard, 1941), 11.

69. ASA (located in the ASAP though they were addressed to the Société d'Autopsie), Notes biographique de Jeanne Véron, December 17, 1906.

70. These included *Le chat* (1880), *Le chien* (1880), *Le petit cousin Charles* (1880), *L'âne* (1881), and *Histoires enfantines* (1881). The publishing enterprise was launched by Eugène Véron, Hovelacque, Lefèvre, Letourneau, and Barodet, all fellow anthropologists.

71. ASA, Jeanne Véron, December 17, 1906. No one ever did read these notes; they were sealed when I found them.

72. ASA, Testament: Vve Eugène Véron, neé Jeanne Guillaud, December 15, 1906.

73. Thulié, "Sur l'autopsie de Louis Asseline," 162.

74. Mathias Duval, "Le cerveau de Louis Asseline," *BSAP* 3d ser., 6 (1883): 260–274.

75. *La Société, l'Ecole et le Laboratoire d'anthropologie de Paris à l'Exposition universelle de 1889* (Paris, 1889), 105.

76. Mathias Duval, Théophile Chudzinski, and Georges Hervé, "Description morphologique du cerveau de Coudereau," *BSAP* 3d ser., 6 (1883): 377–389.

77. As cited in René Rémond, *L'anticléricalisme en France* (Brussels: Complexe, 1985), 176.

78. Gambetta, letter of September 4, 2874, to M. and Mme. Edmond Adam, in Juliette Adam, *Mes souvenirs* vol. 6 (paris: A. Lemerre, 1908), 150.

79. Mathias Duval, "Le poids de l'encéphale de Gambetta," *BSAP* 3d ser., 6 (1883): 400.

80. Chudzinski and Duval, "Description morphologique du cerveau de Gambetta," 129–152.

81. Duval, "Le poids de l'encéphale de Gambetta," 399–417.

82. Rémond, *L'anticléricalisme en France*, 191. Rémond includes several pages of Bert's very colorful anticlerical speeches (188 and 190–193).

83. See chap. 6, below.

84. Duval, "Le poids de l'encéphale de Gambetta," 405.

85. See, for instance, *Le petit bleu*, April 13, 1903.

86. Jean-Baptiste Vincent Laborde, *Léon Gambetta: Biographie psychologique—Le cerveau, la parole, la fonction et l'organe* (Paris, 1898), x.

87. "Cerveau de M. Laborde," *BSAP*, 5th ser., 4 (1903): 422–425.

88. Charles Letourneau, "Adolphe Bertillon," *BSAP*, 3d ser., 6 (1883): 187–192.

89. For a discussion of the use of liturgical models in the modern secular ritual of the court, see Richard K. Fenn, *Liturgies and Trials: The Secularization of Religious Language* (Oxford: Basil Blackwell, 1982).

90. John McManners, *Death and the Enlightenment* (New York: Oxford University Press, 1981), 191–206.

91. Kselman, *Death and the Afterlife*, 95–103.

92. Archives nationales (hereafter AN), F/17/13491–94, Sociétés savantes, Préfecture de police to Ministère de l'instruction publique, August 7, 1880.

93. "Gavarret," *Dictionnaire de biographie française* (Paris: Letouzey et Ané, 1982), 15:875–876.

94. AN, F/17/13491–94, Sociétés savantes, Ministère de l'instruction publique to Gavarret, Inspecteur Général de l'Université, September 21, 1880.

95. AN, F/17/13491–94, Sociétés savantes, Gavarret, Inspecteur général de l'université, to the Ministère de l'instruction publique, November 6, 1880.

96. Brès reports that after having seen Royer (and been moved by the respect with which her much esteemed professor had treated her), she went home and read Darwin the next day (Mme le Docteur [Madeleine] Brès, speech given at the Clémence Royer banquet, March 10, 1897, dossier Clémence Royer, Bibliothèque Marguerite Durand, Paris, cited in Harvey, "Almost a Man of Genius", 104–105).

97. Armande de Quatrefages, L'unité de l'espèce humaine (Paris: Hechette, 1861); and Rapport sur le progrès de l'anthropologie (Paris: Imprimerie Impériale, 1867). Later in the century, his Histoire générale des races humaines (Paris: A. Hennuger, 1887) also became a classic in anthropology.

98. AN, F/17/13491/4, Sociétés savantes, Armande de Quatrefages to the Ministère de l'instruction publique, September 1, 1880.

99. AN, F/17/13491/4, Sociétés savantes, Ministère de l'instruction publique to the Ministère de l'interieur, November 1880.

100. AN, F/17/13491/4, Sociétés savantes, M. Delabarre, Ministère des affaires étrangères, to M. Ferry, Ministère de l'instruction publique, September 1, 1879, including two articles from the Télégraphe dated September 5 and December 7, 1879. The quotation is from the latter article.

101. AN, F/17/13491/4, Sociétés savantes, Ministère de l'instruction publique to the Ministère de l'interieur, November 1880.

102. Freycinet had been complicit in the creation of this decree but apparently had no intention of enforcing it. Indeed, he was already in negotiations for a more amicable settlement when Gambetta encouraged Republican Union ministers to execute the decrees. See, "Constans, Jean," in the Dictionnaire de biographie française, 493–496, and in Yves Benoît, Dictionnaire des ministres: De 1789 à 1989 (Paris: Perrin, 1990), 417–418. See also Jean-Marie Mayeur and Madeleine Rebérioux, The Third Republic from Its Origins to the Great War (Cambridge: Cambridge University Press, 1984) 80.

103. ASAM, préfet de police (Quartier Vivienne), authorization, December 29, 1880.

104. René Fauvelle, "Les desiderata du matérialisme scientifique," L'homme 3 (1886): 107.

105. This issue was a central focus of the freethinkers' movement in general from at least as early as 1879. See Archives of the Préfecture de police, 1,493, Libre pensée: 1879 to 1891, and, in a file with overlapping dates, 1880–1897, both in 193032; Propagation de la foi civil. See also Jacqueline Lalouette, "Science et foi dans l'idéologie libre penseuse (1866–1914)," in Christianisme et science (Paris: Vrin, 1989), 21–54; and idem, La libre pensée en France, 1848–1940 (Paris: Albin Michel, 1997).

106. ASA, letter by Gabriel de Mortillet, February 16, 1886. While cemeteries became religiously neutral under a law of 1881, and secular funeral rights increased thereafter, it was not until a law of 1904 that the church lost its monopoly on pompes funèbres.

107. This prohibition was published in a number of Catholic journals of the time, as well as in L'homme 4 (1887): 382–383.

108. Journal officiel de la République française: Débats parlementaires, Chambre des députés, March 31, 1886, 609–624, cited in Jack D. Ellis, The Physician-Legislators of France: Medicine and Politics in the Early Third Republic, 1870–1914 (Cambridge: Cambridge University Press, 1990), 196. Ellis discusses the cremation law from the point of view of the physician-legislators who,

though also fiercely anticlerical, were primarily attempting to rid the cities of France of their supposedly unhygienic cemeteries.

109. ASAM, Guyot, November 20, 1888; July 28, 1894.

110. Laborde was an important doctor and scientist, an editor of *La tribune médicale,* and a leader of the French Antialcoholic Union.

111. Dias, "La Société d'autopsie mutuelle," 32.

112. ASAM, revised statutes, 1892.

113. Dias, "La Société d'autopsie mutuelle," 32.

114. ASAM, circular, 1903.

115. ASAM, procès-verbal, October 15, 1903. A Dr. Regnault, who was a member of the society, was also present at this meeting, though in what capacity it is difficult to discern. These rare notes were written on loose paper. I have found no book of meeting minutes.

116. "796e Séance, 15 Décembre 1904," *BSAP,* 5th ser., 5 (1904): 648.

117. ASAM, "Questionnaire," 1905–1930.

118. *Revue scientifique* 18, no. 3 (May 6, 1905): 572–573.

119. ASAM, "Questionnaire," Felix Auguste Blachette, Neuilly, February 18, 1905. He was fifty-eight when he completed the questionnaire.

120. ASAM, "Questionnaire," Regnauth Perrier, Paris, March 23, 1905; "Questionnaire," Alfred Guède, Paris, June 1914.

121. ASAM, Testament: Georges Papillault, February 7, 1902.

122. Georges Montandon, "Le squelette du Professeur Papillault," *Bulletins et mémoires de la Société d'anthropologie de Paris,* 8th ser., 6 (1935): 4–22.

123. On Royer's death, see Harvey, *"Almost a Man of Genius."* For another version, in which Royer requested to be buried without a casket, see André Billy, *L'époque 1900* (Paris, 1951), 215.

124. Harvey, *"Almost a Man of Genius,"* 182. See also Albert Milice, *Clémence Royer et sa doctrine de la vie* (Paris: J. Peyronnet, 1926), 210–212.

125. For more on Laborde and Manouvrier see chapter 6. For Manouvrier's quiet insult, see "Cerveau de M. Laborde," *BSAP* 5th ser., 4 (1903): 424.

126. ASAM, Paul Robin, February 14, 1904. Robin's assertion that membership in Freemasonry precluded true atheism was questionable. By the 1890s there was a considerable shift away from belief in a "Supreme Architect" and toward a more radically materialist cosmology within the ranks of French Freemasonry. See Mildred Headings, *French Freemasonry under the Third Republic* (Baltimore: Johns Hopkins Press, 1949).

127. Théophile Chudzinski and Léonce Manouvrier, "Etude sur le cerveau de Bertillon," *BSAP* 3d ser., 10 (1887): 558–590. The offensive remarks were made by Letourneau in the discussion that followed on pages 590–591. For Letourneau's reaction, see "Discussion," *BSAP* 3d ser., 11 (1888): 694–696.

128. ASAM, description of museum.

129. ASAM, "Société d'autopsie-Statuts," 1.

2. Evangelical Atheism and the Rise of French Anthropology

1. England established civil marriage in 1837, Italy in 1866, Spain and Germany after 1870. The term "civil" was explicitly in contrast to confessional or religious. See René Rémond, *Religion and Society in Modern Europe* (Oxford: Blackwell, 1999), 137, 138.

2. Just as the battle was to reach its most fevered pitch, Proudhon was expressing the possibility of reconciliation, though couched in extreme terms: "What difference would I have with you if you would only agree to take up the spiritual side of the Revolution? . . . A Christian, a deist, an antitheist, I am quite as religious as you are-and even in the same terms. I accept all of the categories of Catholicism-with the proviso of interpreting them in the light of this formula which sums up my whole book: *God is the conscience of humanity*" (Pierre-Joseph Proudhon, *De la justice dans la révolution et dans l'église* [Brussels, 1868–1870], end of the 12th étude, cited in Jacques Gadille, "On French Anticlericalism: Some Reflections," *European Studies Review* 2, no. 2 [April 1983]: 138). That was a lot to ask, but it is useful to note that some of those engaged in the matter could see that republicanism and the church were not innately divided.

3. Theodore Zeldin, *France, 1848–1945* (Oxford: Oxford University Press, Clarendon Press, 1977), 998.

4. Maurice Agulhon, *The Republic in the Village: The People of the Var from the French Revolution to the Second Republic*, trans. Janet Lloyd (Cambridge: Cambridge University Press; Paris: Editions de la Maison des sciences de l'homme, 1982).

5. Proudhon, *Correspondance* (1875), 6:110, cited in Zeldin, *France*, 1026.

6. Owen Chadwick, *The Secularization of the European Mind in the Nineteenth Century* (Cambridge: Cambridge University Press, 1975), 155–156.

7. Zeldin, *France*, 1028.

8. Gustave Flaubert, *Madame Bovary*, trans. Geoffrey Wall (London: Penguin, 1992), 61–62.

9. Ernest Renan, *L'avenir de la science: Pensées de 1848* (1890; reprint, Paris: Flammarion, 1995), 75.

10. Renan, *The Future of Science* (London: Watts, 1935), preface.

11. On Dupanloup, see Zeldin, *France*, 1008.

12. See the section "Beliefs and Unbelief" in Jean-Marie Mayeur and Madeleine Rebérioux, *The Third Republic from Its Origins to the Great War, 1871–1914* (Cambridge: Cambridge University Press, 1984), 101–109. See also Chadwick, *The Secularization of the European Mind*; René Rémond, *L'anticléricalisme en France* (Brussels: Complexe, 1985); and the section "Religion and Anticlericalism" in Zeldin, *France*, 983–1039.

13. John McManners, *Church and State in France, 1870–1914* (New York: Harper, 1972), 5–6.

14. "The myth that France was pious and Catholic before the Revolution was invented by modern conservatives idealizing the Middle Ages. In the eighteenth century there was little militant abstention from religion, but only a small and unorganized minority was fervently devout. Most people went to church, but by no means every Sunday" (Zeldin, *France*, 984). John McManners makes the same point: "To understand the dechristianization of rural society, it is important to abandon the myth, dear to conservative churchmen, of an earlier epoch that was 'Christian'" (McManners, *Church and State in France*, 9–10).

15. Gadille, "On French Anticlericalism," 129. Gadille argues that, in this sense, anticlericalism will always be a part of religious history (141).

16. Rémond, *L'anticléricalisme en France*, 219, 220.

17. Jules Michelet, *Le prêtre, la femme et la famille* (1845; reprint, Paris: Michel Lévy and Librairie nouvelle, 1875), 267, 307, cited in Zeldin, *France*, 993.

18. Jean Pommier, "Les idées de Michelet et de Renan sur la confession en 1845," *Journal de psychologie normale et pathologique*, July-October 1936, 520–522; and Zeldin, *France*, 992.

19. Chadwick, *The Secularization of the European Mind*, 124.

20. Cited in the preface of the English translation of his novel *Lourdes*, trans E. A. Vizetelly (Chicago: Neely, 1894), v–vi.

21. McManners, *Church and State in France*, 49.

22. Rémond, *Religion and Society*, 149.

23 G. Hanotaux, cited in Mayeur and Rebérioux, *The Third Republic*, 81.

24. The term was used by Ferdinand Buisson, who, along with Ferry, effectuated the overhaul of the educational system.

25. *Revue pédagogique*, 1882, cited in Mayeur and Rebérioux, *The Third Republic*, 85.

26. Rémond, *L'anticléricalisme en France*, 173.

27. Mayeur and Rebérioux, *The Third Republic*, 126.

28. Having inherited a Napoleonic bureaucracy of some two hundred thousand members, the Third Republic rapidly expanded this corps in order to serve the needs of an increasingly urban, industrial nation. Through its expansion of state monopolies and public service (especially education), the Third Republic employed nearly five times that number by the end of the first decade of the twentieth century. Judith Wishnia has shown that this vast increase in state employees (admittedly, this includes the military) represented "one of every forty living Frenchmen and Frenchwomen, and even more impressive, one of every ten voters received a salary paid out of public funds" (Wishna, *The Proletarianizing of the Fonctionnaires* [Baton Rouge: Louisiana State University Press, 1990], 68).

29. Nancy Leys Stepan, *The Idea of Race in Science* (Hamden, Conn.: Archon, 1982); and idem, "Race and Gender: The Role of Analogy in Science," *Isis* 77, no. 2 (June 1986): 261–277.

30. See George Mosse, *Nationalism and Sexuality* (Madison: University of Wisconsin Press, 1985); Pierre-André Taguieff, *La force du préjugé: Essai sur le racisme et ses doubles* (Paris: Gallimard, 1990); idem, *Face au racisme* (Paris: La Découverte, 1991).

31. See George W. Stocking, *After Tyler: British Social Anthropology, 1888–1951* (Madison: University of Wisconsin Press, 1995); idem, *Victorian Anthropology* (New York: Free, 1987); idem, *Race, Culture and Evolution: Essays in the History of Anthropology* (New York: Free, 1968).

32. Stephen J. Gould, *The Mismeasure of Man* (New York: Norton, 1981); Cynthia Eagle Russett, *Sexual Science: The Victorian Construction of Womanhood* (Cambridge: Harvard University Press, 1989); Robert Nye, *Crime, Madness, and Politics in Modern France: The Medical Conception of National Decline* (Princeton: Princeton University Press, 1984); Susanna Barrows, *Distorting Mirrors: Visions of the Crowd in Late Nineteenth-Century France* (New Haven: Yale University Press, 1981). Key examples of the large literature on anthropology outside France includes George W. Stocking Jr., ed., *Bones, Bodies, Behavior: Essays on Biological Anthropology* (Madison: University of Wisconsin Press, 1988); Renate Bridenthal, Atina Grossmann, and Marion Kaplan, *When Biology Became Destiny* (New York: Monthly Review Press, 1984); Sander L. Gilman, *Freud, Race, and Gender* (Princeton: Princeton University Press, 1993); and Adrian Desmond, *The Politics of Evolution: Morphology, Medicine, and Reform in Radical London* (Chicago: University of Chicago Press, 1989).

33. Claude Blanckaert, "L'anthropologie au féminine: Clémence Royer (1830–1902)," *Revue de synthèse* 105 (1982): 23–38; idem, "Préface: 'L'anthropologie personnifiée'—Paul Broca et la biologie du genre humain," in Paul Broca, *Mémoires d'anthropologie* (Paris: Jean-Michel Place, 1989), i–xliii; and idem, "La science de l'homme entre humanité et inhumanité," in Claude Blanckaert, ed., *Des sciences contre l'homme: Classer, hiérarchiser, exclure* (Paris:

Autrement, 1993), pp. 14–46; Nélia Dias, *Le Musée d'ethnographie du Trocadéro (1878–1908):Anthropologie et musicologie en France* (Paris: CNRS, 1991); Michael Hammond, "Anthropology as a Weapon of Social Combat in Late-Nineteenth-Century France," *Journal of the History of the Behavioral Sciences* 16 (1980): 118–132; Joy Harvey, "Races Specified, Evolution Transformed: The Social Context of Scientific Debates Originating in the Société d'anthropologie de Paris, 1859–1902" (Ph.D. diss., Harvard University, 1983); Elizabeth A. Williams, "The Science of Man: Anthropological Thought and Institutions in Nineteenth-Century France" (Ph.D. diss., Indiana University, 1983).

34. Jennifer Michael Hecht, "Anthropological Utopias and Republican Morality: Atheism and the Mind/Body Problem in France, 1876–1914." (Ph.D. diss., Columbia University, 1995).

35. Philip Nord, *The Republican Moment: Struggles for Democracy in Nineteenth-Century France* (Cambridge: Harvard University Press, 1995), 41–44.

36. Without even entering into the question of data manipulation or the effect of suppositions on experiments, political ideology is inherent in the very choice to study a given relationship scientifically. Importance and meaning are conferred on (and revealed as already socially present in) those questions that we imagine and thus choose to pose. Even when we "find" equality, a given test may reinforce group divisions: there are no research projects on the intellectual variation between middle-class Caucasians in Montana versus Colorado, for example, and neither is there an institute studying moral differences between tall and short people, because we do not see these as legitimate groups. They are not interesting because they have no sociopolitical component; that is to say, no one has an interest.

37. For a discussion of Broca's life and work, see Blanckaert, "Préface." See Blanckaert, *Des sciences contre l'homme.* In English, see Francis Schiller, *Paul Broca: Founder of French Anthropology, Explorer of the Brain* (1979, reprint, Oxford: Oxford University Press, 1992); see also Williams, "The Science of Man"; as well as Harvey, "Races Specified."

38. Letter of March 11, 1948, in Paul Broca, *Correspondance, 1841–1857*, vol. 2, *1848–1857* (Paris: Paul Schmidt, 1886), 12, cited in Jacqueline Lalouette, *La libre pensée en France, 1848–1940* (Paris: Albin Michel, 1997), 27–28.

39. Elizabeth A. Williams, *The Physical and the Moral: Anthropology, Physiology, and Philosophical Medicine in France, 1750–1850* (Cambridge: Cambridge University Press, 1994).

40. The Société de biologie was founded in 1848 by Charles Robin, a close associate of Littré, who would later be honored as a senator for life by the Third Republic. For more on this society, see Eugène Gley, "Histoire de la Société de biologie," *Revue scientifique* 4, no. 13 (1900): 3–11.

41. This is Broca's account of the events. See Paul Broca, "Histoire des travaux de la Société d'anthropologie de Paris (1859–1863)," *Mémoires de la Société d'anthropologie de Paris* 2 (1865): vii–li.

42. Debates on monogenism and polygenism reflected concerns about racial difference and in turn transformed the terms of debate about human difference in general. See Stepan, *The Idea of Race in Science.* For the French case in particular, see Harvey, "Races Specified."

43. The Paris medical profession, in general, was renowned for its republicanism and anticlericalism. See Jack D. Ellis, *The Physician-Legislators of France: Medicine and Politics in the Early Third Republic, 1870–1914* (Cambridge: Cambridge University Press, 1990), 30–51. See also Jacques Léonard, *La médecine entre les saviors et les pouvoirs: Histoire intellectuelle et politique de la médecine française au XIXe siècle* (Paris: Aubier Montaigne, 1981); George Rosen, "The Philoso-

phy of Ideology and the Emergence of Modern Medicine in France," *Bulletin of the History of Medicine* 20 (July 1946): 328–339; and Owsei Temkin, "Materialism in French and German Physiology of the Early Nineteenth Century," *Bulletin of the History of Medicine* 20 (July 1946): 322–327.

44. Henri Thulié, "L'Ecole d'anthropologie de Paris depuis sa fondation," in *L'Ecole d'anthropologie de Paris (1876–1906)* (Paris: Alcan, 1907), 3.

45. Paul Broca, "Discours sur l'homme et les animaux," *Bulletins de la Société d'anthropologie de Paris* (hereafter *BSAP*), 2d ser., 1 (1866): 75. See also his comments in "Discussion sur la religiosité," *BSAP* 2d ser., 12 (1877): 33–36.

46. Cardinals Donnet and Bonnechose led this outcry. See the debates in *Le moniteur*, March 28, 1868, 455–456; May 20, 1868, 689–691; May 21, 1868, 701–704; May 22, 1868, 709; May 23, 1868, 712; May 24, 1868, 718–721. For a discussion of this, see Ellis, *The Physician-Legislators of France*, 39.

47. Schiller, *Paul Broca*, 124.

48. Schiller, *Paul Broca*, 128.

49. For the history of theories of evolution in France, see Yvette Conry, *L'introduction du Darwinisme en France au XIXe siècle* (Paris: Vrin, 1974); Denis Buican, *La révolution de l'évolution* (Paris: PUF, 1984); Thomas F. Glick, ed., *The Comparative Reception of Darwinism* (Austin: University of Texas Press, 1974); Linda Clark, *Social Darwinism in France* (Birmingham: University of Alabama Press, 1984); Peter Bowler, *Theories of Human Evolution: A Century of Debate, 1844–1944* (Baltimore: Johns Hopkins University Press, 1986). For a fascinating discussion of the relationship between Darwinian and Lamarckian evolution and social theory in France, see Robert Nye, "Heredity or Milieu: The Foundations of Modern European Criminological Theory." *Isis* 67, no. 3 (September 1976): 335–355.

50. See, for instance, Paul Broca, "Discours de M. Broca sur l'ensemble de la question," *Congrès international d'anthropologie et d'archéologie préhistoriques* (Paris, 1868), 367–402.

51. Paul Broca, "Sur le transformism: Remarques générales," *BSAP* 2d ser., 5 (1870): 169–170.

52. Nord, *The Republican Moment*, 27–28. For a history of women and the Freemasons in an earlier period, see Dena Goodman, *The Republic of Letters: A Cultural History of the French Enlightenment* (Ithaca: Cornell University Press, 1994), 149–159.

53. For more on this, see Harvey, "Races Specified," 21–24, 38–47.

54. After the society was given its status of "public utility," it was legally able to accept bequests. Several members bequeathed money to the society in order to fund annual essay prizes bearing the name of the benefactor. Many others left a portion of their estates to the society's treasury.

55. *Le Musée d'ethnographie du Trocadéro*, Dias's excellent work on this museum characterizes these various classificatory systems and highlights the significance of museums in the development of ethnographic and anthropological techniques and assumptions.

56. Broca to Vogt, March 8, 1865, Correspondance Carl Vogt, 2188, no. 154, Geneva, cited in Harvey, "Races Specified," 92.

57. Ellis, *The Physician-Legislators of France*, 39.

58. *Official Record of the French Senate*, 3/27/68, 731, cited in Francis Schiller, *Paul Broca*, 273.

59. See the debates in *Le moniteur*, March 28, 1868, 455–456; May 20, 1868, 689–691; May

21, 1868, 701–704; May 22, 1868, 709; May 23, 1868, 712; May 24, 1868, 718–721. See Ellis, *The Physician-Legislators of France*, 39.

60. Ellis, *The Physician-Legislators of France*, 40. In his thesis for Robin, Clemenceau also argued for the possibility of spontaneous generation, which was seen as a necessary starting point for life in the absence of God. See Georges Clemenceau, *De la génération des éléments anatomiques* (Paris: Baillière, 1865), 104–109, 221.

61. Paris, January 17, 1880, Bibliothèque nationale (manuscrit), NAF 24803 ff 89–91, microfilm 1296, doc. 89.

62. Ian Hacking, *The Taming of Chance* (Cambridge: Cambridge University Press, 1990), 3.

63. The only other sustained inquiry into the movement in France (and French-speaking countries) is in a collection of essays entitled *Libre pensée et religion laïque en France: De la fin du Second Empire à la fin de la Troisième République*, intro. J.-M. Mayeur (Strasbourg: Cerdic, 1980). J.-M. Mayeur's piece "La foi laïque de Ferdinand Buisson" (247–257) is particularly insightful on the relationship between philosophy and free thought, as well as on the passionately religious nature of the freethinkers' convictions. The essays in this collection are, however, on very specific subjects and give little sense of the movement overall. They also miss the connection between freethinking and anthropology in France. See also Jacqueline Lalouette, "Science et foi dans l'idéologie libre penseuse (1866–1914)," in *Christianisme et science* (Paris: Vrin, 1989), 21–54. Lalouette chose 1866 as her starting date because that is when the future anthropologists began to publish their journal *La libre pensée*. Through almost all the above-mentioned studies, the importance of this journal is generally recognized, though it goes unnoticed that its authors would all soon become anthropologists.

64. Lalouette, *La libre pensée en France*, 27–28.

65. Pierre et Paul, "Thulié," *Les hommes d'aujourd'hui* 3, no. 144 (1880): 2. For more on this journal, see chap. 3, below.

66. Eugène Véron, *Les associations ouvrières* (Paris: Hachette, 1865).

67. E. Houzé, "Gabriel de Mortillet-Notice nécrologique," in *Extrait du Bulletin de la Société d'anthropologie de Bruxelles*, vol. 17, *1898–1899* (Brussels, 1899), 1; Salomon Reinach, "Gabriel de Mortillet," in *Extrait de La revue historique, 1899* (Paris, 1899), 8–10; *Inauguration du monument de G. de Mortillet-Extrait de L'homme préhistorique, 3e année, no. 11, 1905* (Paris, 1905).

68. For a review of Louis Asseline, *Histoire de l'Autriche depuis la mort de Marie-Thérèse jusqu'à nos jours* (Paris: G. Baillière, 1877), see "L'Autriche contemporaine," *La revue politique et littéraire* 2, no. 15 (1879): 183–185.

69. André Lefèvre, *La renaissance du matérialisme* (Paris: Doin-Bibliothèque matérialiste, 1881), 118.

70. See Robert Fox, "Science, the University, and the State in Nineteenth-Century France," in Gerald L. Geison, ed., *Professions and the French State, 1700–1900* (Philadelphia: University of Pennsylvania Press, 1984), 66–145. Fox notes that the empire deeply angered and frustrated "a network of young, liberally minded philosophers, recent graduates of the Ecole Normale Supérieure" and that "many of them turned from academic life to other pursuits, notably journalism" and cites Véron as typical of this phenomenon (88, 132 n. 83). See also Antoine Prost, *Histoire de l'enseignement en France, 1800–1967* (Paris: Armand Colin, 1968), 80–82. For a personal account, see Francisque Sarcey, *Journal de jeunesse* (Paris: Bibliothèque des annales, n.d.), 44–47.

71. Eugène Véron, "De l'enseignement supérieur en France," *Revue des cours littéraires de la France et de l'étranger* 2 (1865): 401–404, 435–437, 449–452.

72. Lefèvre, *La renaissance du matérialisme*, 120. The "terrible year" was a common term for the French defeat in the Franco-Prussian War.

73. Darwin seems to have been amused by the preface at first, but a few years later another translator was chosen and the text retranslated, at Darwin's request.

74. Peter Bowler, *The Invention of Progress* (Oxford: Basil Blackwell, 1989). See also idem, *The Non-Darwinian Revolution: Reinterpreting a Historical Myth* (Baltimore: Johns Hopkins University Press, 1988).

75. The most complete account of Royer's life is Joy Harvey's *"Almost a Man of Genius": Clémence Royer, Feminism, and Nineteenth-Century Science* (New Brunswick: Rutgers University Press, 1997); see also Claude Blanckaert, "L'anthropologie au féminine: Clémence Royer (1830–1902)," *Revue de synthèse* 105 (1982): 23–38; and Joy Harvey, " 'Strangers to Each Other': Male and Female Relationships in the Life and Work of Clémence Royer," in Pnina Abir-am and Dorinda Outram, eds., *Uneasy Careers and Intimate Lives* (New Brunswick: Rutgers University Press, 1987), 147–171.

76. Clémence Royer, preface to Charles Darwin, *L'origine des espèces*, cited in full in Geneviève Fraisse, *Clémence Royer: Philosophe et femme de science* (Paris: Découverte, 1985), 127.

77. Charles Darwin, *On the Origin of Species* (Cambridge: Harvard University Press, 1964), 488.

78. For a comment on her notion of men's and women's brains see Manouvrier, "L'anthropologie des sexes," *Revue de l'Ecole d'Anthropologie* 9 (1909):41–61.

79. Royer, preface, in Fraisse, *Clémence Royer*, 164.

80. From Royer's unpublished autobiography, "Rectifications biographiques," [1899], 6, in Dossier Clémence Royer, Bibliothèque Marguerite Durand, Paris (hereafter BMD), cited in Harvey, *"Almost a Man of Genius"*, 38.

81. Bert later published some of these articles in book form. See Paul Bert, ed., *Revues scientifiques pour la république française*, 7 vols. (Paris, 1879–1885).

82. M. Mévisse, *Des droits de la femme-Rapport-2ème congrès: Libre pensée* (Brussels, 1893).

83. Chadwick, *The Secularization of the European Mind*, 173.

84. Reinach, "Gabriel de Mortillet," 6–9.

85. Coudereau, "Program," *Libre pensée: Science, lettres, arts, histoire, philosophie* 1 (October 21, 1866): 1–2.

86. Anatole Roujou, "L'anthropologie," *Libre pensée* 6 (November 25, 1866): 42–43. The same issue carried an excerpt of a debate published in the *BSAP* that argued the case for Darwinism (48).

87. Gabriel de Mortillet, "L'homme-singe perfectionné," *Libre pensée* 3 (1866): 23–24.

88. Pierre et Paul, "Thulié," 3.

89. Girard de Rialle, Jules Soury, Paul Lacombe, Emile Leclerq, and P. Sierebois, to name a few.

90. See, for example, Charles Letourneau, "Variabilité des êtres organisées," *La philosophie positive* 3 (1868): 99–121.

91. "Mathias Duval et le dîner du matérialisme scientifique," *L'homme* 3 (1886): 25–28. This

unsigned article appears to have been written by Mortillet, but the quotation comes from Lefèvre's address to the "Dinner," which the article cites at length.

92. Charles Letourneau, "Banquet Tribute," unpublished speech given at the Clémence Royer banquet of 1897, Dossier Clémence Royer, BMD, cited in Harvey, *"Almost a Man of Genius,"* 105.

93. Gabriel de Mortillet, "Au lecteurs," *L'homme* 3 (1886): 1–2.

94. In one of several exceptions, Mortillet spoke out against French colonialism during his term in the Chamber of Deputies (Mortillet addressed the tenth bureau of the chamber). As reported in *L'homme*, he "argued against colonial politics, basing his position on the nonaptitude of the French to acclimatize to Tonkin. To keep this colony is to sentence to death, in full knowledge of the cause, our compatriots forced to go live in that murderous climate" ("Acclimatation," *L'homme* 2 [1885]: 701). Mortillet did not write much on this subject. It would appear that he was more interested in embarrassing Jules Ferry, the initiator of French fin-de-siècle colonialism (whose moderate republicanism drew much anger from the freethinking anthropologists), than he was in championing theories of acclimatization. He was also interested in promoting political anthropology. Earlier in that same volume of *L'homme*, a blurb on "L'anthropologie et la politique" ran as follows: "In a meeting of the Reichstag of March sixteenth, the deputy Virchow combated the colonial politics of Prussia in an argument based on anthropology. It is an excellent example that he has given. Anthropology can furnish precious documents for the solution of the great questions treated in most of our political assemblies. Its introduction into parliamentary practice (*la vie parlementaire*) would make a good number of laws more logical and would be a real act of progress" (222).

95. The badly brought-up man. See Reinach, "Gabriel de Mortillet," 81.

96. Lefèvre, *La renaissance du matérialisme*, 132.

97. Charles Letourneau, *Science et matérialisme* (Paris: Reinwald, 1891), v.

98. Lefèvre, *La renaissance du matérialisme*, 133.

99. Hammond's "Anthropology as a Weapon of Social Combat" and the dissertations of Harvey and Williams ("Races Specified, Evolution Transformed" and "The Science of Man," respectively) all see more of an ideological gulf between Broca and Mortillet and therefore see Broca's death as a more marked turning point in the Société d'anthropologie and its institutions.

100. Coudereau, "De l'influence de la religion sur la civilisation: Réponse à M. Bataillard," *BSAP* 2d ser., 2 (1867): 580.

101. André Lefèvre, "La philosophie devant l'anthropologie," *L'homme* 19 (October 10, 1884): 583.

102. Royer, preface, in Fraisse, *Clémance Royer*, 149.

103. Lefèvre, "La philosophie devant l'anthropologie," 583.

104. René Fauvelle, "Conséquence naturelle de la science libre," *L'homme* 2 (1885): 742.

105. René Fauvelle, "Il faut en finir avec la philosophie," *L'homme* 2 (1885): 139–146.

106. Mortillet, "L'église et la science," *L'homme*, 4 (1887): 609.

107. Letter from Broca to Royer, May 15, 1875, Archives of the Société d'anthropologie de Paris (hereafter ASAP), Musée de l'homme, Paris.

108. The essay appears in its entirety in Harvey, *"Almost a Man of Genius,"* 190, 193–203.

109. Paul Topinard, *La société, l'école, le laboratoire et le musée Broca* (Paris: Chamerot, 1890), 3. Topinard's partial account (he finished his list with "etc.") of the group's membership is as

follows: Asseline, Assézat, Coudereau, Bertillon, Gillet-Vital, de Mortillet (father and son), Duval, Letourneau, Hovelacque, Thulié, Fauvelle, Salmon, Collineau, Hervé, Issaurat, Nicole, Lefèvre, Yves Guyot, and Arthur-Alexandre Bordier (14).

110. Topinard never gave any indication of why he "could not reveal" these facts to Broca. It weakens his case.

111. Photograph taken at the Moscow Anthropological Exhibition, 1897.

112. Charles Issaurat, "Analyse de l'ouvrage de M. A. Hovelacque intitulé: Les nègres de l'Afrique sus-équatoriale," *La tribune médicale* 45 (1890).

113. Topinard, *La société*, 21.

114. See, for instance, Charles Letourneau, "L'origine de l'homme," *Pensée nouvelle*, 1867, reprinted in idem, *Science et matérialisme*.

115. Eugène Dally, comment in "Discussion sur le questionnaire d'ethnographie," *BSAP* 3d ser., 5 (1882): 495.

116. "Suite de la discussion sur l'anthropophagie," *BSAP* 3d ser., 11 (1888): 27–46.

117. Reinach, "Gabriel de Mortillet," 12.

118. These studies—such as *Distorting Mirrors* by Susanna Barrows, Gould's *The Mismeasure of Man*, and *Crime, Madness, and Politics in Modern France* by Robert Nye—all judge Le Bon to have been a synthesizer and popularizer rather than an innovative thinker.

119. ASAP, procès-verbaux, March 27, 1879.

120. Paul Broca, in *Congrès internationale d'anthropologie, 9e session à Lisbonne, 1879* (Lisbon: Académie royale des sciences, 1884).

121. "Godard Prize," *BSAP* 3d ser., 2 (1879): 373–386.

122. Letter from Le Bon, May 22, 1888, ASAP, Box C2 C3, folder C, P2, doc. #4059.

123. Archives nationales (hereafter AN), F/17/17199, "Enquète sur des faits graves imputés à M. Topinard, en sa qualité de Professeur . . . avec pièce justificatives" (printed document).

124. ASAP, procès-verbaux, July 19, 1890.

125. By the end of his life, however, Topinard did have considerable renown. He had written several books especially for an English-speaking audience and had lived for some time in the United States, bringing his work to the attention of the English and American public. His major works, however, were considered by his colleagues to be largely syntheses of works by Broca and others. A significant portion of his *Eléments d'anthropologie*, published in Paris in 1885, was devoted to descriptions and discussions of Broca's final experiments and hypotheses. As Elizabeth Williams notes in her dissertation, "The Science of Man," there were complaints that Topinard did not sufficiently acknowledge his references to the ideas and experiments of his colleagues (238–240). Even Manouvrier, whose equanimity was frequently noted by his colleagues, complained in 1889 that, because of the lack of proper citations in *Eléments d'anthropologie*, Topinard was often credited for work that had actually been done by colleagues (*BSAP* 3d ser., 12 [1889]: 377). Mortillet was more vicious in his critique of the work. Claiming that he did not want to criticize a colleague personally, Mortillet translated and republished—in his journal, *L'homme*—an Italian review of Topinard's work. The review mocked Topinard for his narrow interpretation of anthropology and called *Eléments* a "detailed and rather tedious discussion of results obtained only from craniology and anthropometry" (*L'homme*, 1885, 345–347). The review first appeared in *Rivista di Filosofia scientifica* (Turin), May 25, 1881.

126. *Actes du deuxième congrès d'anthropologie criminelle* (Paris, 1889), 490–496.

127. Léonce Manouvrier, "L'atavisme et le crime," *Revue de l'Ecole d'anthropologie* 1 (1891): 225–240; and idem, "L'anthropologie et le droit," *Revue internationale de sociologie* 2 (1894): 241–273, 351–370. For his discussion of Topinard's position in the latter, see 352–353, 360–370.

128. Paul Broca, *Mémoires d'anthropologie* (Paris, 1871), 1:30, cited in Manouvrier, "L'anthropologie et le droit," 365.

129. "Un coup d'état à l'école d'anthropologie," *La patrie*, February 8, 1890.

130. Marc-Antoine-Marie-François Duilhé, "Le problème anthropologique et les théories évolutionnistes," in *Congrès scientifique international des Catholiques tenu à Paris du 8 au 13 avril 1888* (Paris: Bureaux des Annales de philosophie chrétienne, 1888), 2:621.

131. Charles Jeannolle, "Le positivisme et les sociétés de libre-pensée," *Revue occidentale*, 1st semester (1885): 98–113, 237–257.

132. André Lefèvre, *Religions et mythologies comparées* (Paris: LeRoux, 1877), 328.

133. Abel Hovelacque, *Les débuts de l'humanité* (Paris: Doin, 1881), iii–iv.

134. Abel Hovelacque, *Plus les laiques sont éclairés, moins les prâtres pourront faire du mal* (Paris, 1880), 5.

135. ASAP, folder: année 1916, April 26, 1916.

136. ASAP, box LCO, section C, March 18, 1895.

137. ASAP, box LCO, loose, March 31, 1897.

138. ASAP, box LCO, H3787, February 4, 1895.

139. Jacques Bertillon, "Le musée de l'Ecole d'anthropologie," *La nature* 6 (1878): 39.

140. Jacques Bertillon, "Des monstruosités: Principes généraux de tératologie," *La nature* 1 (1874): 209.

141. For another article that demonstrates this interest in the science of "monsters," see Jacques Bertillon, "Rosa et Josefa: Les deux soeurs tchèques," *La nature* 2, no. 2 (1884): 293–294, in which Bertillon claimed, for instance, that the conjoined girls' back-to-back position conformed to the "theory of self-attraction" put forward by the famous naturalist Etienne Geoffroy Saint-Hilaire.

142. ASAP, folder: année 1916, note dated 1891.

143. Jean-Marie Lanessan, *Revue internationale des sciences biologiques* 1, nos. 1 and 5 (1878): 61–63, 158–159, cited in Ellis, *The Physician-Legislators of France*, 99.

144. Mortillet file, carton 127573, Préfecture de police, Paris.

145. ASAP, box LCO, H3787, February 4, 1895.

146. Pierre et Paul, "Hovelacque," *Les hommes d'aujourd'hui* 3, no. 128 (1880): 2–4.

147. Pierre et Paul, "Thulié."

148. Etienne Roc, "Professeur Mathias Duval," *Les hommes d'aujourd'hui* 6, no. 273 (1886): 2–4.

149. "Mathias Duval et le dîner du matérialisme scientifique," 25–26.

150. Assézat seems to have been a friend of Thulié at first; the two cofounded *Le realisme* in the fifties. He wrote a few articles for *Pensée nouvelle* and then pretty much disappeared from the anthropological record, but he must have continued to work with the group in other capacities since they usually list him among their true believers. He is also among the list of freethinkers that Topinard provided (see n. 109, above).

151. He is speaking here of the Ligue d'union républicaine des droits de Paris, an association of moderate republicans who worked to reach a compromise between the leaders of the Paris Commune and Versailles in 1871.

152. Clémence Royer, speech for "Banquet 1897" March 10, 1897, cited in Harvey, *"Almost a Man of Genius,"* 173.

3. Scientific Materialism and the Scholarly Response

1. Paul Broca, *Instructions générales pour les recherches anthropologiques* (Paris, 1865), 2.

2. See, for instance, Charles Letourneau, *Questionnaire de sociologie et d'ethnographie* (Paris, 1882), which also appeared, in a slightly revised form, in the *Bulletins de la Société d'anthropologie de Paris* (hereafter *BSAP*), 3d ser., 6 (1883): 578–597.

3. Charles Letourneau, *La sociologie d'après l'ethnologie*, Bibliothèque des sciences contemporaines (Paris: Reinwald, 1884), cited in Frédéric Paulhan, "Analyses et comptes rendus: Letourneau-*La sociologie d'après l'ethnographie*," *Revue philosophique* 11 (1881): 546.

4. Eugène Dally, comment in "Discussion sur le questionnaire d'ethnographie," *BSAP* 3d ser., 5 (1882): 574.

5. For a more detailed discussion of the questionnaires in general, see Nélia Dias, *Le Musée d'ethnographie du Trocadéro (1878–1908): Anthropologie et muséologie en France* (Paris: CNRS, 1991): 72–89.

6. Michael Hammond, "Anthropology as a Weapon of Social Combat in Late-Nineteenth-Century France," *Journal of the History of the Behavioral Sciences* 16 (1980): 123–125.

7. All societies, everywhere, were believed to go through the exact same stages of development. They believed that "one still encounters today human populations that are the living image of ancient prehistoric races" (Abel Hovelacque, *Les débuts de l'humanité* [Paris: Doin, 1881], i). Thus every society progressively emerges from any given savage behavior, and "primitive" populations were understood to "occupy at the present time the last (or the first) rung on the human ladder" (ii). Since the "human ladder" was essentially consistent in all instances, the anthropological prehistory and history of France could stand in for any other place; there was no need to travel. Or, if one was inclined to travel, there was no need to stay home and dig.

8. Henri Thulié's book *La femme: Essai de sociologie physiologique* (Paris: Delahaye et Lecrosnier, 1885), for instance, complained that the French state hurt women through unequal political, social, and economic policy but also claimed that the only true role for women was maternity.

9. Eugène Véron, "De l'enseignement supérieur en France," *Revue des cours littéraires de la France et de l'étranger* 2 (1865): 403.

10. Mortillet, "L'antisémitisme," *L'homme* 1 (1884): 522–528. Much of his argument was constructed by offering historical explanations for the various prejudices against the Jewish people. Follow-up pieces on the same subject included a "rectification" in which Mortillet softened his claim of anti-Semitism among the Germans by citing the campaign against anti-Semitism waged by fellow anthropologist (and member of the society) Professor Virchow (564–565).

11. Zola, Bibliothèque nationale, don 3988, vol. 81, NAF 10, 315. Among Zola's many notes (including a section called "Differences between Balzac and me," where he confirms, to himself, his intention to write "scientific" novels) are the résumés of only three books, Letourneau's, a study of the Second Empire by Taxile Delord, and *Traité de l'hérédité naturelle* by Prosper Lucas. For a transcription of these notes (abridged but including all Zola's notes on Letourneau), see Henri Massis, *Comment Emile Zola composait ses romans—D'après ses notes personnelles et inédites* (Paris: Charpentier, 1906), 28–35. Letourneau, *Physiologie des passions* (Paris: G. Baillière, 1868).

12. Zola, "Compte Rendu: Letourneau," *Le globe*, January 23, 1868. For mentions of the Letourneau-Zola connection, see Henri Mitterand, *Zola*, vol. 1, *1840–1871* (Paris: Fayard, 1999), 590, 721. Mitterand refers to Letourneau and Lucas as the "bibliothèque souterraine des Rougon-Macquart." See also Philip Walker, *Zola* (London: Routledge and Kegan Paul, 1985), 90.

13. Walker, *Zola*, 90.

14. Henri Troyat, *Zola* (Paris: Flammarion, 1992), 91.

15. Cited in Frederick Brown, *Zola* (New York: Farrar Straus Giroux, 1995), 460.

16. Bibliothèque nationale, don 3988, vol. 81, NAF 10, 315, cited in Walker, *Zola*, 91.

17. On the Bibliothèque anthropologique, see "Livres et journaux," *L'homme* 2 (1885): 403. The article lists seven volumes of this collection. Their authors include Véron, Mortillet, Letourneau, Hervé, Hovelacque, Thulié, and Duval.

18. Vogt to Mortillet, Geneva, April 20, 1880, Musée de l'homme, Paris, Manuscrits: Ms 54 R; emphasis in original.

19. Preface to Abel Hovelacque, Charles Issaurat, André Lefèvre, Charles Letourneau, Gabriel de Mortillet, Henri Thulié, and Eugène Véron, eds., *Dictionnaire des sciences anthropologiques* (Paris, n.d.). The first section seems to have come out in 1881, and the first full volume in 1884.

20. Charles Letourneau, *The Evolution of Marriage and of the Family* (London: Scott, 1891), 227.

21. Hovelacque et al., *Dictionnaire des sciences anthropologiques*, 478.

22. Charles Letourneau, "Banquet de Mme. Clémence Royer, Grand Hìtel," March 19, 1897, Clémence Royer Dossier, 122, Bibliothèque Marguerite Durand, cited in Joy Harvey, "Races Specified, Evolution Transformed: The Social Context of Scientific Debates Originating in the Société d'anthropologie de Paris, 1859–1902" (Ph.D. diss., Harvard University, 1983), 251 n. 38. For more of the speech, see 332 n. 80.

23. Harvey, "Almost a Man of Genius," 1.

24. Hovelacque et al., *Dictionnaire des sciences anthropologiques*, 475.

25. The "prohibition on searching out paternity" refers to a law, originating under Napoleon, forbidding mother or child from seeking out a missing father.

26. His Nobel Prize (1913) was in medicine and physiology and was specifically for his work on anaphylaxis. He worked in many fields and made several important contributions; for instance, scientific study of allergic phenomena dates from his work.

27. "Causerie bibliographique," *Revue scientifique* 21 (1884): 536–537.

28. André Lefèvre, *La philosophie*, Bibliothèque des sciences contemporaines (Paris: Reinwald, 1879), 414–415.

29. Charles Letourneau, *La sociologie d'après l'ethnographie*, Bibliothèque des sciences contemporaines (Paris: Reinwald, 1884), 450. In his *Evolution of Marriage*, Letourneau wrote: "Man is neither a demigod nor an angel; he is a primate more intelligent than the others, and his relationship with the neighboring species of the animal kingdom is more strongly shown in his psychic than his anatomical traits" (341–342). This comment means that though we resemble apes physically, we are even more like them mentally.

30. Letourneau, *La sociologie*, 552.

31. Durkheim, "Sociology in France in the Nineteenth Century (1900)," in *On Morality and*

Society (Selected Writings), ed., Robert Bellah (Chicago: University of Chicago Press, 1973), 3–24.

32. "Causerie bibliographique," *Revue scientifique* 21 (1884): 598.

33. Abel Hovelacque, *La linguistique*, Bibliothèque des sciences contemporaines (Paris: Reinwald, 1892): 23.

34. "La linguistique moderne: D'après M. Hovelacque," *Revue scientifique* 12 (1876): 424–428.

35. Eugène Véron, *L'esthétique*, Bibliothèque des sciences contemporaines (Paris: Reinwald, n.d.), v. This quotation is taken from the English version, dated 1879: *Aesthetics*, trans. W. H. Armstrong (London: Chapman and Hall, 1879), v.

36. See page 60 in the English version and page 72 in the French.

37. Lars Ahnebrink, *The Beginnings of Naturalism in American Fiction: A Study of the Works of Hamlin Garland, Stephan Crane, and Frank Norris with Special Reference to Some European Influences, 1891–1903* (New York: Russell and Russell, 1961), 140.

38. See Véron, *L'esthétique*, 466.

39. From a letter to Eldon C. Hill, February 14, 1939, cited in Ahnebrink, *The Beginnings of Naturalism*, 139.

40. From the manuscript of "Literature of Democracy," chap. 10, cited in Ahnebrink, *The Beginnings of Naturalism*, 142.

41. Régis Michaud, *The American Novel Today: A Social and Psychological Study* (Boston: Little, Brown, 1928), 200.

42. Paul Topinard, *L'anthropologie*, Bibliothèque des sciences contemporaines (Paris: Reinwald, 1876). Quotations in the text have been taken from the English version, *Anthropology*, trans. Robert Bartley (London, 1890).

43. This addition must have been a terrific sore spot for Topinard, especially after he had been kicked out of the Ecole.

44. "Bibliographie scientifique," *Revue scientifique* 12 (1876): 333–334.

45. Insisting that the "human races" are very different from one another had two major and very different political meanings at the time: (a) it could be part of an apology for slavery, since humanity was not an equal brotherhood; and (b) it could be antireligious, since it contradicted the biblical account of creation.

46. Mortillet, *Le préhistorique*, Bibliothèque des sciences contemporaines (Paris: Reinwald, 1883).

47. The Reinachs were an interesting leftist, Jewish family of scholars, authors, and scientists. Salomon is perhaps best known for his arguments against the historical validity of the Christian Gospels. His brothers were Théodore Reinach, the Hellenist scholar and archaeologist, best known for his now classic study of ancient Jewish coins and for having built, in France, a remarkable re-creation of an ancient Greek villa (now an official historic landmark); and Joseph Reinach, a publicist and lawyer who had a long parliamentary career: he was an associate of Gambetta and waged a campaign against General Boulanger in the journal *République française* in 1889. He was elected a deputy that year but lost his seat in 1898 because he was an early and vociferous Dreyfusard, publishing articles to this effect in Yves Guyot's *Le siècle*. His pro-Dreyfus campaign was understood as particularly courageous because Reinach was Jewish himself. He was again a deputy from 1906 until 1914. He wrote a history of the Drey-

fus Affair in seven volumes, and during World War I he wrote celebrated military articles for *Le figaro*.

48. Reinach, "Gabriel de Mortillet," 31.

49. Mortillet's system is described in most general treatments of archaeology. See, for example, Warwick Bray and David Trump, *The Penguin Dictionary of Archaeology* (Middlesex: Penguin, 1970), 152–153.

50. Glyn E. Daniel, *A Hundred Years of Archaeology* (London: Duckworth, 1952), 99. Daniel cites Gabriel de Mortillet, "Silex taillés de l'époque tertiaire du Portugal," *BSAP* 3d ser., 1 (1878): 428; and idem, "L'homme quaternaire à l'Exposition," *Revue d'anthropologie* 2d ser., no. 2 (1879): 116; Abel Hovelacque, "La linguistique et le precurseur de l'homme," in *Association française pour l'avancement des sciences—Compte rendu*, vol. 2 (Lyon, 1873).

51. Here is a translation of Daniel's Mortillet quotation: "It would be impossible to put into doubt the great law of human progress. Stone shaped by blows, polished stone, bronze, iron, they are the great stages that all humanity has traversed in order to arrive at our civilization." Daniel does not give the citation, but the quotation comes from Mortillet, "Promenades préhistoriques à l'Exposition universelle," *Matériaux pour l'histoire positive et philosophique de l'homme* 3 (1867): 181–283, 285–368.

52. Mortillet, *Le préhistorique*, 665. After Mortillet's death, this edition was revised by his son, Adrien de Mortillet.

53. "Bibliographie scientifique," *Revue scientifique* 18 (1881): 59.

54. "Bibliographie scientifique," *Revue scientifique* 23 (1886): 535–536. See Charles Letourneau, *L'évolution de la morale* (Paris: Delahaye et Lecrosnier, 1887).

55. "Bibliographie scientifique: La physiologie des passions," *Revue scientifique* 14 (1878): 1167–1170. See Charles Letourneau, *La physiologie des passions*, 2d ed. (Paris: Reinwald, 1878).

56. Schopenhauer disagreed, arguing that our internal thoughts are intelligible only in chronological series; therefore they are possessed of time and consequently not noumenal.

57. Isaac Benrubi, *Contemporary Thought of France* (London: Williams and Norgate Limited, 1926), 38.

58. Frédéric Paulhan, "Analyses et comptes rendus: Letourneau—*La sociologie d'après l'ethnographie*," *Revue philosophique* 11 (1881): 546.

59. Frédéric Paulhan, "Analyses et comptes rendus: Letourneau—*L'évolution de la morale*," *Revue philosophique* 23 (1887): 80.

60. Frédéric Paulhan, "Analyses et comptes rendus: Letourneau—*L'évolution du mariage et de la famille*," *Revue philosophique* 26 (1888): 184.

61. Paulhan, "Analyses et comptes rendus: Letourneau—*La sociologie d'après l'ethnographie*," 550.

62. Georges Belot, "Analyses et comptes rendus: Letourneau—*L'évolution de la propriété*," *Revue philosophique* 28 (1889): 645.

63. G.S., "Analyses et comptes rendus: Dr. H. Thulié, *La femme: Essai de sociologie physiologique*," *Revue philosophique* 20 (1885): 538–540.

64. H. Dereux, "Analyses et comptes rendus: André Lefèvre, *La philosophie*," *Revue philosophique* 7 (1879): 457–458.

65. Frédéric Paulhan, "Analyses et comptes rendus: E. Véron-*La morale*," *Revue philosophique* 18 (1884): 475.

66. See Léonce Manouvrier, "Société de psychologie physiologique: Les premières circon-

volutions temporales droite et gauche chez un sourd de l'oreille gauche (Bertillon)," *Revue philosophique* 25 (1888): 330–335; idem, "Société de psychologie physiologique: Mouvements divers et sueur palmaire consécutifs à des images mentales," *Revue philosophique* 26 (1886): 203–207. See chap. 6, below, for more on these articles.

67. Léonce Manouvrier, "Etude comparative sur les cerveaux de Gambetta et de Bertillon," *Revue philosophique* 25 (1888): 453–461.

68. For another Manouvrier monograph for the *Revue philosophique*, see "La fonction psycho-motrice," *Revue philosophique* 17 (1884): 503–525, 638–651.

69. P.-G. Mahoudeau, "Analyses et comptes rendus: Manouvrier, Sur l'interprétation de la quantité dans l'encéphale et du poids de cerveau en particulier," *Revue philosophique* 24 (1887): 321.

70. Léonce Manouvrier, "Analyses et comptes rendus: Duval—*Le Darwinisme*," *Revue philosophique* 21 (1886): 398.

71. Léonce Manouvrier, "Analyses et comptes rendus: Quatrefages—Introduction à *L'étude des races humaines*," *Revue philosophique* 24 (1887): 322.

72. Léonce Manouvrier, "Analyses et comptes rendus: Hovelacque et Hervé—*Précis d'anthropologie*," *Revue philosophique* 24 (1887): 325.

73. For more of Manouvrier's reviews for *Revue philosophique*, see Léonce Manouvrier, "Analyses et comptes rendus: Broca, *Mémoires sur le cerveau de l'homme et des primates*," *Revue philosophique* 27 (1889): 405–409; idem, "Analyses et comptes rendus: Hervé, *La circonvolution de Broca*," *Revue philosophique* 27 (1889): 409–411.

74. *La critique philosophique*, August 8, 1872, 1.

75. *La critique philosophique*, April 15, 1875, p. 166.

76. "Darwin et le bon Dieu," *La critique philosophique* 5 (1876): 205.

77. François Pillon, "La lutte contre le cléricalisme, ce qu'elle ne doit pas être et ce qu'elle doit être: Il ne faut pas que la politique anticléricale soit une politique d'irréligion," *La critique philosophique* 9 (1880): 113–123.

78. F. Grindelle, "Bibliographie: *La renaissance du matérialisme*," *La critique philosophique* 10 (1881): 397–399.

79. F. Grindelle, "Bibliographie: *Les débuts de l'humanité*," *La critique philosophique* 10 (1881): 399–400.

80. F. Grindelle, "Bibliographie: *La morale*," *La critique philosophique* 13 (1884): 236–239.

81. See, for instance, François Pillon, "A propos du substantialisme de Mme. Clémence Royer et de M. Roisel," *La critique philosophique* 11 (1882): 81–89; F. Grindelle, "Bibliographie: *Le bien et la loi morale*," *La critique philosophique* 13 (1884): 204–208.

82. François Pillon, "Revue bibliographique: Léfevre, *L'histoire: Entretiens sur l'évolution historique*," *L'année philosophique* 8 (1897): 261–262; idem, "Letourneau: *L'évolution de l'éducation dans les diverses races humaines*," *L'année philosophique* 9 (1898): 262–263; idem, "Letourneau: *L'évolution de l'esclavages dans les diverses races humaines*," *L'année philosophique* 8 (1897): 266–267; idem, "Letourneau: *L'évolution du commerce dans les diverses races humaines*," *L'année philosophique* 8 (1897): 267–268; idem, "*Letourneau: L'évolution religieuse dans les diverses races humaines*," *L'année philosophique* 3 (1892): 260–261.

83. Pillon, "Letourneau: *L'évolution religieuse*," 263.

84. Adrien Arcelin, "Matériaux pour l'histoire primitive et naturelle de l'homme," *Revue des questions scientifiques* 6 (1879): 319.

85. Adrien Arcelin, "Le cerveau de Gambetta," *Revue des questions scientifiques* 21 (1887): 271–272.

86. Abbé E. Vacandard, "Le nouvel homme préhistorique de Menton," *Revue des questions scientifiques* 20 (1886): 74–122.

87. Lucien Lévy-Bruhl, "La morale de Darwin," *Revue politique et littéraire* 5 (1883): 169–175.

88. Alphonse Bertillon, "Questions des récidivistes: L'identité des récidivistes et la loi de relégation," *Revue politique et littéraire* 3, no. 17 (1883): 513–521.

89. Jean Réville, "Une histoire des religions par un adversaire de la religion: M. Eugène Véron," *Revue politique et littéraire* 5 (1883): 14, 15, 17.

90. Observed by Robert Alun Jones in his "Religion and Science in the Elementary Forms," in N. J. Allen, W. S. F. Pickering and W. Watts Miller, eds., *On Durkheim's Elementary Forms of Religious Life* (NewYork: Routledge, 1998), 39–52. Réville is one of the most frequently cited historians of religion in Durkheim's *Elementary Forms*.

91. Réville, "Une histoire des religions," 14.

92. Jacqueline Lalouette, "Science et foi dans l'idéologie libre penseuse (1866–1914)," in *Christianisme et science* (Paris: Vrin, 1989), 38–39.

93. André Lefèvre, *Philosophy: Historical and Critical*, trans. A. H. Keane (London: Chapman and Hall; Philadelphia: Lippincott, 1879).

94. Annick Chauvière, "Les hommes d'aujourd'hui," in Jean-Michel Place and André Vasseur, eds., *Bibliographie des revues et journaux littéraires des XIXe et Xxe siècles* (Paris: Jean-Michel Place, 1974), 90–97.

95. Etienne Roc, "Professeur Mathias Duval," *Les hommes d'aujourd'hui* 6, no. 273 (1886): 1.

96. Pierre et Paul (the journal's generic pseudonym), "Yves Guyot," *Les hommes d'aujourd'hui* 1, no. 46 (1878): 1.

97. Pierre et Paul, "Hovelacque," *Les hommes d'aujourd'hui* 3, no. 128 (1880): 1.

98. Pierre et Paul, "Thulié," *Les hommes d'aujourd'hui* 3, no. 144 (1880): 1.

99. Pierre et Paul, "Clémence Royer," *Les hommes d'aujourd'hui* 4, no. 170 (1881).

4. Careers in Anthropology and the Bertillon Family

1. "Abel Hovelacque," *La grande encyclopédie*, 3d ed. (Paris, n.d.), cited in Joy Harvey, "Races Specified, Evolution Transformed: The Social Context of Scientific Debates Originating in the Société d'anthropologie de Paris, 1859–1902" (Ph.D. diss., Harvard University, 1983), 66. The freethinkers' attitude toward Ferry fluctuated over the years, but he was never sufficiently anticlerical for them. I do not know to what extent the anecdote about the lockout should be accepted or rejected; Mortillet was one of the contributors to *La grande encyclopédie* and may have embellished his friend's biography.

2. See Jacques Léonard, *La médecine entre les savoirs et les pouvoirs: Histoire intellectuelle et politique de la médecine française au XIXe siècle* (Paris, Aubier, 1981), 274.

3. Henri Thulié, "Sur l'autopsie de Louis Asseline, membre de la Société d'anthropologie et de la Société d'autopsie mutuelle," *Bulletins de la Société d'anthropologie* (hereafter BSAP) 3d ser., 10 (1878): 162.

4. Henry T. F. Rhodes, *Alphonse Bertillon: Father of Scientific Detection* (New York: Abelard-Schuman, 1956), 50.

5. See, for instance, Yves Guyot, *La science economique et ses lois inductives* (Paris, 1881); idem,

Les principes de 1889 et le socialisme (Paris, 1894); idem, *Sophismes socialiste et faits economiques* (Paris, 1908).

6. Archives of the Société d'anthropologie de Paris (hereafter ASAP), procès-verbaux, January 15, 1891. See also Henri Vallois, "Le laboratoire Broca," *Bulletins et mémoires de la Société d'anthropologie* 9th ser., 11 (1940): 3.

7. Archives nationales (hereafter AN), F/17/17199, "Ecole d'anthropologie, 22 mai 1880." These numbers are for 1872. See Antoine Prost, *Histoire de l'enseignement en France, 1800–1967* (Paris: Armand Colin, 1968), 74, 75. It is also useful to compare Vacher de Lapouge's librarian salary over several years (see chap. 5, below). In 1887 he earned 2,000 francs as a librarian, in 1893 his salary was 3,000 francs, and it reached 4,000 francs in 1900. AN, F/17/22640, documents 98–170.

8. AN, F/17/13140.

9. "Excursions préhistoriques," *L'homme* 2 (1885): 276.

10. AN, F/17/2994, "Missions-Mortillet."

11. AN, F/17/2994, September 12, 1872; August 1874.

12. AN, F/17/2994, letter from Mortillet to Ferry, Ministère de l'instruction publique, June 17, 1879.

13. AN, F/17/2994, letter from M. Delabarre, Ministère des affaires étrangères, to M. Ferry, Ministère de l'instruction publique, September 1, 1879, including two articles from *Télégraphe* dated September 5 and December 7, 1879.

14. AN, F/17/2994, letter from M. Laboulaye to Jules Ferry, Ministère de l'instruction publique, Lisbon, September 30, 1880.

15. André Lefèvre, "La vie philosophique," *La vie littéraire* 52 (December 28, 1876); idem, "L'histoire (sonnet)," *La vie littéraire* 6 (February 8, 1877); idem, "Paléontologie intellectuelle," *La vie littéraire* 46 (November 16, 1876). Lefèvre was a "principal collaborator" on *La jeune France*, and his work was featured in a series called "Les hommes de la jeune France"; see Emmanuel des Essarts, "André Lefèvre," *La jeune France*, no. 35 (March 1, 1881). For Lefèvre's many articles and poetry publications, see Jean-Michel Place and André Vasseur, eds., *Bibliographie des revues et journaux littéraires des XIX et XX siècles* (Paris: Editions Jean-Michel Place, 1974).

16. André Lefèvre, "Sonnets philosophiques," *La jeune France*, no. 21 (1880–1881): 38–40.

17. André Lefèvre, "Un vice de la littérature enfantine," *La jeune France*, no. 33 (1880–1881): 404–406.

18. André Lefèvre, "Voltaire et les religions," *La jeune France*, no. 10 (February 1, 1879): 367.

19. André Lefèvre, "L'homme, d'après les découvertes de l'anthropologie," *La jeune France*, no. 1 (October 1, 1878): 214–221.

20. Testament: Clémence Royer, May 5, 1895, Bibliothèque Marguerite Durand, cited in Harvey, *"Almost a Man of Genius": Clémence Royer, Feminism, and Nineteenth-Century Science* (New Brunswick: Rutgers University Press, 1997), 168–169.

21. Edward Hallett Carr, *The Romantic Exiles* (1933; reprint, Cambridge, Mass.: MIT Press, 1981), 349–362.

22. In 1876 Monod founded *Revue historique*, a journal dedicated to "strictly scientific" history, which transformed historical studies in modern France. See his introduction to the journal: *Revue historique* 1 (January–June 1876): 36. For a discussion of Monod's scientific history,

see William R. Keylor, *Academy and Community: The Foundation of the French Historical Profession* (Cambridge: Harvard University Press, 1975).

23. Harvey, *"Almost a Man of Genius,"* 80–81, 222–223.

24. On criminal deportation, see Robert Nye, *Crime, Madness, and Politics in Modern France: The Medical Conception of National Decline* (Princeton: Princeton University Press, 1984).

25. For more on forced sterilization in the United States and elsewhere, see Daniel J. Kevles, *In the Name of Eugenics: Genetics and the Uses of Human Heredity* (Berkeley: University of California Press, 1985).

26. Michel Foucault, *The History of Sexuality:Volume One-An Introduction* (NewYork: Vintage, 1990); and "Omnes et Singulatum:Towards a Criticism of 'Political Reason,'" in *The Turner Lectures on HumanValues*, (Salt Lake City: Sterling M. McMurrin, 1981), vol. 2.

27. Republican students had a medal made for Michelet, who was at the time being harassed by the government for his republicanism, and Bertillon was chosen to award the gift. Thus there were informal meetings that led to friendship. See Rhodes, *Alphonse Bertillon*, 17–18. Rhodes borrowed a great deal from Suzanne Bertillon, *Vie d'Alphonse Bertillon: Inventeur de l'anthropométrie* (Paris: Gallimard, 1941). (Much of his book is simply uncredited translations of her text.)

28. Michel Dupaquier, "La famille Bertillon et la naissance d'une nouvelle science sociale: La démographie," *Annales de démographie historique*, 1983, 294.

29. Rhodes, *Alphonse Bertillon*, 24.

30. Dupaquier, "La famille Bertillon" 294.

31. *La vie et les oeuvres du docteur L.-A. Bertillon, professeur de démographie à l'Ecole d'anthropologie, chef des travaux de la statistique municipale de la ville de Paris* (Paris, 1883), 25–26.

32. Rhodes, *Alphonse Bertillon*, 29.

33. Louis-Adolphe Bertillon, "La biologie," *La pensée nouvelle* 22 (October 11, 1868): 170–172.

34. Dupaquier, "La famille Bertillon," 295.

35. Joshua Cole, *The Power of Large Numbers* (Ithaca: Cornell University Press, 2000), 116 and throughout.

36. See, for instance, Gisela Bock and Pat Thane, eds., *Maternity and Gender Policies:Women and the Rise of the EuropeanWelfare States, 1880–1950s* (NewYork: Routledge, 1991); Susan Pedersen, *Family, Dependence, and the Origins of the Welfare State: Britain and France, 1914–1945* (Cambridge: Cambridge University Press, 1993); and Seth Kovin and Sonya Michel, eds., *Mothers of a NewWorld: Maternalist Politics and the Origins of Welfare States* (NewYork: Routledge, 1933).

37. Louis-Adolphe Bertillon, *La démographie figurée de la France* (Paris: G. Masson, 1874), 2.

38. Louis-Adolphe Bertillon, "Des diverses manières de mesurer la durée de la vie humaine," *Journal de la Société de statistiques de Paris* 7 (1866): 45–64; idem, "Méthode pour calculer la mortalité d'une collectivité pendant son passage dans un milieu déterminé . . . ," *Journal de la Société de statistiques de Paris* 10 (1869), 29–40, 57–65, cited in Theodore M. Porter, *Trust in Numbers:The Pursuit of Objectivity in Science and Public Life* (Princeton: Princeton University Press, 1995), 82–83.

39. Dupaquier, "La famille Bertillon," 308.

40. Suzanne Bertillon, *Vie de Alphonse Bertillon*, 79. See also Rhodes, *Alphonse Bertillon*, 95.

41. Sometimes Bertillon offered some explanation of the numbers he reported, sometimes not. See, for instance, *Le temps*, July 19, 1889; July 27, 1889; August 2, 1889; August 9, 1889.

42. Jacques Bertillon, "The International System of Nomenclature of Diseases and Causes of Death (Bertillon Classification)," *Public Health Reports* 15, no. 49 (December 7, 1900): 42, 40.

43. Jacques Bertillon, *Cours élémentaire de statistique administrative, élaboration des statistiques, organisation des bureaux de statistique, éléments de démographie* (Paris: Société d'éditions scientifiques, 1895), 262.

44. World Health Organization Website, "Library and Information Networks for Knowledge, Disease Classifications and Nomenclature Documents," 2002, www.who.int (August 2, 2002).

45. See Jean-Marie Mayeur and Madeleine Rebérioux, *The Third Republic from Its Origins to the Great War, 1871–1914* (Cambridge: Cambridge University Press, 1984), 330; and Cole, *The Power of Large Numbers*, 2. On the belly strike (or womb strike), also see Francis Ronsin, *La grève des ventres: Propagande néo-malthusienne et baisse de la natalité en France, 19e-20e siècles* (Paris: Aubier Montaigne, 1980).

46. See, for instance, Cheryl A. Koos, "Gender, Anti-Individualism, and Nationalism: The Alliance Nationale and the Pronatalist Backlash Against the Femme moderne, 1933–1940," *French Historical Studies* 19 (1996): 699–723.

47. Cole, *The Power of Large Numbers*, 198.

48. Jacques Bertillon, *De la dépopulation de la France et des remèdes à apporter* (Paris, 1896); idem, *Le problème de la dépopulation* (Paris, 1897); idem, *La dépopulation de la France, ses conséquences, ses causes, mesures à prendre pour la combattre* (Paris: Alcan, 1911).

49. Joseph J. Spengler, *France Faces Depopulation* (Durham, North Carolina: Duke University Press, 1938), 234.

50. Karen Offen, "Depopulation, Nationalism, and Feminism in Fin-de-Siècle France," *American Historical Review* 89 (1984): 648–676.

51. Madame Jacques Bertillon, *La femme médecin au XXe siècle* (thèse de médecine, Faculté de médicine, Paris, 1888), cited by Jacques Léonard, *La médecine entre les savoirs et les pouvoirs: Histoire intellectuelle et politique de la médecine française au XIXe siècle* (Paris: Aubier, 1981), 33. Madeleine Brès became the first French woman to work as a doctor.

52. Bertillon, *Vie d'Alphonse Bertillon*.

53. Jacques Bertillon, "Le problème de la dépopulation: Le programme de l'Alliance nationale pour l'accroissement de la population française," *Revue politique et parlementaire*, 4th year, no. 12 (April-June 1897), 538–539. See also Spengler, *France Faces Depopulation*, 123.

54. Spengler, *France Faces Depopulation*, 127.

55. See Miranda Pollard, *Reign of Virtue: Mobilizing Gender in Vichy France* (Chicago: University of Chicago Press, 1998); and Robert Owen Paxton, *Vichy France: Old Guard and New Order, 1940–1944* (New York: Columbia University Press, 1972), 166–167.

56. *Revue de l'alliance nationale* 359 (September 1943): 143. It was hoped that the sentence would act as a deterrent to others. See Pollard, *Reign of Virtue*, 237 n. 24.

57. Suzanne Bertillon, *Vie d'Alphonse Bertillon*, 22. See also Rhodes, *Alphonse Bertillon*, 33.

58. Rhodes, *Alphonse Bertillon*, 35.

59. Suzanne Bertillon, *Vie d'Alphonse Bertillon*, 33.

60. François Favre, "Enseignement professionel des femmes," *Le monde maçonnique*, November 1862, 385, cited in Philip Nord, *The Republican Moment: Struggles for Democracy in Nineteenth-Century France* (Cambridge: Harvard University Press, 1995), 26.

61. Rhodes, *Alphonse Bertillon*, 33. Suzanne Bertillon related a similar version of this story: see her *Vie d'Alphonse Bertillon*, 22–24.

62. Suzanne Bertillon, *Vie d'Alphonse Bertillon*, 79–80.

63. Alphonse Bertillon, *Les races sauvages: Les peuples de l'Afrique, les peuples de l'Amérique, les peuples de l'Océanie, quelques peuples de l'Asie et des regions bovéales* (Paris, 1882).

64. On recidivism, see Nye, *Crime, Madness, and Politics*.

65. Rhodes, *Alphonse Bertillon*, 87, 89.

66. Head measurements were first divided up into small, medium, and large, and these were physically separated in the file cabinet, so that the measurement card of a criminal with a big head was sure to be in a single area. The secondary measurement narrowed the location further, and tertiary measurements still more, until the card could only be in one place. It was essentially like finding a word in the dictionary: to spell *cat* you first eliminate the twenty-five letters that are not *c*, then you do the same for *a*, and *t*, and you have your definition.

67. Rhodes, *Alphonse Bertillon*, 103.

68. Document archives de la Préfecture de police, as reproduced in Ronsin, *La grève des ventres*.

69. Ronsin tells us the phrase was a favorite among anarchists. Ronsin, *La grève des ventres*, illustrations, Robin caption.

70. Ida M. Tarbell, "Identification in Criminals: The Scientific Method in Use in France," *New McClure's Magazine* 2, no. 4 (March, 1894): 355–356.

71. *Alphonse Bertillon's Instructions for Taking Descriptions for the Identification of Criminals and Others by the Means of Anthropometric Indications*, trans. and intro. Gallus Muller (Chicago: American Bertillon Prison Bureau, 1889), 11–12.

72. Ian Hacking, *The Taming of Chance* (Cambridge: Cambridge University Press, 1990), 187.

73. R. Heindl, cited in Rhodes, *Alphonse Bertillon*, 193.

74. Rhodes, *Alphonse Bertillon*, 146–152.

75. Rhodes has it that bloody fingerprints marked the woman as having committed a double homicide, but an account closer to the sources identifies the crime as infanticide. See ibid., 149; and Kristin Ruggiero, "Fingerprinting and the Argentine Plan for Universal Identification in the Late Nineteenth and Early Twentieth Centuries," in Jane Caplan and John Torpey, eds., *Documenting Individual Identity: The Development of State Practices in the Modern World* (Princeton: Princeton University Press, 2001), 191. Caplan and Torpey's book contains nineteen essays covering a great range of topics relevant to the matter of identity information, spanning several centuries.

76. The bill was passed on July 18 and became law on July 20. Vucetich was appointed director on August 3, 1916.

77. Rhodes, *Alphonse Bertillon*, 151.

78. Ruggiero, "Fingerprinting and the Argentine Plan," 193.

79. Mayeur and Rebérioux, *The Third Republic*, 180.

80. The note he had allegedly written is generally understood to have been forged by a Major Esterhazy in an attempt to pin recent intelligence leaks on Dreyfus.

81. Matt K. Matsuda, *The Memory of the Modern* (Oxford: Oxford University Press, 1996), 134–139.

82. Reported in *Le matin* (December 17, 1903), collected by the Préfecture de police, DB 47, cited in Matsuda, *The Memory of the Modern*, 136.

83. Musacchio, "Bertillonnades," L'assiette au beurre, July 3, 1909, as reproduced in Rhodes, Alphonse Bertillon, 113.

84. Arthur Conan Doyle, The Annotated Sherlock Holmes, ed. and intro. William S. Baring-Gould (New York: Clarkson and Potter, 1967), 2:202.

85. Rhodes, Alphonse Bertillon, 108, 218.

86. Tarbell, "Identification in Criminals," 356.

87. Edmond Locard, Alphonse Bertillon: L'homme, le savant, la pensée philosophique (Lyon: A. Rey, 1914), 7.

5. No Soul, No Morality:Vacher de Lapouge

1. Georges Vacher de Lapouge, "L'anthropologie et la science politique," Revue d'anthropologie 16 (1887): 1422. He had some reservations about this equation of terms, largely because other anthropologists sometimes depicted Aryans as brachycephalic. In his later works, he suggested that the term "Homo Europaeus" should replace "Aryan," but before and after this clarification he largely used "Aryan" to denote the dolichocephalic. On the idea of the Aryan people, see Leon Poliakov, The Aryan Myth: A History of Racist and Nationalist Ideas in Europe (New York: Basic, 1974).

3. See, for instance, his "Questions aryennes," Revue d'anthropologie 18 (1889): 183.

4. One of his clearest explanations of this point is in L'Aryen: Son rôle social (Paris: A. Fontemoing, 1899), 352n.

5. Georges Vacher de Lapouge, preface to Ernst Haeckel, Le monisme: Lien entre la religion et la science, trans. Georges Vacher de Lapouge (Paris: Schleicher, 1897), 1–8.

6. Lapouge, L'Aryen, 467.

7. Ferdinand Brunetière, "Après une visite au Vatican," Revue des deux mondes 7, no. 127 (1895): 97–118.

8. Harry Paul, "The Debate Over the Bankruptcy of Science in 1895," French Historical Studies 3 (1968): 299–327.

9. Brunetière, "Après une visite," 100.

10. André Lefèvre, La religion (Paris: A. Costes, 1921), 572–573, cited in Brunetière, "Après une visite," 98; emphasis mine.

11. Brunetière, "Après une visite," 104.

12. Elected in 1881, he also served as inspector general of higher education, minister of public instruction, and minister of foreign affairs.

13. Marcellin Berthelot, Science et morale (Paris, 1897), 28, 34, 43.

14. Marcellin Berthelot, "La science et la morale," Revue de Paris, February 1, 1895, 461.

15. Clémence Royer, La constitution du monde: Natura rerum, dynamique des atomes, nouveaux principes de philosophie naturelle (Paris: Schleicher Frères, 1900), xx.

16. Charles Richet, "La science a-t-elle fait banqueroute?" Revue scientifique 3 (January 12, 1895): 33–39.

17. Lapouge, L'Aryen, 513.

18. Georges Vacher de Lapouge, Les sélections sociales (Paris: A. Fontemoing, 1896), 307.

19. Lapouge, L'Aryen, 508.

20. Lapouge, "L'anthropologie et la science politique," 143.

21. Lapouge, Sélections sociales, 306–307.

22. Lapouge, preface, 1–8. On Haeckel, see Daniel Gasman, The Scientific Origins of Na-

tional Socialism: Social Darwinism in Ernst Haeckel and the German Monist League (London: Macdonald, 1971).

23. Lapouge, *Sélections sociales*, 306–307.

24. André Béjin, "Le sang, le sens, et le travail: Georges Vacher de Lapouge, darwiniste social, fondateur de l'anthroposociologie," *Cahiers internationaux de sociologie* 73 (1982): 335–336.

25. Charles Letourneau, "Adolphe Bertillon," *Bulletins de la Société d'anthropologie* (hereafter *BSAP*) 3d ser., 6 (1883): 192.

26. Lapouge, *Sélections sociales*, 190–191.

27. Lapouge, *L'Aryen*, ix.

28. Several other early eugenists wrote from this pessimistic Darwinian stance and were concerned with combating "degeneration." It seems that none of these other authors also rejected the Lamarckian evolutionary mechanism. See William Schneider, *Quality and Quantity: The Quest for Biological Regeneration in Twentieth-Century France* (Cambridge: Cambridge University Press, 1990), 11 and 56–63.

29. Lapouge illustrated this by describing an academician and a fish in a row boat on the high seas: if the boat springs a leak, the fish is the fittest. Progress, as such, is not guaranteed (*L'Aryen*, 503).

30. In his *Politics of Evolution: Morphology, Medicine, and Reform in Radical London* (Chicago: University of Chicago Press, 1989), Adrian Desmond wrote of "Red Lamarckians" in England in the 1830s, who "used Lamarckism explicitly to legitimate a cooperative society, female emancipation, and an equal education program" (329–330). Also see his "Lamarckism and Democracy: Corporations, Corruption, and Comparative Anatomy in the 1830s," in James R. Moore, ed., *History, Humanity, and Evolution: Essays in Honor of John C. Green* (Cambridge: Cambridge University Press, 1989), 99–130.

31. Lapouge, *L'Aryen*, 511.

32. Lapouge, *Sélections sociales*, 472–473.

33. Lapouge, *L'Aryen*, 394.

34. As he wrote to his son Claude in 1916: "Concerning your marriage, it is important to me that you continue my effort, but that does not make it my business to choose your wife for you. I only ask, in the name of the race, that she be capable of giving children of my worth or of yours and that she have enough good qualities that it will not be out of the question for them to turn their merits to the best account" (Georges Vacher de Lapouge to Claude Vacher de Lapouge, Fonds Vacher de Lapouge, Université de Montpellier [hereafter FVL/UM], A068–50). The fact that he felt he had to write such a disavowal suggests that he had exerted pressure on his son in the past, and in the rest of the letter Lapouge is plainly responding to considerable anger from his son.

35. Lapouge, *Les sélections sociales*, 262. The statement is an echo of Gambetta's famous dictum that the republic would be scientific or not at all.

36. Some, especially Toussenel, made anti-Semitism a cornerstone of their socialism. To a degree, one can identify anti-Semitism as a leftist phenomenon in its origins and follow its crossover to the right along a path similar to that of nationalism, which also began on the left (when pride in the nation was in opposition to obedience to the monarchy). For a concise discussion of the history of anti-Semitism on the left, see Paul Bénichou, "Sur quelques sources françaises de l'antisémitisme moderne," *Commentaire* 1 (1978): 67–79.

37. Lapouge, *Race et milieu social* (Paris: Rivière, 1909).

38. Edouard Drumont, "Napoleon antisemite," *La libre parole*, March 26, 1900: 1.

39. Lapouge, *L'Aryen*, 464.

40. *L'Aryen*, 376–377.

41. For a detailed treatment of Liard's role in remaking the French university system, see George Weisz, *The Emergence of Modern Universities in France, 1863–1914* (Princeton: Princeton University Press, 1983).

42. See Raymond Lenoir, "L'oeuvre sociologique d'Emile Durkheim," *Europe* 22 (1930): 294, cited in Terry Clark, "Emile Durkheim and the Institutionalization of Sociology in the French University System," *European Journal of Sociology* 9 (1968): 37–71.

43. Henri Bégouen, "Vacher de Lapouge-Père de l'Aryenisme," *Journal des débats politiques et littéraires*, August 22, 1936, 3. See also Georges Vacher de Lapouge, "Souvenirs," in Henri H. de La Haye Jousselin, *Georges Vacher de Lapouge: Essai de bibliographie* (self-published, 1986), 15. This essay was written on request for the German journal *Hammer* in 1929, on the occasion of Lapouge's seventy-fifth birthday.

44. Archives nationales (hereafter AN), F/17/22640, doc. 98.

45. Lapouge, "Souvenirs," 11.

46. AN, F/17/25839, personnel file, Louis Liard, renseignements confidentiels, Académie de Bordeaux, July 1870.

47. AN, F/17/25839, personnel file, Louis Liard, Liard to the Ministre de l'instruction publique, Académie de Poitiers, October 24, 1972.

48. AN, F/17/25839, personnel file, Louis Liard, renseignements confidentiels, Poitiers, August 29, 1974.

49. Ibid. If Contejean published, the work does not seem to have survived.

50. AN, F/17/22640, doc. 99.

51. AN, BB/6(II)419, February 17, 1879.

52. AN, F/17/22640, docs. 96 and 97.

53. AN, BB/6(II)419, March 29, 1879.

54. For example: "The closer the moment approaches when I will have to begin functioning, the more I find myself struck by the heavy moral responsibility that weighs upon the magistrature. There cannot be a magistrate so attentive and so hard-working that he doesn't occasionally condemn the innocent and ruin, in civil affairs, honest men who are in the right" (AN, BB/6(II)419, May 18, 1879).

55. AN, BB/6(II)419, March 29, 1879. Lapouge published articles in the town's republican newspaper under the name Verax. See the following articles by him in *L'avenir de la Vienne*: "Liberté, liberté, chérie," May 22, 1879, 2; "La liberté des fils de famille," May 23, 1879, 1; "Paroles pour la liberté," July 1 and 2, 1879, 1; "La messe de corvée," January 20, 1880, 1; "La réforme de l'organisation judiciaire," February 24 and 25, 1880, 1–2; February 26, 1880, 1; February 27, 1880, 2.

56. Told to Henri Bégouen (who was writing an obituary of Lapouge in 1936) by the people of Niort; see Bégouen, "Vacher de Lapouge," 3.

57. The very existence of the law is a reminder that this deconsecration project had a great deal in common with the revolutionary interest in deism and with destroying the contemporary vestments of hierarchy. When Lapouge revived the law, he did it for different particular reasons, but the law's two incarnations were both about manipulating people and things in order to revise the ideological terrain in which people live.

58. AN, BB/6(II)419, 1881

59. Another report indicated that he was creating embarrassing situations and needed to be in a much more supervised position (AN, BB/6[II]419, n.d.).

60. "Magistrats de M. Cazot-Danton," *Journal du centre*, August 5, 1881. It seems Lapouge had been quite vocal at several political banquets at Blanc and Mezières, and the *Journal du centre* hypothesized that this had "troubled the sleep" of his newly appointed superior.

61. AN, BB/6(II)419, 1883.

62. AN, BB/6(II)419, divers. On March 22, 1883, Lapouge walked into an arms shop, chose, and was handed a gun. He opened the barrel, looked inside, closed the barrel, and pointed the gun at a wall. Two seconds later one of the proprietors had a bullet in his neck. As all witnesses would later attest, the revolver had gone off on its own, "for no discernible reason," except that it was rather old. The victim, a M. Aufort, did not press charges and was soon up and about (AN, BB/6(II)419, April 6, 1883).

63. He also studied at the Ecole des hautes études, the Louvre, and the Ecole des langues orientales. See Lapouge, "Souvenirs," 13.

64. Hovelacque to Lapouge, FVL/UM, A 047–3 to A 047–4. See also Topinard to Lapouge, FVL/UM, no. 5, A099–1 through A099–37.

65. On his friendship with Topinard, see Lapouge, "Souvenirs," 14, as well as the correspondence between Topinard and Lapouge in the FVL/UM. The idea that Topinard might be seen as an influence on Lapouge was noted by Etienne Patte in 1937, a year after Lapouge's death. "Certain French historians have previously entertained some degree of racial determinism; Topinard, who was one of Vacher de Lapouge's teachers, had echoed this but was much less enthusiastic about it, it seems to me" ("Georges Vacher de Lapouge," *Revue générale du centre-ouest de la France* 46 [1937]: 775). When Lapouge finally retired from anthropometrical research, only a few years before his death, he gave his vast collection to Patte. The two men did not know each other very well, but Lapouge did not know of anyone else to whom he could make the gift and Patte wanted the skulls for his own research. This helps to explain why Patte wrote the article.

66. During his years in Paris, Lapouge wrote several letters to Liard asking him about possible job opportunities, keeping him apprised of his intellectual development, and thanking him for having given direction to his studies. See Lapouge to Louis Liard, FVL/UM, no. 8, A068–95, 96.

67. Lapouge, "Souvenirs," 14.

68. AN, F/17/22640, doc. 17.

69. For his own explanation of why he left Paris, see Lapouge, "Souvenirs," 15.

70. AN, F/17/22640, doc. 97.

71. AN, F/17/22640, docs. 86 and 90.

72. AN, F/17/22640, doc. 167.

73. Topinard to Lapouge, July 30, 1887, FVL/UM, no. 5, A099–16.

74. Bégouen, "Vacher de Lapouge," 3. See also Lapouge, "Souvenirs," 15.

75. AN, F/17/22640, doc. 164. As noted in a report of 1888, "he still sees—and doesn't try to hide it—the functions of a librarian as beneath him, and he only does them so that he need not worry about his daily bread while he engages in his advanced and original scientific research."

76. AN, F/17/22640, doc. 83.

77. AN, F/17/22640, docs. 81, 67, and 83. Lapouge also asked for permission to teach his anthroposociological courses in the Faculté de Droit. He had already asked a committee of the law professors if they would allow this, but they had declined. Liard intervened for him, award-

ing him a salary of 500 francs for his courses. According to Lapouge, the law faculty was favorably disposed to his teaching law; however, they asked him to set aside his specialty and teach a "preliminary course" that the other professors did not want to teach.

78. Lapouge, "Souvenirs," 15.

79. Letter of Paul Valéry to Henri Bégouen (1936), cited in Bégouen, "Vacher de Lapouge," 3. In light of this quotation, and others like it, Eugen Weber's comment that Valéry admired Lapouge is a bit misleading (*My France* [Cambridge: Harvard University Press, 1991], 34).

80. Letter from Vacher de Lapouge to Collignon, October 5, 1892, cited in Bégouen, "Vacher de Lapouge," 3.

81. Lapouge, *L'Aryen*, ix.

82. Lapouge, *Sélections sociale*, 472–473. An anonymous source asserts that Lapouge's mistress received the package and telegenetically conceived a child (interview with the author, Paris, 1993).

83. AN, F/17/22640, doc. 141.

84. AN, F/17/22640, docs. 62–64. An inquiry ensued and soon found the five girls in question, who all attested that the accusations were true. The mother of one of the girls, a Madame Guillemin, went to confront Lapouge in his laboratory and appropriated a nude picture of her daughter and several other girls. One of the girls apparently offered the further evidence that Monsieur Lapouge had a scar on his "membre viril." When confronted with this, Lapouge agreed that he did have such a scar, but at the time of the alleged events it was still a painful wound. He had undergone surgery, later attested to by his doctor (who agreed that it could not have yet healed), and had offered this as evidence in his favor when confronted by the furious mother. He suggested to the inquisitor that she herself had later coached the girls to mention the scar. The girls were then questioned again, and some eventually retracted all accusations regarding sexual acts, saying they had been put up to it by the young Guillemin. Perhaps Madame Guillemin was so angry about the nude photographs that she embellished the girls' reports in an effort to ensure a conviction. Alternatively, the accusations may have been true.

85. AN, F/17/22640, doc. 64. With his usual inability to understand his own reputation, Lapouge requested that if the minister of public instruction should see the necessity of transferring him, he should be promoted in the process, because a simple parallel move would seem like a condemnation (doc. 62).

86. AN, F/17/22640, doc. 62.

87. See, for instance, Nancy Stepan, "Race and Gender: The Role of Analogy in Science," *Isis* 77, no. 2 (June 1986): 261–277; and Ann Laura Stoler, *Race and the Education of Desire: Foucault's History of Sexuality and the Colonial Order of Things* (Durham, N.C.: Duke University Press, 1995).

88. AN, F/17/22640, doc. 136.

89. Letter, Lapouge to Ecole, September 22, 1897, Archives de la Société d'anthropologie, box: "Correspondance: 1905."

90. From 1900 to 1923 Lapouge also served as librarian for the Ecole préparatoire de médecine et de pharmacie, also located at Poitiers. (AN, F/17/22640, doc. 35, among others).

91. *Badische Press*, August 11, 1901, as quoted in Jean Boissel, "Autour du Gobinisme: Correspondance inédite entre L. Schemann et G. Vacher de Lapouge," *Annales du CESERE* 4 (1981): 114.

92. AN, F/17/22640, doc. 124.

93. AN, F/17/22640, doc. 121.

94. Lapouge to Liard, August 1902, FVL/UM, no. 8, A068.

95. The chair had been occupied by Ernst Hamy. See AN, F/17/22640, doc. 57, for details on the librarian application. For the museum chair, see Patte, "Georges Vacher de Lapouge," 782.

96. Muffang to Lapouge, May 20, 1909, FVL/UM, A067–83.

97. Jennifer Michael Hecht, "Vacher de Lapouge and the Rise of Nazi Science," *Journal of the History of Ideas* 61, no. 2 (April 2000): 285–304.

98. Often Lapouge is mixed in with these figures. Consider Karl Dietrich Bracher on "the European background": "The second half of the nineteenth century has been called the Darwinian Age. Otto Ammon, Georges Vacher de Lapouge, Madison Grant, and the great sociologist Ludwig Gumplowicz, in his early writings, all sought to apply biological considerations to sociohistorical developments. While Gumplowicz ultimately discarded this line of inquiry, one of his students, Ludwig Woltmann, developed an extreme form of Social Darwinism which later was incorporated into the ideology of National Socialism" (Bracher, *The German Dictatorship: The Origins, Structure, and Effects of National Socialism*, trans. Jean Steinberg, intro. Peter Gay [New York: Praeger, 1970], 14).

99. Correspondence: Schemann to Lapouge, April 24, 1900, FVL/UM. In this same letter, Schemann promised once again to do everything in his power to make Lapouge's work known and appreciated in his country.

100. Correspondence: Ammon to Lapouge, May 17, 1892, FVL/UM.

101. Correspondence: Ammon to Lapouge, February 4, 1893, FVL/UM. A letter later that same month merits quotation:

> Your news that you lost the struggle struck me as a disaster that has happened to science in all the civilized countries! I understand well that the rulers of your country would not be enchanted by your theories, but I was naive enough to believe that science is free, especially in a republican country. I see that I was wrong. Having recovered my countenance, I told myself that the war is never decided by one battle alone. You will be the winner in the end, that is to say: the *real* winner. If I was in your place, I know what I would have to do. I do not dare give you counsel, recognizing your superiority, but if I possessed your knowledge and your faculties, I would be sure of my success. (February 19, 1893, FVL/UM)

Their correspondence continued until Ammon's death in 1916.

102. Hans F. K. Günther, *Racial Elements of European History*, trans. G. C. Wheeler (New York, 1927). Ernst Haeckel, whose influence on Nazi doctrine is insisted upon in Gasman, *The Scientific Origins of National Socialism*), was cited once, and without significance. Houston Stewart Chamberlain is absent entirely.

103. Correspondence: Lapouge/Günther, FVL/UM.

104. Correspondence: Lapouge to Schemann, May 28, 1931, FVL/UM.

105. Karl Saller, *Die Rassenlehre des Nationalsozialismus in Wissenschaft und Propaganda* (Darmstadt, 1961), 27.

106. Lapouge to Ludwig Plate, March 20, 1930, FVL/UM. The fact that Günther later dedicated one of his books to Plate further suggests that Plate came to his aid.

107. Saller, *Die Rassenlehre*, 27–28. This event is mentioned in many studies. See, for example, Robert Proctor, "From *Anthropologie* to *Rassenkunde*," in George Stocking, ed., *Bones, Bodies, Behavior: Essays on Biological Anthropology* (Wisconsin: University of Wisconsin Press, 1988),

158; Bracher, *The German Dictatorship*, 166; and Eugene Davidson, *The Trial of the Germans: Nuremberg, 1945–1946* (New York: Macmillan, 1966), 265.

108. The German term for *anthroposociologie* is *Sozialanthropologie*. See Hans Fabricius, *Reichsinnenminister Dr. Frick: Der revolutionaere Staatsmann* (Berlin, 1939), 44. Günther was often cited above Rosenberg as the single most important racial theorist. See, for example, Robert Proctor's study "Nazi Medicine and the Politics of Knowledge," in Sandra Harding, ed., *The "Racial" Economy of Science: Toward a Democratic Future* (Bloomington: Indiana University Press, 1993), 352, which discusses Günther's "widely recognized status as father of German Rassenkunde and the Nordic movement" (323). Saller attributes great importance to Günther, arguing a direct relationship between Nazi theories and Günther's work. Hans-Jürgen Lutzhöft, *Der Nordische Gedanke in Deutschland, 1920–1940* (Stuttgart: Klett, 1971), on the other hand, sees the ideology of Günther and other Nordicists as significantly different from that of the Nazis, describing the Nazi use of Günther as highly opportunistic. The truth, I would argue, is a combination of the two: Günther's work may well have inspired the young Hitler, and it was certainly taught to a generation of schoolchildren, and yet that does not mean that it was a blueprint for later events. Practice and doctrine both changed over time (and place) and were never in more than loose alignment.

In September 1933 "racial science" became a compulsory subject in German schools, and there arose a sudden, acute need for a textbook on the subject. Teachers met this problem by giving their students selections from the works of Günther and of Alfred Rosenberg. See Wolfgang Wippermann, "Das Berliner Schulwesen in der NS-Zeit: Fragen, Thesen und Methodische Bemerkungen," in Benno Schmoldt, ed., *Schule in Berlin* (Berlin: Colloquium, 1989), 57–73; and Michael Burleigh and Wolfgang Wippermann, *The Racial State: Germany, 1933–1945* (New York, 1991), 213. According to one school principal: "In our school, a thorough course in racial studies and hereditary studies was enacted. Special emphasis was put on racial studies of the Jews following Günther and his skull measurements" (cited in Wippermann, "Das Berliner Schulwesen in der NS-Zeit," 65). In a parallel effort, for purposes both pedagogical and classificatory, the children's heads were measured, and their cephalic indexes calculated. That Günther was an avid head measurer is evident in all his works, but it is rarely mentioned in historical accounts of the period. In general, in modern studies, the details of racial science are simply omitted. In Bracher's study, the first chapter is devoted to historico-social anti-Semitism, but while he recognizes the profound influence of early scientific racist doctrine and that of Lapouge in particular, the subject is dispatched in two pages (14 and 15). Ammon, Grant, Gumplowicz, and Woltmann are also briefly mentioned. A few sentences in the rest of the work mention scientific doctrine during the Nazi period (see, for instance, 252).

When mention of these measuring tasks is made, it takes on an oddly comical tone, as if the practitioners were crackpots, outside the official doctrine. In fact, they were the official doctrine. One modern scholar reports, for example, that, "there were men like . . . Hans F. K. Guenter [*sic*], who conducted an investigation in Dresden that showed the streetcar motormen to have more Northern blood than the conductors" (Davidson, *The Trial of the Germans*, 40). Another states that "skull measurements were used by the Nazis in an attempt to sort out those with Jewish ancestry" (Steve Jones, *The Language of Genes* [New York: Doubleday, 1993], 201).

109. Fabricius, *Reichsinnenminister Dr. Frick*, 44.

110. Lapouge to Grant, March 23, 1919, FVL/UM. For the book's first publication, see Madison Grant, *The Passing of the Great Race* (New York: Scribners, 1916). Assire's French trans-

lation was published as *Le déclin de la grande race* in 1926. Lapouge continued to write instructive (and very friendly) letters to Grant. See, in particular, Lapouge to Grant, April 4, 1929, FVL/UM.

111. Lapouge to Assire, April 2, 1932, FVL/UM.

112. For Hitler's reliance on Günther, consider, for example, Karl Dietrich Bracher on *Mein Kampf*: "The book borrowed from the *Rassenkunde des deutschen Volkes* (Munich, 1922) by the anthropologist Hans F. K. Günther and his theories of 'Nordification.' " He adds that "Chamberlain, Fritsch, Spengler, Lagarde, Schopenhauer, Wagner, in addition to numerous obscure, pseudo-scientific works, are the sources which the author, naturally without direct citation or annotation, used and vulgarized" (*The German Dictatorship*, 128). See also Joachim C. Fest's *The Face of the Third Reich: Portraits of the Nazi Leadership* (New York: Pantheon, 1970), which cites Günther as the source of Hitler's race theory, particularly the notion of the blond "Nordic type" (99–100). According to Ernst Nolte, "Hitler was probably not familiar with Vacher de Lapouge, but the ideas which Lapouge was one of the first to express were well known to him" (*Three Faces of Fascism*, trans. Leila Vennewitz [New York: Rinehart and Winston, 1966], 515 n. 4).

113. Georges Vacher de Lapouge, *Der Arier und seine Bedeutung für die Gemeinschaft* (Frankfurt A. M.: Moritz Diesterweg, 1939). The preface to the German translation places the book as a major work in racial science: "Much of what here appears as an admonition is now the practical politics of today. This translation was encouraged by the high governmental minister Dr. Ruttke and Miss Kietta, and it comes with the clear permission of the publisher and the estate of the author." Falk Ruttke went on to publish several works under the title *Rasse und Recht* in 1937 and 1938.

114. Edgar Tatarin-Tarnheyden, "Georges Vacher de Lapouge: Visionnaire française de l'avenir européen," *Cahiers franco-allemands* 9 (October-December 1942): 336–346.

115. Peter Viereck, *Metapolitics: The Roots of the Nazi Mind* (New York, 1961), 254. This is the revised edition of *Metapolitics: From Romanticism to Hitler* (New York, 1941).

116. Hans F. K. Günther, "Zum Tode des Grassen Georges Vacher de Lapouge," *Rasse: Monatschrift der Nordischen Bewegung* 3 (1936): 95–98.

117. Dr. Werner Kulz, "Marquis de Lapouge zum Gedenken!" *Volke und Rasse* 6 (June 1936): 255.

118. Lapouge to Schemann, December 10, 1934, in Boissel, "Autour du Gobinisme," 109.

119. Lapouge to Madame DuPont, May 12, 1935, FVL/UM. Lapouge was writing to Madame DuPont because her husband had just died. The letter is, however, full of anthroposociological discussions.

120. See Lapouge's extensive collection of correspondence in the Fonds Vacher de Lapouge, housed in the Paul Valéry library of the University of Montpellier, as well as his government personnel files, located in the Archives nationales de Paris, especially BB/6(II)419 and F/17/22640; and the Archives of the Société d'anthropologie de Paris, located at the Musée de l'homme in Paris.

121. I will here note one major work from each: Grant, *The Passing of the Great Race*; Margaret Sanger, *Woman and the New Race* (New York: Blue Ribbon, 1920); Charles Davenport, *Heredity in Relation to Eugenics* (New York: Holt, 1911); Lothrop Stoddard, *The Rising Tide of Color Against White World-Supremacy*, intro. Madison Grant (New York: Scribner, 1920); William Ripley, *The Races of Europe* (London: Trench, Trübner, 1897); John Beddoe, *The Races of Britain: A Contribution to the Anthropology of Western Europe* (Bristol: J. W. Arrowsmith, 1885); Ernst

Haeckel, *The Riddle of the Universe at the Close of the Nineteenth Century*, trans. Joseph McCabe (New York: Harper, 1901); Francis Galton. *Hereditary Genius: An Inquiry into Its Laws and Consequences* (London: Macmillan, 1869); Gustave Le Bon, *La psychologie des foules* (Paris, 1895); Charles Robert Drysdale, *The Cause of Poverty: A Paper Read at the National Liberal Club on 21st October, 1890* (London: Standring, 1891); Bessie Ingman Drysdale, *Labour Troubles and Birth Control* (London: Heinemann, 1920); Luis Huerta, *La doctrina eugénica* (Madrid: Editorial Instituto Samper, 1933); Angelo Crespi, *Contemporary Thought of Italy* (London: Williams and Norgate, 1926); Georges Chatterton-Hill, *La physiologie morale* (Paris: P. V. Stock, 1904); Carl Closson, "Further Data of Anthropo-Sociology," *Journal of Political Economy* 7 (March 1899), 243–252; Jean-Richard Bloch, *L'anoblissement en France au temps de François Ier: Essai d'une définition de la condition juridique et sociale de la noblesse au début du 16e siècle* (Paris: Alcan, 1934); Charles Ujfalvy, *Les Aryens au nord et au sud de l'Hindou-Kouch* (Paris: G. Masson, 1896).

122. Closson to Lapouge, April 24, 1896, FVL/UM.

123. Lapouge to Haeckel, December 22, 1896, cited in Daniel Gasman, *Haeckel's Monism and the Birth of Fascist Ideology* (New York: Lang, 1998), 140–141.

124. Lapouge to unknown, November 26, 1926, FVL/UM.

125. Lapouge to Madame Albertine Lapouge, September 23, 1921, and September 28, 1921, FVL/UM, no. 068–50 and no number. See also Guy Thuillier, "Un anarchiste positiviste: Georges Vacher de Lapouge," in Pierre Guiral and Emile Temine, eds., *L'idée de race dans la pensée politique française contemporaine* (Paris: CNRS, 1977), 59.

126. Lapouge to Sanger, April 24, 1925, FVL/UM, no. 10, A068–110.

127. Margaret Sanger, *Margaret Sanger: An Autobiography* (New York: Norton, 1938), 372. The scalding anecdote runs as follows:

> The next morning the Hotel McAlpin, where the convention was to be held, called me up to report that Dr. Lapouge had been severely burned, and an interpreter was needed. Dr. Drysdale hurried off to find the poor little man of seventy in excruciating pain but carrying on a dissertation, highly amusing, about the hazards of America's much advertised plumbing. Without understanding how to regulate a shower he had stood under it and turned on the hot water. The skin fairly peeled off his chest. Nevertheless, bandaged and oiled, he undauntedly attended all the sessions." (372–373)

128. Even in *L'Aryen*, published in 1899, Lapouge was much heartened by the eugenics movement in the United States and the work that had been done there to control births according to eugenic concerns (504–507).

129. Lapouge to Charles Davenport, February 20, 1921, FVL/UM, no. 8, A068–70.

130. Calvin Coolidge, "Whose Country Is This?" *Good Housekeeping* 72 (February 1921): 14, cited in Daniel J. Kevles, *In the Name of Eugenics: Genetics and the Uses of Human Heredity* (Berkeley: University of California Press, 1985), 97.

131. Buck v. Bell, 274 U.S. 205, 207 (1927), cited in Kevles, *In the Name of Eugenics*, 111.

132. See chap. 7, "Eugenic Enactments," in Kevles, *In the Name of Eugenics*, 96–112. See also Edward Larson, *Sex, Race, and Science: Eugenics in the Deep South* (Baltimore: Johns Hopkins University Press, 1995).

133. Including a grand rabbi. See Lapouge to unknown, November 26, 1926, FVL/UM.

134. Gaultier would later serve as general editor of the Bibliothèque de philosophie scientifique published by Flammarion in the thirties and forties.

135. Lapouge to Gaultier, June 15, 1915, FVL/UM.

136. Lapouge to Grant, April 1929, FVL/UM.

137. The "zealous assistant" was Lapouge disciple Du Pont, who published under the pseudonym Warren Kincade. At this point, Du Pont was the European correspondent for *Review of Reviews*. See Correspondence: Lapouge/Du Pont, FVL/UM.

138. See Philippe Bernard and Henri Dubief, *The Decline of the Third Republic, 1914–1938* (Cambridge, 1985), 131, 168, 224, 226. Marthe Hanau "made extravagant promises and went bankrupt at the end of 1928" (131). *Le quotidien* "did not survive . . . the rash connection with Madam Hanua" (168).

139. By the very beginning of the century, Lapouge's theories had become well-established aspects of the study of sociology in the United States. In 1905 Thomas Nixon Carver, professor of political economy at Harvard, compiled a sourcebook "for students of sociology" for a series called "Selections and Documents in Economics." The text, entitled *Sociology and Social Progress: A Handbook for Students of Sociology* (Boston: Ginn, 1906), included a significant section by Lapouge, as well as one by the racialist writer William Ripley, then professor of economics at Harvard and the editor of the series. Pitirim Sorokin, professor of sociology at Harvard, published *Contemporary Sociological Theories* with Harper and Brothers in 1928 and included a large section on Lapouge (219–308). It was not utterly accepting of anthroposociological principles—indeed, it argued against them on several points—but it explained them extensively and took them seriously. In this context, Günther's *Rassenkunde des Deutschen Volkes* (1924) was praised as "a very valuable work" (263).

6. Body and Soul: Léonce Manouvrier and the Disappearing Numbers

1. Léonce Manouvrier, "Conclusions générales sur l'anthropologie des sexes et applications sociales," *Revue de l'Ecole d'Anthropologie* 13 (1903): 406.

2. The only study dedicated to Manouvrier until now, that I know of, is my article "A Vigilant Anthropology: Léonce Manouvrier and the Disappearing Numbers," *Journal of the History of the Behavioral Sciences* 33, no. 3 (summer 1997): 221–240.

3. On the history of statistics, see Theodore Porter, *The Rise of Statistical Thinking, 1820–1900* (Princeton: Princeton University Press, 1986); Ian Hacking, *The Taming of Chance* (Cambridge: Cambridge University Press, 1990); and idem, *The Probabilistic Revolution*, vol. 1, ed. Lorenz Kruger, Lorraine J. Daston, and Michael Heidelberger; vol. 2, ed. Lorenz Kruger, Gerd Gigerenzer, and Mary Morgan (Cambridge, Mass.: MIT Press, 1987).

4. On eugenics, see William Schneider, *Quality and Quantity: The Quest For Biological Regeneration in Twentieth-Century France* (Cambridge: Cambridge University Press, 1990); Daniel J. Kevles, *In the Name of Eugenics: Genetics and the Uses of Human Heredity* (Berkeley: University of California Press, 1985); and Linda Clark, *Social Darwinism in France* (Birmingham: University of Alabama Press, 1984).

5. See, for instance, "Académie de médecine (1er Octobre)," *Le temps*, October 3, 1889. The article reported Laborde's argument that even the vapors of absinthe could have fatal effects.

6. This took over twenty years because "preparateur" had a salary to it, and "director" did not, and this was Manouvrier's main employment. Until Laborde suddenly wanted use of his personal office at the lab—which had been Manouvrier's work space for two decades—Manouvrier did not mind the situation. When Laborde insisted (despite explanations of the difficulty involved), Manouvrier contacted the minister of public instruction who sided with Manouvrier and gave him the official power he had lacked until then. AN F/17/23860.

7. Stephen J. Gould, *The Mismeasure of Man* (New York: Norton, 1981), 73–113.

8. Archives of the Société d'anthropologie de Paris (hereafter ASAP), Musée de l'homme, Paris, Bibliothèque.

9. ASAP, procès-verbaux.

10. Paul Broca, "Sur le volume et la forme du cerveau suivant les individus et suivant les races," *Bulletins de la Société d'anthropologie de Paris* (hereafter *BSAP*), 1st ser., 2 (1861): 139–207, 301–321, 441–446.

11. For a discussion of the wide influence of Broca's data on women's inferiority, see Cynthia Eagle Russett, *Sexual Science: The Victorian Construction of Womanhood* (Cambridge: Harvard University Press, 1989), 35–36.

12. Gustave Le Bon, "Recherches anatomiques et mathématique sur les lois des variations du volume du cerveau et sur leur relations avec l'intelligence," *Revue d'anthropologie* 2, no. 2 (1879): 60.

13. Léonce Manouvrier, "Recherches sur le développement quantitatif comparé de l'encéphale et de diverses parties du squelette," *Bulletin de la Société zoologique de France* 6 (1881): 77–94.

14. Léonce Manouvrier, "Sur la grandeur du front et des principales régions du crâne chez l'homme et chez la femme," *BSAP* 3d ser., 6 (1883): 694–698.

15. Léonce Manouvrier, "Variétés: L'internat en mèdicine des femmes," *Revue scientifique* 3, no. 19 (1884): 592–597.

16. Léonce Manouvrier, "Indications anatomiques et physiologiques relatives aux attributions naturelles de la femme," *Congrès français et international du droit des femmes* (Paris, 1889), 41–51.

17. Madame Conta, "Quelques considérations d'ordre social concernant l'homme et la femme," *Congrès français et international du droit des femmes* (Paris, 1889), 131–135.

18. Manouvrier, "Indications anatomiques," 49.

19. *Congrès français et international du droit des femmes* (Paris, 1889), 226, 40, and 146–163.

20. Manouvrier, "Indications anatomiques," 49.

21. Manouvrier, "Conclusions générales," 405.

22. Condorcet's feminism may not be as well-known as Mills's. Consider, for example, his discussion of the "tenth stage" of human development:

> Among the causes of the progress of the human mind that are of the utmost importance to the general happiness, we must number the complete annihilation of the prejudices that have brought about an inequality of rights between the sexes, an inequality fatal even to the party in whose favor it works. It is vain for us to look for a justification of this principle in any differences of physical organization, intellect, or moral sensibility. This inequality has its origin solely in an abuse of strength, and all the later sophisticated attempts that have been made to excuse it are in vain. (Antoine-Nicolas de Condorcet, *Sketch for a Historical Picture of the Progress of the Human Mind* [1795], trans. June Barraclough [New York: Noonday, 1955], 193)

23. Maria Montessori, *Pedagogical Anthropology*, trans. F. T. Cooper (New York: Fredrick A. Stokes, 1913). For a brief discussion of this work, see Gould, *Mismeasure of Man*, 107, 123.

24. Jacqueline Lalouette, *La libre pensée en France, 1848–1940* (Paris: Albin Michel, 1997), 331.

25. Montessori, *Pedagogical Anthropology*, 257–258.

26. "Intelligence," in Charles Turgeon, *Le féminisme français* (Paris: Larose, 1907), 1:131–134.

27. Russett, *Sexual Science*, 186–187.

28. The observation is Gould's. See his *Mismeasure of Man*, 123. "Nordau" refers to Max Nordau, author of *Degeneration* (New York, 1895).

29. "Metaphysical criminology," was a term utilized by its opponents rather than by anyone claiming to practice it. See, for example, E. Ferri, "Various Short Contributions to Criminal Sociology," *Internationaler Kongress der Kriminalanthropologie* 7 (1911): 251.

30. Gould, *Mismeasure of Man*, 26, 106, 107, 139; Robert Nye, *Crime, Madness, and Politics in Modern France: The Medical Conception of National Decline* (Princeton: Princeton University Press, 1984), 106, 108, 116, 118, and 125.

31. Both Nye and Gould mention Manouvrier as one of the most dedicated French critics of Lombrosian theory. A thorough examination of Manouvrier's critique, however, lay outside the projects outlined by both Nye and Gould. Perhaps because of this, neither appears to appreciate Manouvrier's dominant role in formulating the French position. In particular, Nye gives inordinate weight to the influences of Paul Topinard and Gabriel Tarde, both of whom fashioned far less rigorous critiques of Lombroso than did Manouvrier. Nye and Gould also mistakenly exaggerate the exceptional nature of Manouvrier's egalitarianism. Neither seems to have been aware that the Société d'anthropologie de Paris was, at this time, dominated by left-wing egalitarians.

32. Gould, *Mismeasure of Man*, 26; Nye, *Crime, Madness, and Politics*, 108.

33. *Actes du deuxième congrès d'anthropologie criminelle* (Paris, 1889).

34. Manouvrier discusses this and similar metaphors in his "Les aptitudes et les actes dans leurs rapports avec la constitution anatomique et avec le milieu extérieur" (paper delivered at the Septième Conférence Broca), *BSAP* 4th ser., 1 (1890): 918–939.

35. *Actes du deuxième congrès*, 31–35.

36. See Thomas Wilson, "Criminal Anthropology," *Annual Report of the Smithsonian Institution* (Washington: U.S. Government Printing Office, 1891).

37. *Actes du deuxième congrès*, 92–106.

38. Tarde also discussed his belief that there exists "a residue of incorrigible criminals, real antisocial monsters" (104).

39. See Nye, *Crime, Madness, and Politics*, 106, for a similar assessment of the importance of this congress.

40. See his "Questions préalables dans l'étude comparative des criminels et des honnàtes gens," in *Actes du troisième congrès d'anthropologie criminelle de Bruxelles* (Brussels, 1893), 171–182.

41. Ida M. Tarbell, "Identification in Criminals: The Scientific Method in Use in France," *New McClure's Magazine* 2, no. 4 (March 1894): 369.

42. *Actes du deuxième congrès*, 490–496.

43. Léonce Manouvrier, "L'anthropologie et le droit," *Revue internationale de sociologie* 2 (1894): 367.

44. Manouvrier, "La genèse normale du crime," *BSAP* 4th ser., 4 (1893): 405–458. The conference referred to evolution as "species transformism" so as not to imply a bias toward the Darwinian model. Manouvrier used the term instead of "evolution," but their meaning here is identical, and I paraphrase it as such for clarity.

45. Emile Durkheim, *L'année sociologique* 1 (1896–1897): 519.

46. Muffang to Lapouge, Fonds Vacher de Lapouge, Université de Montpellier (hereafter FVL/UM), A067–1 through A067–91.

47. Muffang to Lapouge, February 6, 1900, FVL/UM, A51–6.

48. Topinard to Lapouge, May 14, 1887, no. 5, FVL/UM, A099–11.

49. Léonce Manouvrier, "L'indice céphalique et la pseudo-sociologie," *Revue de l'Ecole d'anthropologie* 9 (1899): 233–259.

50. Gustave Rouanet, "Les théories aristocratique devant la science," *La petit republique*, January 2, 1900, 1.

51. Salomon Reinach, "Compte rendu: L'Aryen," *Revue critique d'histoire et de littérature*, February 12, 1900, 121–125.

52. Emile Durkheim, "Notre siècle: La sociologie en France au XIXe siècle," *Revue bleue*, 4th ser., 13, no. 20 (May 19, 1900): 651.

53. Muffang to Lapouge, January 1901, FVL/UM, A060–27.

54. Muffang to Lapouge, May 31, 1900, FVL/UM, A060–55.

55. Manouvrier, "L'anthropologie et le droit," 265.

56. Manouvrier, "L'individualité de l'anthropologie," *Revue de l'Ecole d'anthropologie* 14 (1904): 397–410.

57. See Prokopec, "Hrdlička," 57–61; Erik Trinkaus, "A History of Homo erectus and Homo sapiens Paleontology in America," in Frank Spencer, ed., *A History of American Physical Anthropology, 1930–1980* (New York: Academic, 1982), 261–280; and Thelma S. Baker and Phyllis B. Eveleth, "The Effects of Funding Patterns on the Development of Physical Anthropology," in *A History of American Physical Anthropology*, 31–48.

58. Trinkaus, "A History of Homo erectus," 261; Aleš Hrdlička, *The Skeletal Remains of Early Man* (Washington D.C.: The Smithsonian institution, 1930).

59. Baker and Eveleth, "The Effects of Funding Patterns," 33.

60. Royer to Ghénia Avril de Sainte-Croix, n.d. [1900], Bibliothèque Marguerite Durand, cited in Joy Harvey, *"Almost a Man of Genius": Clémence Royer, Feminism, and Nineteenth-Century Science* (New Brunswick: Rutgers University Press, 1997), 179.

61. Gilles de La Tourette, "Le professeur J.-M. Charcot," *Revue hebdomadaire* 15 (1893): 608–622. Tourette's association with the syndrome named for him was due to a long article he wrote on the tic disorder in 1885. After Charcot died, Tourette switched professions and took up forensic medicine. He was, by the way, such an unbearable personality that his obituary demanded a rebuttal. See Christopher G. Goetz, Michel Bonduelle, and Toby Gelfand, *Charcot: Constructing Neurology* (New York: Oxford University Press, 1995), 322.

62. Charles Richet, "Les démoniaques d'aujourd'hui: L'hystérie et le somnambulisme," *Revue des deux mondes* 37 (1880): 340.

63. Tourette, "Le professeur J.-M. Charcot," 612.

64. Jan Goldstein, *Console and Classify: The French Psychiatric Profession in the Nineteenth Century* (Cambridge: Cambridge University Press, 1987), 322–377.

65. Richet, "Les démoniaques d'aujourd'hui," 363.

66. Tourette, "Le professeur J.-M. Charcot," 611.

67. Reproduced in Goetz, Bonduelle, and Gelfand, *Charcot*, 92–93. The tableau is just the stage: the men here are the select group allowed to come in close, while a large audience outside the frame watches from the seats of an amphitheater.

68. Richet, "Les démoniaques d'aujourd'hui"; Charles Richet, "Les démoniaques d'autrefois: Les sorcières et les possédées," *Revue des deux mondes* 37 (1880): 552–583.

69. Richet, "Les démoniaques d'aujourd'hui," 341.

70. Richet, "Les démoniaques d'autrefois," 583.

71. Goldstein, *Console and Classify*, 361–377.

72. Jean-Marie Charcot, "Leçon d'ouverture," cited in ibid., 368.

73. Paul Bert, ed., *Revues scientifiques pour la république française publiée par le journal "La république française" sous la direction de M. Paul Bert* (Paris: G. Masson, 1879), 1:4, cited in Goldstein, *Console and Classify*, 368.

74. Sigmund Freud, "Charcot," trans. J. Strachey, in *Selections-Collected Papers*, authorized translation under the supervision of Joan Riviere, trans. and ed. James Strachey with Alix Strachey (New York: Basic, 1959), 1:10, 11. See also Goldstein, *Console and Classify*, 383–384.

75. Theodore Zeldin wrote that in 1960 about two-thirds of French nuns were engaged in hospital or other social services and one-third in education. "It is no wonder the lay republic could not expel them *en masse*" (*France, 1848–1945* [Oxford: Oxford University Press, Clarendon, 1977), 2:1015, see also 1010–1015.

76. Archives of the Société d'autopsie mutuelle, Musée de l'homme, Paris, clipped article, "Le cerveau des grands hommes: Au musée d'anthropologie," *Le petit bleu*, no. 329 (April 13, 1903).

77. Jack D. Ellis, *The Physician-Legislators of France: Medicine and Politics in the Early Third Republic, 1870–1914* (Cambridge: Cambridge University Press, 1990), 195.

78. Désiré-Magloire Bourneville and Dr. Voulet, *De la contracture hystérique permanente; ou, Appréciation scientifique des miracles de Saint Louis et de Saint Médarde* (Paris, 1872).

79. Désiré-Magloire Bourneville, *Louise Lateau; ou, La stigmatisée belge* (Paris, 1875).

80. See Joan Jacobs Brumberg, *Fasting Girls: The Emergence of Anorexia Nervosa as a Modern Disease* (Cambridge: Harvard University Press, 1988); Caroline Walker Bynum, *Holy Feast and Holy Fast: The Religious Significance of Food to Medieval Women* (Berkeley: University of California Press, 1987).

81. Désiré-Magloire Bourneville and P. Renarde, *Iconographie photographique de la Salpêtrière*, 3 vols. (Paris: Progrès médical, 1876–1880).

82. Jean Weir, *Histoires disputes et discours des illusions et impostures des diables*, pref. Desiré-Magloire Bourneville (Paris, 1885).

83. Goldstein, *Console and Classify*, 371.

84. Jean-Marie Charcot, "Faith-Cure," *New Review* 8 (January-June 1893): 18–31; idem, "La foi qui guérit," *Revue hebdomadaire* 1, no. 5 (December 3, 1892): 122–132. I take my quotations from the original English.

85. Léonce Manouvrier, "Etude comparative sur les cerveaux de Gambetta et de Bertillon," *Revue philosophique* 25 (1888): 453–461; idem, "Société de psychologie physiologique: 'Mouvements divers et sueur palmaire consécutifs à des images mentales.' Séance du 29 mars 1886 (présidence de M. Charcot)," *Revue philosophique* 22 (1886): 203–207; idem, "Les premières circonvolutions temporales droite et gauche chez un sourd de l'oreille gauche (Bertillon)," *Revue philosophique* 26 (1888): 330–335.

86. See Goldstein, *Console and Classify*; Steven Lukes, *Emile Durkheim: His Life and Work* (Stanford: Stanford University Press, 1985), 372–373; W. Paul Vogt, "Political Connections, Professional Advancement, and Moral Education in Durkheimian Sociology," *Journal of the History of*

the *Behavioral Sciences* 27 (1991): 56–75; Ellis, *The Physician-Legislators of France*, 39; George Rosen, "The Philosophy of Ideology and the Emergence of Modern Medicine in France," *Bulletin of the History of Medicine* 20 (July 1946): 328–339; Gerald L. Geison, ed., *Professions and the French State, 1700–1900* (Philadelphia: University of Pennsylvania Press, 1984); H. Tint, "The Search for a Laic Morality Under the French Third Republic: Renouvier and the 'Critique Philosophique,'" *Sociological Review* 5, no. 1 (July 1957): 5–26.

7. The Leftist Critique of Determinist Science

1. The first section of this chapter grew out of my article: "The Solvency of Metaphysics: The Debate Over Racial Science and Moral Philosophy in France, 1890–1914," *Isis: Journal of the History of Science Society* 90 (spring 1999): 1–24.

2. Tolstoy and Dostoyevsky had novels translated into French between 1884 and 1888.

3. "Introduction," Revue de métaphysique et de morale (hereafter *RMM*) 1 (1893): 1. See also the unsigned article "La philosophie au Collège de France," *RMM* 1 (1893): 369–381.

4. The "terrible forces" probably referred to the rise of scientific racism (see chap. 5 for a discussion of this in *RMM*) as well as the anarchist violence that raged between 1892 and 1894.

5. Alphonse Darlu, "Réflexions d'un philosophe sur les questions du jour: Science, morale et religion," *RMM* 3 (1895): 249.

6. For a discussion of solidarism, see J. E. S. Hayward, "The Official Philosophy of the French Third Republic: Léon Bourgeois and Solidarism," *International Review of Social History* 9 (1961): 22–25; and idem, "Solidarity: The Social History of an Idea in Nineteenth-Century France," *International Review of Social History* 4 (1959): 261–284. See also Célestin Bouglé, *Le solidarisme* (Paris, 1907).

7. The relationship between Fouillée's philosophy and Bourgeois's political career has been well established; see Hayward, "Solidarity," as well as John A. Scott, *Republican Ideas and the Liberal Tradition in France, 1870–1914* (New York: Columbia University Press, 1951); and Linda Clark, *Social Darwinism in France* (Birmingham: University of Alabama Press, 1984), 54–57, 157–186.

8. As William Logue has characterized it, solidarism was a kind of neoliberalism that asserted that the maximum liberty for all could only be attained through an organized social action that ran counter to classic liberalism's concept of freedom. See William Logue, "Sociologie et politique: Le libéralisme de Célestin Bouglé," *Revue française de sociologie* 20 (1979): 141–161. In a different context, Hayward had said as much when he cited the growing nineteenth-century notion that "in the inegalitarian economic sphere, it was laissez-faire that oppressed and social intervention that liberated" ("The Official Philosophy," 33). Neither Hayward nor Logue, however, take into account the themes of natural history and evolution that dominated nineteenth- and early-twentieth-century discussions of solidarism.

9. Several studies have demonstrated that French scientists (and in some cases the English as well) held on to Lamarckianism long after Darwinian theory had replaced it elsewhere. See Yvette Conry, *L'introduction du Darwinisme en France au XIXe siècle* (Paris: Vrin, 1974); Thomas F. Glick, ed., *The Comparative Reception of Darwinism* (Austin: University of Texas, 1974); and Robert Nye, "Heredity or Milieu: The Foundations of Modern European Criminological Theory," *Isis* 67, no. 3 (1976): 335–355; Peter Bowler, *Theories of Human Evolution: A Century of Debate, 1844–1944* (Baltimore: Johns Hopkins University Press, 1986). As for the prominence of Darwinian theory in political discourse, the works examined in the present study all ignore

Lamarck (though, in some cases, he would have aided their arguments) and use the terms "Darwinism," "Darwinian," "struggle for life" ("lutte pour la vie" or "lutte pour l'existence"), and "natural selection" in their references to evolution—as is clear in the quotations used herein. See also Bourgeois, *Essai d'une philosophie de la solidarité: Conférences et discussions* (Paris, 1902). This conference was attended by twenty-one prominent French philosophers, academics, lawyers, and politicians. In the discussions, no one brought up Lamarck or his "inheritance of acquired characteristics." This is not to say that they discussed Darwinism in detail, either, or that they never assumed a gradual human progress that might be construed as a sort of Lamarckian improvement. My point is that the French wrote copiously about the meaning and consequences of the struggle for existence. Our discovery of the French romance with Lamarck should not blind us to their obsession with the Darwinian "struggle" and "violent natural selection" (for specific references to these terms, see, for instance, Bourgeois, *Essai d'une philosophie*, 3, 29, and 78).

10. Hayward, "The Official Social Philosophy of the French Third Republic," 27.

11. Alfred Fouillée, "La psychologie des peuples et l'anthropologie," *Revue des deux mondes* 128 (March 15, 1895): 365. Fouillée also quoted Lapouge's warning of "copious exterminations"—without attribution—on this page. His reference to Manouvrier predated the latter's specific critique of Lapouge.

12. See, for instance, Alfred Fouillée, "As Others See Us," *The Living Age* 212 (1899): 67–72.

13. Alfred Fouillée, *Psychologie du peuple français* (Paris, 1898), 281.

14. Muffang to Lapouge, undated, and Fouillée to Lapouge, undated, Fonds Vacher de Lapouge, Université de Montpellier, A 91–6 and A 42–1 through A 42–3, respectively.

15. Alfred Fouillée, *Esquisse psychologique des peuples* (Paris, 1903), 529–530.

16. Bourgeois, *Essai d'une philosophie*, 6.

17. Célestin Bouglé, "Anthropologie et démocratie," *RMM* 5 (1897): 443–461.

18. He correctly named Otto Ammon as Lapouge's German counterpart and referred to other authors as "their disciples" (443 and throughout).

19. Of course, Lapouge did not agree. He held that members of the superior race were fewer in number and sometimes constitutionally delicate, such that they required and deserved privilege.

20. Topinard, in particular, was singled out (444).

21. Bouglé cited Manouvrier's "Les aptitudes et les actes" (1890) and his "Genèse normale du crime" (1893). See chap. 6 for a discussion of these works.

22. Bouglé to Brunetière, no date, Bibliothèque nationale de Paris, Manuscrit, NAF 25033 ff 79–80.

23. Célestin Bouglé, *Essais sur le régime des castes* (Paris: Alcan, 1908). See esp. the section "Race," 129–156.

24. Don Martindale, *The Nature and Types of Sociological Theory* (Boston: Houghton Mifflin, 1981), 265.

25. Bouglé, *La démocratie devant la science* (Paris: Alcan, 1904), 18.

26. Célestin Bouglé, *Qu'est-ce que la sociologie?* (Paris: Alcan, 1907), 143.

27. Georges Vacher de Lapouge, *L'Aryen: Son rôle social* (Paris: A. Fontemoing, 1899), 514.

28. Bouglé, *La démocratie*, 286.

29. Later called *La revue* and then *La revue mondiale*. Finot wrote many of the journal's articles. According to his son, in the few years before he became a French citizen he used ten different pseudonyms in order to write freely and fill the journal's pages. See Jean-Louis Finot, "Mon père," *La revue mondiale* 33 (May 15, 1922): 143–150.

30. Bibliothèque nationale de Paris, Manuscrit: 24494 (1) doc. 264; microfilm 2278/NAF 24519 (174–180); NAF 24530 ff 380–382; NAF 25038 ff 292–293.

31. René Worms, "Jean Finot, sociologue," *La revue mondiale* 33 (May 15, 1922): 228.

32. The only figure Worms mentioned was Gobineau, but he spoke at length about Finot's attack on the cephalic index.

33. Jean Finot, *Le préjugé des races*, 2d ed. (Paris, 1905–1906); and idem, *Le préjugé et problème des sexes* (Paris: Alcan, 1912).

34. Finot, *Le préjugé des races*, 103.

35. Jean Finot, *Race Prejudice*, trans. Florence Wade-Evans (New York, 1906), vi.

36. Finot, *Le préjugé des races*, 505.

37. He also listed Fouillée's *Tempérament et caractère* in this context.

38. Henri Bergson, "Rapport sur 'Progrès et bonheur' de J. Finot," in *Mélanges*, ed. André Robinet (Paris: PUF, 1972), 1090–1094.

39. According to Finot's son, this book was translated into fifteen languages. For the English version, see *The Science of Happiness*, trans. Mary Stafford (New York: Putnam, 1914). All quotations in the text are drawn from this edition.

40. Paul Bourget, *Essais de psychologie contemporaine: Baudelaire, M. Renan, Flaubert, M. Taine, Stendhal* (Paris: A. Lemerre, 1883), quoted in Finot, *The Science of Happiness*, 97 n. 1.

41. Finot, *The Science of Happiness*, 72.

42. Georges Vacher de Lapouge, *Race et milieu social* (Paris: Rivière, 1909), xi, xx.

43. Madison Grant, *The Passing of the Great Race* (New York: Scribners, 1916).

44. Lapouge to Madison Grant, March 23, 1919, Fonds Vacher de Lapouge, Université de Montpellier, A 068–80.

45. Charles Péguy, *Oeuvres en prose, 1898–1908* (Paris: Gallimard, 1959), 483.

46. Henri Bergson, *Creative Evolution*, trans. A. Mitchell (New York: Holt, 1911), 270–271.

47. *The Times*, October 28, 1911, EB VII, 38–41, reprinted in Henri Bergson, *Mélanges*, ed. André Robinet (Paris: PUF, 1972), 951–959.

48. Jacques Chevalier, *Henri Bergson*, trans. Lilian Clare (1928; reprint, New York: AMS, 1969), 168–169.

49. Robert Owen Paxton, *Vichy France: Old Guard and New Order, 1940–1944* (New York: Columbia University Press, 1972), 151, 159, 266, 272, 344.

50. Steven Lukes, *Emile Durkheim: His Life and Work* (Stanford: Stanford University Press, 1985), 44.

51. Emile Durkheim, "La philosophie dans les universités allemandes," *Revue internationale de l'enseignement* 13 (1887): 439–440, cited in W. Paul Vogt, "Political Connections, Professional Advancement, and Moral Education in Durkheimian Sociology," *Journal of the History of the Behavioral Sciences* 27 (1991): 64.

52. Vogt, "Political Connections," 60–61.

53. Emile Durkheim, *Education and Sociology* (New York: Free, 81.

54. Arguably, Tarde shared leadership with Réne Worms.

55. Emile Durkheim, *Les règles de la méthode sociologique* (Paris, 1894), 127.

56. Alphonse Darlu, "Réflexion d'un philosophe sur les questions du jour: La solidarité," *Revue de métaphysique et de morale* 5 (1897): 126.

57. Emile Durkheim, "Représentations individuelles et représentations collectives," *RMM* 6 (1898), translated as "Individual and Collective Representations," in *Sociology and Philosophy*, trans. D. F. Pocock (New York: Free, 1974), 33.

58. Durkheim, *Les règles de la méthode sociologique*, 6, 8.

59. Célestin Bouglé, "Sociologie et démocratie," *RMM* 4 (1896): 119.

60. Charles Andler, "Sociologie et démocratie," *RMM* 4 (1896): 246.

61. Bouglé, "Sociologie, psychologie, et histoire," *RMM* 4 (1896): 362–371.

62. Emile Durkheim, "Sociologie," in *La science française* (Paris: Imprimerie nationale, 1915), 382.

63. Emile Durkheim, *The Elementary Forms of Religious Life*, trans. and intro. by Karen E. Fields (New York: Free, 1995).

64. A relevant passage reads:

Nevertheless, it will be said, no matter how religions are explained, they have certainly erred about the true nature of things: The sciences have demonstrated that. So the modes of action they encourage or imposed upon man could only rarely have had useful effects: It is not with purifications that sicknesses are cured, or with sacrifices or songs that the crop is made to grow. . . . But . . . let us suppose that religion answers a need quite different from adapting us to tangible things: There will be no risk of its being weakened solely because it satisfies this need poorly or not at all. . . . But for that to occur, religious ideas must not draw their origin from a feeling that is disturbed by the setbacks of experience, for otherwise, where would their resilience come from? (80–81)

8. Coda

1. "Considérations présentées par M. L. Manouvrier à l'appui de sa canditature," Collège de France Archives, G.IV.f-45F. In detailing these struggles, Manouvrier cited his studies showing that the height of the French people had not diminished since prehistoric times, contrary to the opinion universally accepted, and thus asserted that he had proven invalid the supposed degeneration of the French population. He further noted that he had defused both of the "two opposed camps" to which the feminist movement had given rise. One of these groups, he explained, attributed to women mental incapacities corresponding to a plethora of anatomical inferiorities. "The majority of these inferiorities, cerebral or otherwise, do not exist, and the others have been wrongly interpreted." As for the other camp, which considered social competition between the sexes to be possible and desirable, Manouvrier announced that "physiologically and anatomically, I can demonstrate the inanity and the danger of these aspirations." This was more critical than his usual take on feminism, perhaps because of the constraints of trying to get a job. As for the Lombrosian theories of atavistic criminality, Manouvrier asserted that they contain "such an immense number of errors of every sort" that one can hardly begin to critique them (ibid., companion essay).

2. Ibid., main application.

3. Papillault, "Discours de M. G. Papillault aux obsèques de M. Manouvrier," *BSAP* 7th ser., 6–8 (1927–1927): 6.

4. *Journal officiel*, Sénat, March 26, 1892, 286–287.

5. Collège de France Archives, G.IV

6. The most extensive account of this event can be found in George Sarton, "Paul, Jules, and Marie Tannery (with a note on Grégoire Wyrouboff)," *Isis* 38 (November 1947): 33–51. See also Pierre Duhem, *Paul Tannery* (Montligeon [Orne]: Librairie de Montligeon, 1905), 14; and

Harry Paul, "The Debate Over the Bankruptcy of Science in 1895," *French Historical Studies* 3 (1968): 325–327. See also Paul's study of the origins of twentieth-century history of science through an analysis of the Collège de France chair: "Scholarship and Ideology: The Chair of the General History of Science at the Collège de France, 1892–1913," *Isis* 67, no. 238 (1976): 376–387. There are several recent contributions to the history of the science and religion debates in the Third Republic. Fritz Ringer has explored the question in terms of educational practice and ideology; see *Fields of Knowledge: French Academic Culture in Comparative Perspective, 1890–1920* (Cambridge: Cambridge University Press, 1992), esp. pp. 207–225. Also important is Herman Lebovics, *True France: The Wars Over Cultural Identity, 1900–1945* (Ithaca: Cornell University Press, 1992). Chapter 1 of Lebovics's book considers the anthropologist and politician Louis Marin in a study of the activity and eventual decline of the "old right" vision of France in the twentieth century (12–50).

7. Paul, "The Debate Over the Bankruptcy of Science," 322. See also Antonin Eymieu, *La part des croyants dans les progrès de la science au XIXe siècle*, 2 vols. (Paris: Perrin, 1920–1935).

8. Sarton, "Paul, Jules, and Marie Tannery," 40. In his "Debate over the Bankruptcy of Science," Paul argues that Sarton was overstating the point a bit but agrees that Tannery was considerably more qualified.

9. See Emmanuel de Margerie, "Albert de Lapparent," *Annales de géographie* 17 (1908): 344–347, esp. 346; and Charles Barrois, "Albert de Lapparent et sa carrière scientifique," *Revue des questions scientifiques*, 3d ser., 16 (1909)): 9–44, esp. 18–19, cited in Robert Fox, "Science, the University and the State in Nineteenth-Century France," in Gerald L. Geison, ed., *Professions and the French State, 1700–1900*, (Philadelphia: University of Pennsylvania Press, 1984), 66–145, esp. 119, 145 n. 198.

10. See Hélène Pierre-Duhem, *Un savant français: Pierre Duhem* (Paris, 1936), 95–157, cited in Fox, "Science, the University and the State," 119, 145 n. 198.

11. The marquis protested that the republic denied favors to those who were not in complete accord with it on questions of politics and religion. Apparently, in 1877 he was removed from his prefecture in Indre-et-Loire because of his oppositional stance. See his "Foi et science," *Le correspondant*, no. 179 (1895), 801–835, cited in Paul, "The Debate Over the Bankruptcy of Science," 316–317. Interestingly, in the context of the Brunetière controversy, Paul here notes de Nadaillac's opposition to "the narrow and outmoded rationalism of the encyclopedists of the eighteenth century," without, it would seem, any knowledge of the specific censorship de Nadaillac had experienced at the hands of the freethinking anthropologists.

12. "Banquet offert à M. Berthelot," *Revue scientifique*, April 13, 1895, 466–474, cited in Paul, "The Debate Over the Bankruptcy of Science," 320.

13. "Considérations présentées par M. L. Manouvrier," Collège de France Archives, G.IV.g–13 Q.

14. Thomas A. Kselman, *Death and the Afterlife in Modern France* (Princeton: Princeton University Press, 1993), 158–159.

15. Charles Richet, "Du somnambulism provoqué," *Revue philosophique* 10 (1880): 337–374, 462–484.

16. Charles Richet, *Dans cent ans* (Paris, 1892).

17. Ruth Brandon, *The Spiritualists: The Passion of the Occult in the Nineteenth and Twentieth Centuries* (New York: Knopf, 1983), 130, 132.

18. Charles Richet, *Metapsychics* (London, 1905); idem, *Our Sixth Sense* (London: Rider, 1929).

19. Sir Oliver Lodge, *Past Years: An Autobiography* (New York: Scribner's, 1932), 296.

20. Sigmund Freud, *On Aphasia: A Critical Study* (New York: International Universities Press, 1953). Some Freud scholars call this the first Freudian book, some see it as a neurological study without much significance for psychoanalysis.

21. Frank Sulloway, *Freud: Biologist of the Mind* (New York: Basic, 1979), 271.

22. Freud, "Hysteria," in *The Standard Edition of the Complete Psychological Works of Sigmund Freud*, trans. and ed. James Strachey, with Anna Freud, Alix Strachey, and Alan Tyson (London: Hogarth, 1953), 1:41.

23. William J. McGrath, *Freud's Discovery of Psychoanalysis: The Politics of Hysteria* (Ithaca: Cornell University Press, 1986), 166.

24. Sigmund Freud, "A Case of Successful Treatment by Hypnotism," in *The Standard Edition*, 1:126–127.

25. See McGrath, *Freud's Discovery of Psychoanalysis*, 165–169.

26. For a brilliant discussion of this and a number of other ways that Freud fits into the context I have been describing, see Philip Reiff, *Freud: The Mind of the Moralist* (Chicago: University of Chicago Press, 1979), 257–299.

27. Christopher G. Goetz, Michel Bonduelle, and Toby Gelfand, *Charcot: Constructing Neurology* (New York: Oxford University Press, 1995), 294.

28. Edward R. Tannenbaum, *The Action Française: Die-hard Reactionaries in Twentieth-Century France* (New York: Wiley, 1962), 92.

29. Léon Daudet, *Devant la douleur* (Paris: Nouvelle librairie nationale, 1915), 176; and idem, *Paris vécu* (Paris: Gallimard, 1930), 97, as cited in Goetz, Bonduelle, and Gelfand, *Charcot: Constructing Neurology*, 278.

30. On Guyau the philosopher, see Geoffrey C. Fidler, "On Jean-Marie Guyau, Immoraliste," *Journal of the History of Ideas* 55, no. 1 (1994): 75–97.

31. Jean-Marie Mayeur and Madeleine Rebérioux, *The Third Republic from Its Origins to the Great War, 1871–1914* (Cambridge: Cambridge University Press, 1984), 87.

32. *Inauguration de monument de G. de Mortillet-Extrait de L'homme préhistorique, 3e année, no. 11, 1905* (Paris, 1905), 1.

33. "Discours de M. A. Chervin, Président de la Société des Conférences Anthropologique," in ibid., 15–17.

34. "Discours de M. H. Thulié, Directeur de l'Ecole d'anthropologie de Paris," in ibid., 12–15.

35. Raoul Anthony, "Rapport du secrétaire général pour l'année 1936," *Bulletins de la Société d'anthropologie de Paris* (hereafter *BSAP*), 8th ser., 1–11 (1930–1940): 70–71.

36. Anthony, "Discours de M. R. Anthony aux obsèques de M. Manouvrier," *BSAP* 7th ser., 6–8 (1925–1927): 2–4.

37. *Bulletins et mémoires de la Société d'anthropologie de Paris*, 13th ser., 1 (1974).

38. See, for instance, Bernard Chopineaux, "Etude de l'articulation tino-tarsienne chez des populations du 'mésolithique,'" *Bulletins et mémoires de la Société d'anthropologie de Paris*, 8th ser., 1 (1974): 1; and Pranab Ganguly, "Variation in Physique in North India in Relation to Urbanization and Economic Status," *Bulletins et mémoires de la Société d'anthropologie de Paris*, 8th ser., 1 (1974): 6.

Conclusion

1. I am, of course, paraphrasing the Durkheim quotation cited above. See page 291.

2. Robert Owen Paxton, *Vichy France: Old Guard and New Order, 1940–1944* (New York: Columbia University Press, 1972).

3. Detlev J. K. Peukert, "The Genesis of the 'Final Solution' from the Spirit of Science," in Jane Caplan and Thomas Childers, eds., *Reevaluating the Third Reich*, intro. Charles S. Maier (New York: Holmes and Meier, 1993), 234–252.

4. Ann Laura Stoler, "Racial Histories and Their Regimes of Truth," *Political Power and Social Theory* 11 (1997): 183–206.

6. Emile Durkheim, *The Elementary Forms of Religious Life*, trans. and intro. by Karen E. Fields (New York: Free, 1995), 267.

Bibliography

∞

ARCHIVES

Musée de l'homme

Archives of the Société d'anthropologie de Paris, boxes located in metal cabinets in the Musée de l'homme in Paris. Though many of these documents were given codes or numbers when they were originally filed, they are now more or less unclassified. As such, they have been cited in descriptive terms along with any identifying code or number that could be found. Some of the boxes have labels, but these are generally entirely misleading. Most frequently cited herein are box: "Correspondance 1905"; folder (tan): "Correspondance à classer: 1906–07"; and folder (tan): "I. 1888–1902."

Archives of the Société d'autopsie mutuelle, in a box in same cabinets as above in the Musée de l'homme, uncataloged.

Manuscrits, in the library of the Musée de l'homme. Fully cataloged collection of letters, manuscript drafts, and anthropometrical tables. Includes Vogt's letters to Mortillet, Ms 54 R, Geneva.

Archives nationales de Paris

AN, F/17/2994, Missions—Mortillet.

AN, F/17/13491-4, Ministère de l'instruction publique: Sociétés savantes, Société d'autopsie mutuelle.

AN, F/17199, Ministère de l'instruction publique: Ecole d'anthropologie (trial of Topinard, applications for money, general correspondence).

AN, F/17/25839, Ministère de l'instruction publique: personal file, Louis Liard, renseignements confidentiels: reports on Liard and letters from Liard.

AN, F/17/22640, Vacher de Lapouge-Ministère de l'instruction publique: administrative reports on Lapouge and letters from Lapouge to the administration during his career as a librarian. One hundred and seventy numbered documents (numbers are not always perfectly sequential with dates).

AN, BB/6(II)419, Vacher de Lapouge, Ministère de la justice: administrative reports on Lapouge and letters from Lapouge to the administration during his career as a magistrate. Twenty numbered documents.

AN, F/17/13140, Ministère de l'instruction publique: Sociétés savantes.

AN, F/17/23860, Ministère de l'instruction publique: Personal file, Léonce Manouvrier.

Archives de la préfecture de police, Paris

1,493 *Libre pensée*: 1879 to 1891 and, in a file with overlapping dates, 1880 to 1897, both in 193032; Propagation de la foi civil.

Mortillet file, carton 127573.

Université de Montpellier

Fonds Vacher de Lapouge. At the time of my research, this large collection of documents was housed at the Université de Montpellier's Paul Valéry Library, under the personal control of Jean Boissel, who discovered them while doing research on Schemann and Gobineau. They are on microfilm and cataloged as such (according to spool).

Bibliothèque nationale (Manuscrit: Nouvelles acquisitions françaises)

NAF 24803 ff 89–91 (Victor Hugo letters: Broca).

NAF 25033 ff 79–80 (Ferdinand Brunetière letters: Bouglé).

NAF 24519 ff 174–180 (Emile Zola letters: Finot).

NAF 10,315 Don 3988, vol. 81. (Emile Zola).

Collège de France archives

G.II.10. (137): "Assemblée des professeurs—procès-verbaux, November 5, 1899."

G.IV.f-45F: "Considérations présentées par M. L. Manouvrier à l'appui de sa canditature."

G. IV.g-13 Q: "Considérations présentées par M. L. Manouvrier à l'appui de sa canditature pour la chaire d'histoire générale des sciences."

G. VI.f 45: Letter from Charles Lévêque to Henri Bergson, November 2, 1899.

BOOKS AND ARTICLES

"Acclimatation." *L'homme* 2 (1885): 701.

Actes du deuxième congrès d'anthropologie criminelle. Paris, 1889.

Agulhon, Maurice. *The Republic in the Village: The People of the Var from the French Revolution to the Second Republic.* Trans. Janet Lloyd. Cambridge: Cambridge University Press; Paris: Editions de la Maison des sciences de l'homme, 1982.

Andler, Charles. "Sociologie et démocratie," *Revue de métaphysique et de morale* 4 (1896): 243–256.

Anthony, Raoul. "Discours de M. R. Anthony aux obsèques de M. Manouvrier." *Bulletins de la Société d'anthropologie de Paris.* 7th ser. (1927–1927): 6–8.

Anthony, Raoul. "Rapport du secrétaire général pour l'année 1936." *Bulletins de la Société d'anthropologie de Paris.* 8th ser. 1–11 (1930–1940): 70–71.

Arcelin, Adrien. "Matériaux pour l'histoire primitive et naturelle de l'homme." *Revue des questions scientifiques* 6 (1879): 319–325.

——. "Le cerveau de Gambetta." *Revue des questions scientifiques* 20 (1887): 271–272.

Ariès, Philippe. *Essais sur l'histoire de la mort en occident.* Paris: Seuil, 1975.

Asseline, Louis. *Histoire de l'Autriche depuis la mort de Marie-Thérèse jusqu'à nos jours.* Paris: G. Baillière, 1877.

Asseline, Louis and A. Lefèvre. Preface and notes to Denis Diderot, *Jacques le fataliste et son maître.* Paris, Charpentier, n.d.

Badone, Ellen. *The Appointed Hour: Death, Worldview, and Social Change in Brittany.* Berkeley: University of California Press, 1989.

Bakan, David. "The Influence of Phrenology on American Philosophy." *Journal of the History of the Behavioral Sciences* 2 (1966): 200–220.

Baker, Thelma S. and Phyllis B. Eveleth. "The Effects of Funding Patterns on the Development of Physical Anthropology." In Frank Spencer, ed., *A History of American Physical Anthropology,* 31–48. New York: Academic, 1982.

Barrows, Susanna. *Distorting Mirrors: Visions of the Crowd in Late Nineteenth-Century France.* New Haven: Yale University Press, 1981.

Beddoe, John. *The Races of Britain: A Contribution to the Anthropology of Western Europe.* Bristol: J. W. Arrowsmith, 1885.

Béjin, André. "Le sang, le sens et le travail: Georges Vacher de Lapouge, darwiniste social, fondateur de l'anthroposociologie." *Cahiers internationaux de sociologie* 73 (1982): 323–343.

Bégouen, Henri. "Vacher de Lapouge-Père de l'Aryenisme." *Journal des débats politiques et littéraires,* August 22, 1936, 3.

Belot, Georges. "Analyses et comptes rendus: Letourneau-L'évolution de la propriété." *Revue philosophique* 28 (1889): 645–653.

Bénichou, Paul. "Sur quelques sources françaises de l'antisémitisme moderne." *Commentaire* 1 (1978): 67–79.

Benoît, Yves. *Dictionnaire des ministres: De 1789 à 1989.* Paris: Perrin, 1990.

Benrubi, Isaac. *Contemporary Thought of France.* London: Williams and Norgate, 1926.

Bergson, Henri. *Matière et mémoire: Essai sur la relation du corps à l'esprit.* Paris: Alcan, 1896.

——. *Creative Evolution.* Trans. A. Mitchell. New York: Holt, 1911.

——. "Rapport sur 'Progrès et bonheur' de J. Finot." In *Mélanges,* ed. André Robinet, 1090–1094. Paris: PUF, 1972.

Bernard, Philippe and Henri Dubief. *The Decline of the Third Republic, 1914–1938.* Cambridge: Cambridge University Press, 1985.

Bert, Paul., ed. *Revues scientifiques pour la république française publiée par le journal "La republique française" sous la direction de M. Paul Bert.* 7 vols. Paris: G. Masson, 1879–1885.

Berthelot, Marcellin. "La science et la morale." *Revue de Paris,* February 1, 1895, 449–469.

——. *Science et morale.* Paris, 1897.

Bertillon, Alphonse. *Les races sauvages: Les peuples de l'Afrique, les peuples de l'Amérique, les peuples de l'Océanie, quelques peuples de l'Asie et des regions bovéales.* Paris, 1882.

——. "Questions des récidivistes: L'identité des récidivistes et la loi de relégation." *Revue politique et littéraire* 3, no. 17 (1883): 513–521.

———. *Notice sur le fonctionnement du service d'identification de la préfecture de police*. Paris: Masson, 1889.

———. *L'anthropométrie judiciaire à Paris en 1889*. Lyon: Storch, 1890.

———. *La photographie judiciaire*. Paris, 1890.

Bertillon, Jacques. "Des monstruosités: Principes généraux de tératologie," *La nature* 1 (1874): 209–212, 243–246, 273–276, 338–342.

———. "Le Musée de l'Ecole d'anthropologie." *La nature* 6 (1878): 39–42.

———. "Rosa et Josefa: Les deux soeurs tchèques." *La nature* 2, no. 2 (1884): 293–294.

———. *Cours élémentaire de statistique administrative, élaboration des statistiques, organisation des bureaux de statistique, éléments de démographie*. Paris: Société d'éditions scientifiques, 1895.

———. *De la dépopulation de la France et des remèdes à apporter*. Paris, 1896.

———. "Le problème de la dépopulation: Le programme de l'Alliance nationale pour l'accroissement de la population française." *Revue politique et parlementaire*, 4th year, no. 12 (April–June 1897): 531–574.

———. *Le problème de la dépopulation*. Paris, 1897.

———. "The International System of Nomenclature of Diseases and Causes of Death (Bertillon Classification)." *Public Health Reports* 15, no. 49 (December 7, 1900).

———. *La dépopulation de la France, ses conséquences, ses causes, mesures à prendre pour la combattre*. Paris: Alcan, 1911.

Bertillon, Louis-Adolphe. "Des diverses manières de mesurer la durée de la vie humaine." *Journal de la Société de statistiques de Paris* 7 (1866): 45–64.

———. "La biologie." *La pensée nouvelle* 22 (October 11, 1868): 170–172.

———. *La démographie figurée de la France*. Paris: G. Masson, 1874.

Bertillon, Suzanne. *Vie d'Alphonse Bertillon: Inventeur de l'anthropométrie*. Paris: Gallimard, 1941.

Billy, André. *L'époque 1900*. Paris: J. Tallandier, 1951.

Blanckaert, Claude. "L'anthropologie au féminine: Clémence Royer (1830–1902)." *Revue de synthèse* 105 (1982): 23–38.

———. "Préface: 'L'anthropologie personnifée'-Paul Broca et la biologie du genre humain." In Paul Broca, *Mémoires d'anthropologie*, i–xliii. Paris: Jean-Michel Place, 1989.

———. "La science de l'homme entre humanité et inhumanité." In Claude Blanckaert, ed., *Des sciences contre l'homme: Classer, hiérarchiser, exclure*, 14–46. Paris: Autrement, 1993.

Bloch, Jean-Richard. *L'anoblissement en France au temps de François Ier: Essai d'une définition de la condition juridique et sociale de la noblesse au début du 16e siècle*. Paris: Alcan, 1934.

Blumenberg, Hans. *The Legitimacy of the Modern Age*. Trans. Robert M. Wallace. Cambridge, Mass.: MIT Press, 1983.

Bock, Gisela and Pat Thane, eds. *Maternity and Gender Policies: Women and the Rise of the European Welfare States, 1880–1950s*. New York: Routledge, 1991.

Boissel, Jean. "Autour du Gobinisme: Correspondance inédite entre L. Schemann et G. Vacher de Lapouge." *Annales du CESERE* 4 (1981): 91–120.

Bouglé, Célestin. "Sociologie et démocratie." *Revue de métaphysique et de morale* 4 (1896): 118–128.

———. "Sociologie, psychologie, et histoire." *Revue de métaphysique et de morale* 4 (1896): 362–371.

———. "Anthropologie et démocratie." *Revue de métaphysique et de morale* 5 (1897): 443–461.

———. *La démocratie devant la science*. Paris: Alcan, 1904.

——. *Le solidarisme*. Paris, 1907.

——. *Qu'est-ce que la sociologie?* Paris: Alcan, 1907.

——. *Essais sur le régime des castes*. Paris: Alcan, 1908.

Bourgeois, Léon. *Essai d'une philosophie de la solidarité: Conférences et discussions*. Paris, 1902.

Bourget, Paul. *Essais de psychologie contemporaine: Baudelaire, M. Renan, Flaubert, M. Taine, Stendhal*. Paris: A. Lemerre, 1883.

Bourneville, Désiré-Magloire. *Louise Lateau; ou, La stigmatisée belge*. Paris, 1875.

Bourneville, Désiré-Magloire and P. Renarde. *Iconographie photographique de la Salpêtrière*. 3 vols. Paris: Progrès médical, 1876–1880.

Bourneville, Désiré-Magloire and Dr. Voulet. *De la contracture hystérique permanente; ou, Appréciation scientifique des miracles de Saint Louis et de Saint Médarde*. Paris, 1872.

Bowler, Peter. *Theories of Human Evolution: A Century of Debate, 1844–1944*. Baltimore: Johns Hopkins University Press, 1986.

——. *The Non-Darwinian Revolution: Reinterpreting a Historical Myth*. Baltimore: Johns Hopkins University Press, 1988.

——. *The Invention of Progress*. Oxford: Basil Blackwell, 1989.

Bracher, Karl Dietrich. *The German Dictatorship: The Origins, Structure, and Effects of National Socialism*. Trans. Jean Steinberg. Intro. Peter Gay. New York: Praeger, 1970.

Brandon, Ruth. *The Spiritualists: The Passion of the Occult in the Nineteenth and Twentieth Centuries*. New York: Knopf, 1983.

Bray, Warwick and David Trump. *The Penguin Dictionary of Archaeology*. Middlesex: Penguin, 1970.

Bridenthal, Renate, Atina Grossmann, and Marion Kaplan. *When Biology Became Destiny*. New York: Monthly Review Press, 1984.

Broca, Paul. "Sur le volume et la forme du cerveau suivant les individus et suivant les races." *Bulletins de la Société d'anthropologie*, 1st ser., 2 (1861): 139–207, 301–321, 441–446.

——. "Histoire des travaux de la Société d'anthropologie de Paris (1859–1863)." *Mémoires de la Société d'anthropologie de Paris* 2 (1865): vii–li.

——. *Instructions générales pour les recherches anthropologiques*. Paris, 1865.

——. "Discours de M. Broca sur l'ensemble de la question." In *Congrès international d'anthropologie et d'archéologie préhistoriques*, 367–402. Paris, 1868.

——. "Sur le transformism: Remarques générales," *Bulletin de la Société d'anthropologie de Paris* 2d ser., 5 (1870): 169–239.

——. "Discours sur l'homme et les animaux." *Bulletins de la Société d'anthropologie de Paris* 2d ser., no. 1 (1866): 68–85.

——. *Mémoires d'anthropologie*. 2 vols. Paris, 1871.

——. "Discussion sur la religiosité." *Bulletins de la Société d'anthropologie de Paris* 2d ser., no. 12 (1877): 33–36.

——. *Congrès internationale d'anthropologie, 9e session à Lisbonne, 1879*. Lisbon: Académie royale des sciences, 1884.

——. *Correspondance, 1841–1857*. Vol. 2, *1848–1857*. Paris: Paul Schmidt, 1886.

Brown, Frederick. *Zola*. New York: Farrar, Straus, 1995.

Brumberg, Joan Jacobs. *Fasting Girls: The Emergence of Anorexia Nervosa as a Modern Disease*. Cambridge: Harvard University Press, 1988.

Brunetière, Ferdinand. "Après un visite au Vatican." *Revue des deux mondes* 7, no. 127 (1895): 97–118.

Buican, Denis. *La révolution de l'évolution*. Paris: PUF, 1984.

Burleigh, Michael and Wolfgang Wippermann. *The Racial State: Germany, 1933–1945*. New York: Cambridge University Press, 1991.

Bynum, Caroline Walker. "Bodily Miracles and the Resurrection of the Body in the High Middle Ages." In Thomas Kselman, ed., *Belief in History*, 68–106. Notre Dame: University of Notre Dame Press, 1990.

———. *Holy Feast and Holy Fast: The Religious Significance of Food to Medieval Women*. Berkeley: University of California Press, 1987.

Caplan, Jane and John Torpey, eds. *Documenting Individual Identity: The Development of State Practices in the Modern World*. Princeton: Princeton University Press, 2001.

Carr, Edward Hallett. *The Romantic Exiles*. 1933. Reprint, Cambridge, Mass.: MIT Press, 1981.

Carver, Thomas Nixon, ed. *Sociology and Social Progress: A Handbook for Students of Sociology*. Boston: Ginn, 1906.

"Causerie bibliographique." *Revue scientifique* 21 (1884): 535–536, 536–537, 598.

Céard, Henry. "Sapeck l'incomparable." *Le siècle*. October 15, 1889.

"Cerveau de M. Laborde." *Bulletins de la Société d'anthropologie de Paris* 5th ser., 4 (1903): 422–425.

Chadwick, Owen. *The Secularization of the European Mind in the Nineteenth Century*. Cambridge: Cambridge University Press, 1975.

Charcot, Jean-Marie. "Faith-Cure." *New Review* 8 (January-June 1893): 18–31.

———. "La foi qui guérit." *Revue hebdomadaire* 1, no. 5 (December 3, 1892): 122–132

Chatterton-Hill, Georges. *La physiologie morale*. Paris: P. V. Stock, 1904.

Chauvière, Annick. "Les hommes d'aujourd'hui." In Jean-Michel Place and André Vasseur, eds., *Bibliographie des revues et journaux littéraires des XIXe et Xxe siècles*, 90–97. Paris: Jean-Michel Place, 1974.

Chevalier, Jacques. *Henri Bergson*. Trans. Lilian Clare. 1928. Reprint, New York: AMS, 1969.

Chudzinski, Théophile and Mathias Duval. "Description morphologique du cerveau de Gambetta." *Bulletins de la Société d'anthropologie* 3d ser., 6 (1883): 129–152.

Chudzinski, Théophile and Léonce Manouvrier. "Etude sur le cerveau de Bertillon." *Bulletins de la Société d'anthropologie* 3d ser., 10 (1887): 558–590.

Clark, Linda. *Social Darwinism in France*. Birmingham: University of Alabama Press, 1984.

Clark, Terry. "Emile Durkheim and the Institutionalization of Sociology in the French University System." *European Journal of Sociology* 9 (1968): 37–71.

Clemenceau, Georges. *De la génération des éléments anatomiques*. Paris: Baillière, 1865.

Closson, Carl. "Further Data of Anthropo-Sociology." *Journal of Political Economy* 7 (March 1899): 243–252.

Cole, Joshua. *The Power of Large Numbers*. Ithaca: Cornell University Press, 2000.

Compte-rendu: Congrès de la libre pensée à Rome, 1904. Ghent, 1905.

Conan Doyle, Arthur. *The Annotated Sherlock Holmes*. Ed. and intro. William S. Baring-Gould. 2 vols. New York: Clarkson and Potter, 1967.

Condorcet, Antoine-Nicolas de. *Sketch for a Historical Picture of the Progress of the Human Mind* (1795). Trans. June Barraclough. New York: Noonday, 1955.

Conry, Yvette. *L'introduction du Darwinisme en France au XIXe siècle*. Paris: Vrin, 1974.

Conta, Madame. "Quelques considérations d'ordre social concernant l'homme et la femme." *Congrès français et international du droit des femmes*, 131–135. Paris, 1889.

Coolidge, Calvin. "Whose Country Is This?" *Good Housekeeping* 72 (February 1921): 13–14, 106–109.

Coudereau, Auguste. "Program." *Libre pensée: Science, lettres, arts, histoire, philosophie* 1 (October 21, 1866): 1–2.

——. "De l'influence de la religion sur la civilisation: Réponse à M. Bataillard." *Bulletins de la Société d'anthropologie* 2d ser., no. 2 (1867): 580–591.

——. "L'autopsie mutuelle." *La pensée libre* 17 (November 13, 1880): 1.

"Un coup d'état à l'école d'anthropologie." *La patrie*. February 8, 1890.

Crespi, Angelo. *Contemporary Thought of Italy*. London: Williams and Norgate, 1926.

Dally, Eugène. Comment in "Discussion sur le questionnaire d'ethnographie." *Bulletins de la Société d'anthropologie de Paris* 3d ser., no. 5 (1882): 557–578.

——. "Suite de la discussion sur l'anthropologie." *Bulletins de la Société d'anthropologie de Paris* 3d ser., no. 11 (1888): 27–46.

Daniel, Glyn E. *A Hundred Years of Archaeology*. London: Duckworth, 1952.

Darlu, Alphonse. "Réflexion d'un philosophe sur les questions du jour: Science, morale et religion." *Revue de métaphysique et de morale* 3 (1895): 239–251.

——. "Réflexion d'un philosophe sur les questions du jour: La solidarité." *Revue de métaphysique et de morale* 5 (1897): 120–128.

Darwin, Charles. *On the Origin of Species*. Cambridge: Harvard University Press, 1964.

"Darwin et le bon Dieu." *La critique philosophique* 5 (1876): 205.

Daudet, Léon. *Devant la douleur*. Paris: Nouvelle librairie nationale, 1915.

——. *Paris vécu*. Paris: Gallimard, 1930.

Davenport, Charles. *Heredity in Relation to Eugenics*. New York: Holt, 1911.

Davidson, Eugene. *The Trial of the Germans: Nuremberg, 1945–1946*. New York: Macmillan, 1966.

Demaison, André. *Faidherbe*. Paris, 1932.

Demeulenaere-Douyère, Christiane. *Paul Robin: Un militant de la liberté et du bonheur*. Paris: Publisud, 1994.

Dereux, H. "Analyses et comptes rendus: André Lefèvre, La philosophie." *Revue philosophique* 7 (1879): 455–458.

Desmond, Adrian. "Lamarckism and Democracy: Corporations, Corruption, and Comparative Anatomy in the 1830s." In James R. Moore, ed., *History, Humanity, and Evolution: Essays in Honor of John C. Green*, 99–130. Cambridge: Cambridge University Press, 1989.

——. *The Politics of Evolution: Morphology, Medicine, and Reform in Radical London*. Chicago: University of Chicago Press, 1989.

Dias, Nélia. *Le Musée d'ethnographie du Trocadéro (1878–1908): Anthropologie et muséologie en France*. Paris: CNRS, 1991.

——. "La Société d'autopsie mutuelle; ou, Le dévouement absolu aux progrès de l'anthropologie." *Gradhiva* 10 (1991): 26–35.

"Un diner pour la vulgarisation de la dissection mutuelle." *Morning News*, February 7, 1884.

Drysdale, Bessie Ingman. *Labour Troubles and Birth Control*. London: Heinemann, 1920.

Drysdale, Charles Robert. *The Cause of Poverty: A Paper Read at the National Liberal Club on 21st October, 1890*. London: Standring, 1891.

Duhem, Pierre. *Paul Tannery*. Montligeon (Orne): Librairie de Montligeon, 1905.

Duilhé, Marc-Antoine-Marie-François. "Le problème anthropologique et les théories évolutionnistes." In *Congrès scientifique international des Catholiques tenu à Paris du 8 au 13 avril 1888*, 2:621. Paris: Bureaux des Annales de philosophie chrétienne, 1888.

Dupaquier, Michel. "La famille Bertillon et la naissance d'une nouvelle science sociale: La démographie." *Annales de démographie historique*, 1983, 293–311.

Durkheim, Emile. "La philosophie dans les universités allemandes." *Revue internationale de l'enseignement* 13 (1887): 439–440.

———. *Les règles de la méthode sociologique*. Paris, 1894.

———. "Représentations individuelles et représentations collectives." *Revue de métaphysique et de morale* 6 (1898): 273–302.

———. "Notre siècle: La sociologie en France au XIXe siècle." *Revue bleue* 4th ser., 13, no. 20 (May 19, 1900): 609–652.

———. "Sociologie." In *La science française*. Paris: Imprimerie nationale, 1915.

———. *Education and Sociology*. New York: Free, 1956.

———. "Sociology in France in the Nineteenth Century (1900)." In *On Morality and Society (Selected Writings)*, ed. Robert Bellah, 3–24. Chicago: University of Chicago Press, 1973.

———. *Sociology and Philosophy*. Trans. D. F. Pocock. New York: Free, 1974.

———. *The Elementary Forms of Religious Life*. Trans. and intro. Karen E. Fields. New York: Free, 1995.

Duval, Mathias. "Le cerveau de Louis Asseline." *Bulletins de la Société d'anthropologie* 3d ser, 6 (1883): 260–274.

———. "Le poids de l'encéphale de Gambetta." *Bulletins de la Société d'anthropologie* 3d ser., 6 (1883): 399–417.

———. "L'aphasie depuis Broca." *Bulletins de la Société d'anthropologie* 3d ser., 10 (1887): 743–771.

Duval, Mathias, Théophile Chudzinski, and Georges Hervé. "Description morphologique du cerveau de Coudereau." *Bulletins de la Société d'anthropologie* 3d ser., 6 (1883): 377–389.

Echerac, Jean d.' "André Lefèvre." *Revue de l'Ecole d'anthropologie de Paris* 14 (1904): 386.

Ellis, Jack D. *The Physician-Legislators of France: Medicine and Politics in the Early Third Republic, 1870–1914*. Cambridge: Cambridge University Press, 1990.

Eymieu, Antonin. *La part des croyants dans les progrès de la science au XIXe siècle*. 2 vols. Paris: Perrin, 1920–1935.

Fabricius, Hans. *Reichsinnenminister Dr. Frick: Der revolutionaere Staatsmann*. Berlin, 1939.

Fauvelle, Jean-Louis. "Conséquence naturelle de la science libre." *L'homme* 2 (1885): 737–742.

———. "Il faut en finir avec la philosophie." *L'homme* 2 (1885): 139–146.

———. "Les desiderata du matérialisme scientifique." *L'homme* 3 (1886): 103–108.

Favre, François. "Enseignement professionel des femmes." *Le monde maçonnique*, November 1862, 385.

Fenn, Richard K. *Liturgies and Trials: The Secularization of Religious Language*. Oxford: Basil Blackwell, 1982.

Ferri, E. "Various Short Contributions to Criminal Sociology." *Internationaler Kongress der Kriminalanthropologie* 7 (1911): 251.

Fest, Joachim C. *The Face of the Third Reich: Portraits of the Nazi Leadership*. New York: Pantheon, 1970.

Fidler, Geoffrey C. "On Jean-Marie Guyau, Immoraliste." *Journal of the History of Ideas* 55, no. 1 (1994): 75–97.

Finot, Jean. *La philosophie de la longévité*. Paris: Schleicher, 1901.

———. *Le préjugé des races*. 2d ed. Paris, 1905–1906.

——. *Race Prejudice*. Trans. Florence Wade-Evans. New York, 1906.

——. *Le préjugé et problème des sexes*. Paris: Alcan, 1912.

——. *The Science of Happiness*. Trans. Mary Stafford. New York: Putnam, 1914.

Finot, Jean-Louis. "Mon père." *La revue mondiale* 33 (May 15, 1922): 143–150.

Flaubert, Gustave. *Madame Bovary*. Trans. Geoffrey Wall. London: Penguin, 1992.

Foucault, Michel. *Discipline and Punish*. Trans. A. Sheridan. New York: Pantheon, 1977.

——. "Omnes et Singulatum: Towards a Criticism of 'Political Reason.' " In *The Turner Lectures on Human Values*, vol. 2. Salt Lake City: Sterling M. McMurrin, 1981.

——. *The History of Sexuality: Volume One-An Introduction*. New York: Vintage, 1990.

Fouillée, Alfred. "La psychologie des peuples et l'anthropologie." *Revue des deux mondes* 128 (March 15, 1895): 365–396.

——. *Psychologie du peuple français*. Paris, 1898.

——. "As Others See Us." *The Living Age* 212 (1899): 67–72.

——. *Esquisse psychologique des peuples*. Paris, 1903.

Fox, Robert. "Science, the University, and the State in Nineteenth-Century France." In Gerald L. Geison, ed., *Professions and the French State, 1700–1900*, 66–145. Philadelphia: University of Pennsylvania Press, 1984.

Fraisse, Geneviève. *Clémence Royer: Philosophe et femme de science*. Paris: Découverte, 1985.

Freud, Sigmund. "A Case of Successful Treatment by Hypnotism." In *The Standard Edition of the Complete Psychological Works of Sigmund Freud*, trans. and ed. James Strachey, with Anna Freud, Alix Strachey, and Alan Tyson, 1:126–127. London: Hogarth, 1953.

——. *On Aphasia: A Critical Study*. New York: International Universities Press, 1953.

——. "Hysteria." In *The Standard Edition of the Complete Psychological Works of Sigmund Freud*. Trans. and ed. James Strachey, with Anna Freud, Alix Strachey, and Alan Tyson, 1:39–59. London: Hogarth, 1953.

——. "Charcot." In *Selections-Collected Papers*. Authorized translation under the supervision of Joan Riviere, trans. and ed. James Strachey with Alix Strachey, 1:9–23. New York: Basic, 1959.

G.S. "Analyses et comptes rendus: Dr. H. Thulié, *La femme: Essai de sociologie phusiologique*." *Revue philosophique* 20 (1885): 538–540.

Gadille, Jacques. "On French Anticlericalism: Some Reflections." *European Studies Review* 2, no. 2 (April 1983): 127–143.

Galton, Francis. *Hereditary Genius: An Inquiry into Its Laws and Consequences*. London: Macmillan, 1869.

Gasman, Daniel. *The Scientific Origins of National Socialism: Social Darwinism in Ernst Haeckel and the German Monist League*. London: Macdonald, 1971.

——. *Haeckel's Monism and the Birth of Fascist Ideology*. New York: Lang, 1998.

Geison, Gerald L., ed. *Professions and the French State, 1700–1900*. Philadelphia: University of Pennsylvania Press, 1984.

Gilman, Sander L. *Freud, Race, and Gender*. Princeton: Princeton University Press, 1993.

Gley, Eugène. "Histoire de la Société de biologie." *Revue scientifique* 4, no. 13 (1900): 3–11.

Glick, Thomas F., ed. *The Comparative Reception of Darwinism*. Austin: University of Texas Press, 1974.

Goetz, Christopher G., Michel Bonduelle, and Toby Gelfand. *Charcot: Constructing Neurology*. New York: Oxford University Press, 1995.

Goldstein, Jan. *Console and Classify, The French Psychiatric Profession in the Nineteenth Century.* Cambridge: Cambridge University Press, 1987.

Goodman, Dena. *The Republic of Letters: A Cultural History of the French Enlightenment.* Ithaca: Cornell University Press, 1994.

Gould, Stephen J. *The Mismeasure of Man.* New York: Norton, 1981.

Grant, Madison. *The Passing of the Great Race.* New York: Scribners, 1916.

Grindelle, F. "Bibliographie: Les débuts de l'humanité." *La critique philosophique* 10 (1881): 399–400.

———. "Bibliographie: La renaissance du matérialisme." *La critique philosophique* 10 (1881): 397–399.

———. "Bibliographie: Le bien et la loi morale." *La critique philosophique* 13 (1884): 204–208.

———. "Bibliographie: La morale." *La critique philosophique* 13 (1884): 236–239.

Günther, Hans F. K. *Racial Elements of European History.* Trans. G. C. Wheeler. New York, 1927.

———. "Zum Tode des Grassen Georges Vacher de Lapouge." *Rasse: Monatschrift der Nordischen Bewegung* 3 (1936): 95–98.

Guyot, Yves. *La science économique.* Bibliothèque des sciences contemporaines. Paris: Reinwald, 1881.

———. *Les principes de 1889 et le socialisme.* Paris, 1894.

———. *Sophismes socialiste et faits economiques.* Paris, 1908.

Hacking, Ian. *The Taming of Chance.* Cambridge: Cambridge University Press, 1990.

Haeckel, Ernst. *The Riddle of the Universe at the Close of the Nineteenth Century.* Trans. Joseph McCabe. New York: Harper, 1901.

Hammond, Michael. "Anthropology as a Weapon of Social Combat in Late-Nineteenth-Century France." *Journal of the History of the Behavioral Sciences* 16 (1980): 118–132.

Harvey, Joy. "Races Specified, Evolution Transformed: The Social Context of Scientific Debates Originating in the Société d'anthropologie de Paris, 1859–1902." Ph.D. diss., Harvard University, 1983.

———. " 'Strangers to Each Other': Male and Female Relationships in the Life and Work of Clémence Royer." In Pnina Abir-am and Dorinda Outram, eds., *Uneasy Careers and Intimate Lives,* 147–171. New Brunswick: Rutgers University Press, 1987.

———. *"Almost a Man of Genius": Clémence Royer, Feminism, and Nineteenth-Century Science.* New Brunswick: Rutgers University Press, 1997.

Haye Jousselin, Henri H. de La. *Georges Vacher de Lapouge: Essai de bibliographie.* Self-published, 1986.

Hayward, J. E. S. "Solidarity: The Social History of an Idea in Nineteenth-Century France." *International Review of Social History* 4 (1959): 261–284.

———. "The Official Social Philosophy of the French Third Republic: Léon Bourgeois and Solidarism." *International Review of Social History* 6 (1961): 19–48.

Headings, Mildred. *French Freemasonry Under the Third Republic.* Baltimore: Johns Hopkins Press, 1949.

Hecht, Jennifer Michael. "L'anthropologie et la république: Les réactions socialiste et républicaine face aux théories de Vacher de Lapouge." Paper presented at "La race: Idées et pratiques dans les sciences et dans l'histoire," Colloque CNRS, Paris, June 1–2, 1993.

———. "Anthropological Utopias and Republican Morality: Atheism and the Mind/Body Problem in France, 1876–1914." Ph.D. diss., Columbia University, 1995.

———. "A Vigilant Anthropology: Léonce Manouvrier and the Disappearing Numbers." *Journal of the History of the Behavioral Sciences* 33, no. 3 (summer 1997): 221–240.

———. "French Scientific Materialism and the Liturgy of Death: The Invention of a Secular Version of Catholic Last Rites (1876–1914)." *French Historical Studies* 20, no. 4 (fall 1997): 703–735.

———. "The Solvency of Metaphysics: The Debate Over Racial Science and Moral Philosophy in France, 1890–1914." *Isis: Journal of the History of Science Society* 90 (spring 1999): 1–24.

———. "Vacher de Lapouge and the Rise of Nazi Science." *Journal of the History of Ideas* 61, no. 2 (April 2000): 285–304.

Houzé, E. "Gabriel de Mortillet-Notice nécrologique." In *Extrait du Bulletin de la Société d'anthropologie de Bruxelles*, vol. 17, *1898–1899*, 1–3. Brussels, 1899.

Hovelacque, Abel. "La linguistique et le precurseur de l'homme." In *Association française pour l'avancement des sciences—Compte rendu*, vol. 2. Lyon, 1873.

———. *Plus les laîques sont éclairés, moins les prêtres pourront faire du mal*. Paris, 1880.

———. *Les débuts de l'humanité*. Paris: Doin, 1881.

———. *La linguistique*. Bibliothèque des sciences contemporaines. Paris: Reinwald, 1892.

Hovelacque, Abel, Charles Issaurat, André Lefèvre, Charles Letourneau, Gabriel de Mortillet, Henri Thulié, and Eugène Véron, eds. *Dictionnaire des sciences anthropologiques*. Paris, n.d. (1881–).

Huerta, Luis. *La doctrina eugénica*. Madrid: Editorial Instituto Samper, 1933.

Inauguration du monument de G. de Mortillet-Extrait de L'homme préhistorique, 3e année, no. 11, 1905. Paris, 1905.

Issaurat, Charles. "Analyse de l'ouvrage de M. A. Hovelacque intitulé: Les nègres de l'Afrique sus-équatoriale." *La tribune médicale* 45 (1890).

Jeannolle, Charles. "Le positivisme et les sociétés de libre-pensée." *Revue occidentale*, 1st semester (1885): 98–113, 237–257.

Jones, Robert Alun. "Religion and Science in the Elementary Forms." In N. J. Allen, W. S. F. Pickering and W. Watts Miller, eds., *On Durkheim's Elementary Forms of Religious Life*, 39–52. New York: Routledge, 1998.

Jones, Steve. *The Language of Genes*. New York: Doubleday, 1993.

Kevles, Daniel J. *In the Name of Eugenics: Genetics and the Uses of Human Heredity*. Berkeley: University of California Press, 1985.

Keylor, William R. *Academy and Community: The Foundation of the French Historical Profession*. Cambridge: Harvard University Press, 1975.

Koos, Cheryl A. "Gender, Anti-Individualism, and Nationalism: The Alliance Nationale and the Pronatalist Backlash Against the Femme moderne, 1933–1940." *French Historical Studies* 19 (1996): 699–723.

Kovin, Seth and Sonya Michel, eds. *Mothers of a New World: Maternalist Politics and the Origins of the Welfare States*. New York: Routledge, 1933.

Kselman, Thomas A. *Death and the Afterlife in Modern France*. Princeton: Princeton University Press, 1993.

Kulz, Werner. "Marquis de Lapouge zum Gedenken!" *Volke und Rasse* 6 (June 1936): 255–258.

Laborde, Jean-Baptiste Vincent. *Léon Gambetta: Biographie psychologique-Le cerveau, la parole, la fonction et l'organe*. Paris, 1898.

Lalouette, Jacqueline. "Science et foi dans l'idéologie libre penseuse (1866–1914)." In *Chris-

tianisme et science, 21–54. Etudes réunies par l'Association française d'histoire religieuse contemporaine. Paris: Vrin, 1989.

———. *La libre pensée en France, 1848–1940*. Paris: Albin Michel, 1997.

Lapouge, Georges Vacher de (as Verax). "Liberté, liberté, chérie." *L'avenir de la Vienne*, May 22, 1879, 2.

— (as Verax). "La liberté des fils de famille." *L'avenir de la Vienne*, May 23, 1879, 1.

— (as Verax). "Paroles pour la liberté." *L'avenir de la Vienne*, July 1 and 2, 1879, 1.

— (as Verax). "La messe de corvée." *L'avenir de la Vienne*, January 20, 1880: 1.

— (as Verax). "La réforme de l'organisation judiciaire." *L'avenir de la Vienne*, February 24 and 25, 1880, 1–2; February 26, 1880, 1; February 27, 1880, 2.

———. "L'anthropologie et la science politique." *Revue d'anthropologie* 16 (1887): 136–157.

———. "Questions aryennes." *Revue d'anthropologie* 18 (1889): 181–193.

———. *Les sélections sociales*. Paris: A. Fontemoing, 1896.

———. Preface to Ernst Haeckel, *Le monisme: Lien entre la religion et la science*, trans. Georges Vacher de Lapouge, 1–8. Paris: Schleicher, 1897.

———. *L'Aryen: Son rôle social*. Paris: A. Fontemoing, 1899.

———. *Race et milieu social*. Paris: Rivière, 1909.

———. *Der Arier und seine Bedeutung für die Gemeinschaft*. Frankfurt A. M.: Moritz Diesterweg, 1939.

———. "Souvenirs." In Henri H. de La Haye Jousselin, *Georges Vacher de Lapouge: Essai de bibliographie*, 9–16. Self-published, 1986.

Larson, Edward. *Sex, Race, and Science: Eugenics in the Deep South*. Baltimore: Johns Hopkins University Press, 1995.

Le Bon, Gustave. "Recherches anatomiques et mathématique sur les lois des variations du volume du cerveau et sur leur relations avec l'intelligence." *Revue d'anthropologie* 2, no. 2 (1879): 27–104.

———. *La psychologie des foules*. Paris, 1895.

Lebovics, Herman. *True France: The Wars Over Cultural Identity, 1900–1945*. Ithaca: Cornell University Press, 1992.

Lefèvre, André. *Religions et mythologies comparées*. Paris: LeRoux, 1877.

———. "L'homme, d'après les découvertes de l'anthropologie." *La jeune France*, no. 1 (October 1, 1878): 214–221.

———. *La philosophie*. Bibliothèque des sciences contemporaines. Paris: Reinwald, 1879.

———. *Philosophy: Historical and Critical*. Trans. and intro. A. H. Keane. London: Chapman and Hall; Philadelphia: Lippincott, 1879.

———. "Voltaire et les religions." *La jeune France*, no. 10 (February 1, 1879): 361–367.

— "Sonnets philosophiques." *La jeune France*, no. 21 (1880–1881): 38–40.

———. "Un vice de la littérature enfantine." *La jeune France*, no. 33 (1880–1881): 404–406.

———. *La renaissance du matérialisme*. Bibliothèque matérialiste. Paris: Doin, 1881.

———. "La philosophie devant l'anthropologie." *L'homme* 19 (October 10, 1884): 577–584.

———. "Louis Asseline." *Contre-poison*, 1900, 259–314.

———. *La religion*. Paris: A. Costes, 1921.

Lenoir, Raymond. "L'oeuvre sociologique d'Emile Durkheim." *Europe* 22 (1930): 292–296.

Léonard, Jacques. *La médecine entre les savoirs et les pouvoirs: Histoire intellectuelle et politique de la médecine française au XIXe siècle*. Paris: Aubier Montaigne, 1981.

Letourneau, Charles. "L'origine de l'homme." *Pensée nouvelle*, 1867. Reprinted in *Science et matérialisme*. Paris: Reinwald, 1891.

——. "Variabilité des àtres organisées." *La philosophie positive* 3 (1868): 99–121.

——. *La physiologie des passions*. (1st ed. Paris: G. Baillière, 1868) 2d ed. Paris: Reinwald, 1878.

——. *Questionnaire de sociologie et d'ethnographie*. Paris, 1882. Reprinted, in a slightly revised form, in *Bulletins de la Société d'anthropologie de Paris* 3d ser., no. 6 (1883): 578–597.

——. "Adolphe Bertillon." *Bulletins de la Société d'anthropologie* 3d ser., 6 (1883): 187–192.

——. *La sociologie d'après l'ethnographie*. Bibliothèque des sciences contemporaines. Paris: Reinwald, 1884.

——. *L'évolution de la morale*. Paris: Delahaye et Lecrosnier, 1887.

——. *The Evolution of Marriage and of the Family*. London: Scott, 1891.

——. *Science et matérialisme*. Paris: Reinwald, 1891.

——. *La condition de la femme*. Paris: Giard/Brière, 1903.

Lévy-Bruhl, Lucien. "La morale de Darwin." *Revue politique et littéraire* 5 (1883): 169–175.

Libre pensée et religion laîque en France: De la fin du Second Empire à la fin de la Troisième République. Intro. J.-M. Mayeur. Strasbourg: Cerdic, 1980.

Locard, Edmond. *Alphonse Bertillon: L'homme, le savant, la pensée philosophique*. Lyon: A. Rey, 1914.

Lodge, Sir Oliver. *Past Years: An Autobiography*. New York: Scribner's, 1932.

Logue, William. "Sociologie et politique: Le libéralisme de Célestin Bouglé." *Revue française de sociologie* 20 (1979): 141–161.

Löwith, Karl. *Meaning in History: The Theological Presuppositions of the Philosophy of History*. Chicago: University of Chicago Press, 1949.

Lukes, Steven. *Emile Durkheim: His Life and Work*. Stanford: Stanford University Press, 1985.

Lützhoft, Hans-Jürgen. *Der Nordische Gedanke in Deutschland, 1920–1940*. Stuttgart: Klett, 1971.

McGrath, William J. *Freud's Discovery of Psychoanalysis: The Politics of Hysteria*. Ithaca: Cornell University Press, 1986.

McManners, John. *Church and State in France, 1870–1914*. New York: Harper, 1972.

——. *Death and the Enlightenment*. New York: Oxford University Press, 1981.

Mahoudeau, P.-G. "Analyses et comptes rendus: Manouvrier, 'Sur l'interprétation de la quantité dans l'encéphale et du poids de cerveau en particulier.'" *Revue philosophique* 24 (1887): 317–321.

Manouvrier, Léonce. "Recherches sur le développement quantitatif comparé de l'encéphale et de diverses parties du squelette." *Bulletin de la Société zoologique de France* 6 (1881): 77–94.

——. "Sur la grandeur du front et des principales régions du crâne chez l'homme et chez la femme." *Bulletins de la Société d'anthropologie* 3d ser., 6 (1883): 694–698.

——. "La fonction psycho-motrice." *Revue philosophique* 17 (1884): 503–525, 638–651.

——. "Variétés: L'internat en mèdicine des femmes." *Revue scientifique* 3, no. 19 (1884): 592–597.

——. "Analyses et comptes rendus: Duval—*Le Darwinisme*." *Revue philosophique* 21 (1886): 398.

——. "Société de psychologie physiologique: 'Mouvements divers et sueur palmaire consécutifs à des images mentales.' Séance du 29 mars 1886 (présidence de M. Charcot)." *Revue philosophique* 22 (1886): 203–207.

———. "Analyses et comptes rendus: Hovelacque et Hervé-*Précis d'anthropologie.*" *Revue philosophique* 24 (1887): 325–327.

———. "Analyses et comptes rendus: Quatrefages-Introduction à *L'étude des races humaines.*" *Revue philosophique* 24 (1887): 321–325.

———. "Etude comparative sur les cerveaux de Gambetta et de Bertillon." *Revue philosophique* 25 (1888): 453–461.

———. "Les premières circonvolutions temporales droite et gauche chez un sourd de l'oreille gauche (Bertillon)." *Revue philosophique* 26 (1888): 330–335.

———. "Indications anatomiques et physiologiques relatives aux attributions naturelles de la femme." In *Congrès français et international du droit des femmes*, 41–51. Paris, 1889.

———. "Analyses et comptes rendus: Broca-*Mémoires sur le cerveau de l'homme et des primates.*" *Revue philosophique* 27 (1889): 405–409.

———. "Analyses et comptes rendus: Hervé-*La circonvolution de Broca.*" *Revue philosophique* 27 (1889): 409–411.

———. "Les aptitudes et les actes dans leurs rapports avec la constitution anatomique et avec le milieu extérieur" (paper delivered at the Septième Conférence Broca), *Bulletins de la Société d'anthropologie* 4th ser., 1 (1890): 918–939.

———. "L'atavisme et le crime." *Revue de l'Ecole d'anthropologie* 1 (1891): 225–240.

———. "La genèse normale du crime." *Bulletins de la Société d'anthropologie* 4th ser., 4 (1893): 405–458.

———. "Questions préalables dans l'étude comparative des criminels et des honnêtes gens." In *Actes du troisième congrès d'anthropologie criminelle de Bruxelles*, 171–182. Brussels, 1893.

———. "L'anthropologie et le droit." *Revue internationale de sociologie* 2 (1894): 241–273, 351–370.

———. "L'indice céphalique et la pseudo-sociologie." *Revue de l'Ecole d'anthropologie* 9 (1899): 233–259.

———. "Cerveau de M. Laborde." *Bulletins de la Société d'anthropologie* 5th ser., no. 4 (1903): 424.

———. "Conclusions générales sur l'anthropologie des sexes et applications sociales." *Revue de l'Ecole d'anthropologie* 13 (1903): 405–423.

———. "L'individualité de l'anthropologie." *Revue de l'Ecole d'anthropologie* 14 (1904): 397–410.

———. "L'anthropologie des sexes." *Revue de l'Ecole d'anthropologie* 9 (1909): 41–61.

Margerie, Emmanuel de. "Albert de Lapparent." *Annales de géographie* 17 (1908): 344–347.

Martindale, Don. *The Nature and Types of Sociological Theory.* Boston: Houghton Mifflin, 1981.

Massis, Henri. *Comment Emile Zola composait ses romans-D'après ses notes personnelles et inédites.* Paris: Charpentier, 1906.

"Mathias Duval et le dîner du matérialisme scientifique." *L'homme* 3 (1886): 25–28.

Matsuda, Matt K. *The Memory of the Modern.* Oxford: Oxford University Press, 1996.

Mayeur, Jean-Marie and Madeleine Rebérioux. *The Third Republic from Its Origins to the Great War, 1871–1914.* Cambridge: Cambridge University Press, 1984.

Mévisse, M. *Des droits de la femme—Rapport—2ème congrès: Libre pensée.* Brussels, 1893.

Michelet, Jules. *Le prêtre, la femme et la famille.* 1845. Reprint, Paris: Michel Lévy and Librairie nouvelle, 1875.

Milice, Albert. *Clémence Royer et sa doctrine de la vie.* Paris: J. Peyronnet, 1926.

Mitterand, Henri. *Zola*. Vol. 1, *1840–1871*. Paris: Fayard, 1999.

Montandon, Georges. "Le squelette du Professeur Papillault." *Bulletins et mémoires de la Société d'anthropologie de Paris*, 8th ser., 6 (1935): 4–22.

Montessori, Maria. *Pedagogical Anthropology*. Trans. F. T. Cooper. New York: Fredrick A. Stokes, 1913.

Moore, James R., ed. *History, Humanity, and Evolution: Essays in Honor of John C. Green*. Cambridge: Cambridge University Press, 1989.

Mortillet, Gabriel de. "L'homme-singe perfectionné." *Libre pensée* 3 (1866): 23–24.

———. "Promenades préhistoriques à l'Exposition universelle." *Matériaux pour l'histoire positive et philosophique de l'homme* 3 (1867): 181–283, 285–368.

———. "Silex taillés de l'époque tertiaire du Portugal." *Bulletins de la Société d'anthropologie* 3d ser., 1 (1878): 428–429.

———. "L'homme quaternaire à l'Exposition." *Revue d'anthropologie* 2d ser., no. 2 (1879): 114–118.

———. *Le préhistorique*. Bibliothèque des sciences contemporaines. Paris: Reinwald, 1883.

———. "L'antisémitisme." *L'homme* 1 (1884): 522–528.

———. "Au lecteurs." *L'homme* 3 (1886): 1–2.

———. "L'église et la science." *L'homme* 4 (1887): 609–614.

Nagel, Günter. *Georges Vacher de Lapouge (1854–1936): Ein Beitrag zur Geschichte des Sozialdarwinismus in Frankreich*. Freiburg i. Br.: Schulz, 1975.

Nolte, Ernst. *Three Faces of Fascism*. Trans. Leila Vennewitz. New York: Rinehart and Winston, 1966.

Nord, Philip. *The Republican Moment: Struggles for Democracy in Nineteenth-Century France*. Cambridge: Harvard University Press, 1995.

Nye, Robert. "Heredity or Milieu: The Foundations of Modern European Criminological Theory." *Isis* 67, no. 3 (1976): 335–355.

———. *Crime, Madness, and Politics in Modern France: The Medical Conception of National Decline*. Princeton: Princeton University Press, 1984.

Offen, Karen. "Depopulation, Nationalism, and Feminism in Fin-de-Siècle France." *American Historical Review* 89 (1984): 648–676.

Papillault, G. "Discours de M. G. Papillault aux obsèques de M. Manouvrier." *Bulletins de la Société d'anthropologie* 7th ser. (1925–1927): 4–7.

Patte, Etienne. "Georges Vacher de Lapouge." *Revue générale du centre-ouest de la France* 46 (1937): 769–789.

Paul, Harry. "The Debate Over the Bankruptcy of Science in 1895." *French Historical Studies* 3 (1968): 299–327.

———. "Scholarship and Ideology: The Chair of the General History of Science at the Collège de France, 1892–1913." *Isis* 67, no. 238 (1976): 376–387.

Paulhan, Frédéric. "Analyses et comptes rendus: Letourneau-*La sociologie d'après l'ethnographie*." *Revue philosophique* 11 (1881): 546–551.

———. "Analyses et comptes rendus: E. Véron-*La morale*." *Revue philosophique* 18 (1884): 475–478.

———. "Analyses et comptes rendus: Letourneau-*L'évolution de la morale*." *Revue philosophique* 23 (1887): 74–80.

———. "Analyses et comptes rendus: Letourneau-*L'évolution du mariage et de la famille*." *Revue philosophique* 26 (1888): 177–184.

Paxton, Robert Owen. *Vichy France: Old Guard and New Order, 1940–1944.* New York: Columbia University Press, 1972.

Pedersen, Susan. *Family, Dependence, and the Origins of the Welfare State: Britain and France, 1914–1945.* Cambridge: Cambridge University Press, 1993.

Péguy, Charles. *Oeuvres en prose, 1898–1908.* Paris: Gallimard, 1959.

Peukert, Detlev J. K. "The Genesis of the 'Final Solution' from the Spirit of Science." In Jane Caplan and Thomas Childers, eds., *Reevaluating the Third Reich*, intro. Charles S. Maier, 234–252. New York: Holmes and Meier, 1993.

Pierre et Paul (the journal's generic pseudonym). "Yves Guyot." *Les hommes d'aujourd'hui* 1, no. 46 (1878): 1–4.

——. "Hovelacque." *Les hommes d'aujourd'hui* 3, no. 128 (1880): 2–4.

——. "Thulié." *Les hommes d'aujourd'hui* 3, no. 144 (1880): 2–4.

——. "Clémence Royer." *Les hommes d'aujourd'hui* 4, no. 170 (1881): 2, 3.

Pillon, François. "La lutte contre le cléricalisme, ce qu'elle ne doit pas être et ce qu'elle doit être: Il ne faut pas que la politique anticléricale soit une politique d'irréligion." *La critique philosophique* 9 (1880): 113–123.

——. "A propos du substantialisme de Mme. Clémence Royer et de M. Roisel." *La critique philosophique* 11 (1882): 81–89.

——. "Letourneau: *L'évolution religieuse dans les diverses races humaines.*" *L'année philosophique* 3 (1892): 260–261.

——. "Letourneau: *L'évolution de l'esclavages dans les diverses races humaines.*" *L'année philosophique* 8 (1897): 266–267.

——. "Letourneau: *L'évolution du commerce dans les diverses races humaines.*" *L'année philosophique* 8 (1897): 267–268.

——. "Revue bibliographique: Lefèvre, *L'histoire: Entretiens sur l'évolution historique.*" *L'année philosophique* 8 (1897): 261–262.

——. "Letourneau: *L'évolution de l'éducation dans les diverses races humaines.*" *L'année philosophique* 9 (1898): 262–263.

Poliakov, Leon. *The Aryan Myth: A History of Racist and Nationalist Ideas in Europe.* New York: Basic, 1974.

Pollard, Miranda. *Reign of Virtue: Mobilizing Gender in Vichy France.* Chicago: University of Chicago Press, 1998.

Porter, Theodore M. *The Rise of Statistical Thinking, 1820–1900.* Princeton: Princeton University Press, 1986.

——. *Trust in Numbers: The Pursuit of Objectivity in Science and Public Life.* Princeton: Princeton University Press, 1995.

The Probabilistic Revolution. Vol. 1. Ed. Lorenz Kruger, Lorraine J. Daston, and Michael Heidelberger. Cambridge, Mass.: MIT Press, 1987.

The Probabilistic Revolution. Vol. 2. Ed. Lorenz Kruger, Gerd Gigerenzer, and Mary Morgan. Cambridge, Mass.: MIT Press, 1987.

Proctor, Robert. "From *Anthropologie* to *Rassenkunde.*" In George Stocking, ed., 138–180. *Bones, Bodies, Behavior: Essays on Biological Anthropology.* Wisconsin: University of Wisconsin Press, 1988.

——. "Nazi Medicine and the Politics of Knowledge." In Sandra Harding, ed., *The "Racial" Economy of Science: Toward a Democratic Future*, 344–358. Bloomington: Indiana University Press.

Prokopec, Miroslav. "Hrdlička: A Scientist and a Man." In Vladimir Novotny, ed., *Anthropological Congress Dedicated to Aleš Hrdlička*, 57–61. Prague: Academia, 1971.

Prost, Antoine. *Histoire de l'enseignement en France, 1800–1967*. Paris: Armand Colin, 1968.

Proudhon, Pierre-Joseph. *De la justice dans la révolution et dans l'église*. Brussels, 1868–1870.

Quatrefages, Armande de. *L'unité de l'espèce humain*. Paris: Hachette, 1861.

——. *Rapport sur le progrès d'anthropologie*. Paris: Imprémerie impériale, 1867.

——. *Histoire générale des races humaines*. Paris: A. Hennuyer, 1887.

Reiff, Philip. *Freud: The Mind of the Moralist*. Chicago: University of Chicago Press, 1979.

Reinach, Salomon. "Gabriel de Mortillet." In *Extrait de la Revue historique, 1899*, 8–10. Paris, 1899.

——. "Compte rendu: *L'Aryen*." *Revue critique d'histoire et de literature*, February 12, 1900, 121–125.

Rémond, René. *L'anticléricalisme en France*. Brussels: Complexe, 1985.

——. *Religion and Society in Modern Europe*. Oxford: Blackwell, 1999.

Renan, Ernest. *L'avenir de la science: Pensées de 1848*. 1890. Reprint, Paris: Flammarion, 1995.

Réville, Jean. "Une histoire des religions par un adversaire de la religion: M. Eugène Véron." *Revue politique et littéraire* 5 (1883): 13–17.

Rhodes, Henry T. F. *Alphonse Bertillon: Father of Scientific Detection*. New York: Abelard-Schuman, 1956.

Richardson, Ruth. *Death, Dissection, and the Destitute*. London: Routledge and Kegan Paul, 1987.

Richet, Charles. "Du somnambulisme provoqué." *Revue philosophique* 10 (1880): 337–374, 462–484.

——. "Les démoniaques d'aujourd'hui: L'hystérie et le somnambulisme." *Revue des deux mondes* 37 (1880): 340–372.

——. "Les démoniaques d'autrefois: Les sorcières et les possédées." *Revue des deux mondes* 37 (1880): 552–583.

——. *Dans cent ans*. Paris, 1892.

——. "La science a-t-elle fait banqueroute?" *Revue scientifique* 3 (1895): 33–39.

——. *Metapsychics*. London, 1905.

——. *Our Sixth Sense*. London: Rider, 1929.

Ringer, Fritz. *Fields of Knowledge: French Academic Culture in Comparative Perspective, 1890–1920*. Cambridge: Cambridge University Press, 1992.

Ripley, William. *The Races of Europe*. London: Trench, Trübner, 1897.

Roc, Etienne. "Professeur Mathias Duval." *Les hommes d'aujourd'hui* 6, no. 273 (1886): 1–4.

Ronsin, Francis. *La grève des ventres: Propagande néo-malthusienne et baisse de la natalité, 19e–20e siècles*. Paris: Aubier Montaigne, 1980.

Rosen, George. "The Philosophy of Ideology and the Emergence of Modern Medicine in France." *Bulletin of the History of Medicine* 20 (July 1946): 328–339.

——. *Professions and the French State, 1700–1900*. Ed. Gerald L. Geison. Philadelphia: University of Pennsylvania Press, 1984.

Rouanet, Gustave. "Les théories aristocratique devant la science." *La petit republique*, January 2, 1900.

Roujou, Anatole. "L'anthropologie." *Libre pensée* 6 (November 25, 1866): 42–43.

Royer, Clémence. Preface to Charles Darwin, *L'origine des espèces*. Paris, 1862.

——. *La constitution du monde: Natura rerum, dynamique des atomes, nouveaux principes de philosophie naturelle*. Paris: Schleicher Frères, 1900.

Ruggiero, Kristin. "Fingerprinting and the Argentine Plan for Universal Identification in the Late Nineteenth and Early Twentieth Centuries." In Jane Caplan and John Torpey, eds., *Documenting Individual Identity: The Development of State Practices in the Modern World*, 184–196. Princeton: Princeton University Press, 2001.

Russett, Cynthia Eagle. *Sexual Science: The Victorian Construction of Womanhood*. Cambridge: Harvard University Press, 1989.

Saller, Karl. *Die Rassenlehre des Nationalsozialismus in Wissenschaft und Propaganda*. Darmstadt, 1961.

Sanger, Margaret. *Woman and the New Race*. New York: Blue Ribbon, 1920.

——. *Margaret Sanger: An Autobiography*. New York: Norton, 1938.

Sarcey, Francisque. *Journal de jeunesse*. Paris: Bibliothèque des annales, n.d.

Sarton, George. "Paul, Jules, and Marie Tannery (with a note on Grégoire Wyrouboff)." *Isis* 38 (November 1947): 33–51.

Schiller, Francis. *Paul Broca: Founder of French Anthropology, Explorer of the Brain*. 1979. Reprint, Oxford: Oxford University Press, 1992.

Schneider, William. *Quality and Quantity: The Quest For Biological Regeneration in Twentieth-Century France*. Cambridge: Cambridge University Press, 1990.

Scott, John A. *Republican Ideas and the Liberal Tradition in France, 1870–1914*. New York: Columbia University Press, 1951.

La Société, l'Ecole et le Laboratoire d'anthropologie de Paris à l'Exposition universelle de 1889. Paris: Imprimeries réunites, 1889.

"Société d'autopsie." *Revue de l'Ecole d'anthropologie de Paris* 3 (1893): 233–236.

Sorokin, Pitirim. *Contemporary Sociological Theories*. New York: Harper, 1928.

Spengler, Joseph J. *France Faces Depopulation*. Durham, N.C.: Duke University Press, 1938.

"Statuts de la Société d'autopsie mutuelle." *Revue scientifique* 11 (November 25, 1876): 527–528.

Stepan, Nancy Leys. *The Idea of Race in Science*. Hamden, Conn.: Archon, 1982.

——. "Race and Gender: The Role of Analogy in Science." *Isis* 77, no. 2 (June 1986): 261–277.

Stocking, George W. Jr., ed. *Bones, Bodies, Behavior: Essays on Biological Anthropology*. Madison: University of Wisconsin Press, 1988.

Stoddard, Lothrop. *The Rising Tide of Color Against White World-Supremacy*. Intro. Madison Grant. New York: Scribner, 1920.

Stoler, Ann Laura. *Race and the Education of Desire: Foucault's History of Sexuality and the Colonial Order of Things*. Durham, N.C.: Duke University Press, 1995.

——. "Racial Histories and Their Regimes of Truth." *Political Power and Social Theory* 11 (1997): 183–206.

"Suite de la discussion sur l'anthropophagie." *Bulletins de la Société d'anthropologie* 3d ser., 11 (1888): 27–46.

Sulloway, Frank. *Freud: Biologist of the Mind*. New York: Basic, 1979.

Tannenbaum, Edward R. *The Action Française: Die-hard Reactionaries in Twentieth-Century France*. New York: Wiley, 1962.

Tarbell, Ida M. "Identification in Criminals: The Scientific Method in Use in France." *New McClure's Magazine* 2, no. 4 (March 1894): 355–369.

Tatarin-Tarnheyden, Edgar. "Georges Vacher de Lapouge: Visionnaire française de l'avenir européen." *Cahiers franco-allemands* 9 (October-December 1942): 336–346.

Temkin, Owsei. "Materialism in French and German Physiology of the Early Nineteenth Century." *Bulletin of the History of Medicine* 20 (July 1946): 322–327.

Thuillier, Guy. "Un anarchiste positiviste: Georges Vacher de Lapouge." In Pierre Guiral and Emile Temine, eds., *L'idée de race dans la pensée politique française contemporaine*, 48–65. Paris: CNRS, 1977.

Thulié, Henri. "Sur l'autopsie de Louis Asseline, membre de la Société d'anthropologie et de la Société d'autopsie mutuelle." *Bulletins de la Société d'anthropologie* 3d ser., 10 (1878): 161–167.

———. *La femme: Essai de sociologie physiologique; ce qu'elle a été, ce qu'elle est, les théories, ce qu'elle doit être.* Paris: Delahaye et Lecrosnier, 1885.

———. "L'Ecole d'anthropologie de Paris depuis sa fondation." In *L'Ecole d'anthropologie de Paris (1876–1906)*, 1–27. Paris: Alcan, 1907.

Tint, H. "The Search for a Laic Morality Under the French Third Republic: Renouvier and the 'Critique Philosophique.'" *Sociological Review* 5, no. 1 (July 1957): 5–26.

Topinard, Paul. *L'anthropologie.* Bibliothèque des sciences contemporaines. Paris: Reinwald, 1876.

———. *Eléments d'anthropologie.* Paris, 1885.

———. *Anthropology.* Trans. Robert Bartley. London, 1890.

———. *La société, l'école, le laboratoire et le musée Broca.* Paris: Chamerot, 1890.

Tourette, Gilles de la. "Le professeur J.-M. Charcot." *Revue hebdomadaire* 15 (1893): 608–622.

Trinkaus, Erik. "A History of Homo erectus and Homo sapiens Paleontology in America." In Frank Spencer, ed., *A History of American Physical Anthropology, 1930–1980*, 261–280. New York: Academic, 1982.

Troyat, Henri. *Zola.* Paris: Flammarion, 1992.

Turgeon, Charles. *Le féminisme français.* Vol. 1. Paris: Larose, 1907.

Ujfalvy, Charles. *Les Aryens au nord et au sud de l'Hindou-Kouch.* Paris: G. Masson, 1896.

Vacanard, Abbé E. "Le nouvel homme préhistorique de Menton." *Revue des questions scientifiques* 20 (1886): 74–122.

Vallois, Henri. "Le laboratoire Broca." *Bulletins et mémoires de la Société d'anthropologie* 9th ser., 11 (1940): 1–18.

Vasseur, André and Jean-Michel Place, eds. *Bibliographie des revues et journaux littéraires des XIXe et Xxe siècles.* Paris: Jean-Michel Place, 1974.

Véron, Eugène. *Les associations ouvrières.* Paris: Hachette, 1865.

———. "De l'enseignement supérieur en France." *Revue des cours littéraires de la France et de l'étranger* 2 (1865): 401–404, 435–437, 449–452.

———. *Aesthetics.* Trans. W. H. Armstrong. London: Chapman and Hall, 1879.

———. *L'esthétique.* Bibliothèque des sciences contemporaines. Paris: Reinwald, n.d.

Véron, Jeanne. *Le chat.* Paris: Librairie d'éducation laïque, 1880.

———. *Le chien.* Paris: Librairie d'éducation laïque, 1880.

———. *Le petit cousin Charles.* Paris: Librairie d'éducation laïque, 1880.

———. *L'âne.* Paris: Librairie d'éducation laïque, 1881.

———. *Histoires enfantines.* Paris: Librairie d'éducation laïque, 1881.

La vie et les oeuvres du docteur L.-A. Bertillon, professeur de démographie à l'Ecole d'anthropologie, chef des travaux de la statistique municipale de la ville de Paris. Paris, 1883.

Viereck, Peter. *Metapolitics: The Roots of the Nazi Mind.* New York, 1961. Rev. ed. of *Metapolitics: From Romanticism to Hitler.* New York, 1941.

Vogt, W. Paul. "Political Connections, Professional Advancement, and Moral Education in Durkheimian Sociology." *Journal of the History of the Behavioral Sciences* 27 (1991): 56–74.

Vovelle, Michel. *La mort en l'occident de 1300 à nos jours*. Paris: Gallimard, 1983.

Walker, Philip. *Zola*. London: Routledge and Kegan Paul, 1985.

Weber, Eugen. *My France*. Cambridge: Harvard University Press, 1991.

Weir, Jean. *Histoires disputes et discours des illusions et impostures des diables*. Pref. Désiré-Magloire Bourneville. Paris, 1885.

Weisz, George. *The Emergence of Modern Universities in France, 1863–1914*. Princeton: Princeton University Press, 1983.

Williams, Elizabeth A. "The Science of Man: Anthropological Thought and Institutions in Nineteenth-Century France." Ph.D. diss., Indiana University, 1983.

———. *The Physical and the Moral: Anthropology, Physiology, and Philosophical Medicine in France, 1750–1850*. Cambridge: Cambridge University Press, 1994.

Wilson, Thomas. "Criminal Anthropology." *Annual Report of the Smithsonian Institution*. Washington: U.S. Government Printing Office, 1891.

Wippermann, Wolfgang. "Das Berliner Schulwesen in der NS-Zeit: Fragen, Thesen und Methodische Bemerkungen." In Benno Schmoldt, ed., *Schule in Berlin*, 57–73. Berlin: Colloquium, 1989.

Wishnia, Judith. *The Proletarianizing of the Fonctionnaires*. Baton Rouge: Louisiana State University Press, 1990.

Worms, René. "Jean Finot, sociologue." *La revue mondiale* 33 (May 15, 1922): 228–232.

Zeldin, Theodore. *France, 1848–1945*. Vol. 2. Oxford: Oxford University Press, Clarendon, 1977.

Zola, Emile. "Compte Rendu: Letourneau." *Le globe*, January 23, 1868.

———. *Lourdes*. Trans. E. A. Vizetelly. Chicago: Neely, 1894.

INDEX